AF321955

# Navigation: Science and Technology

Volume 17

This series *Navigation: Science and Technology (NST)* presents new developments and advances in various aspects of navigation - from land navigation, marine navigation, aeronautic navigation to space navigation; and from basic theories, mechanisms, to modern techniques. It publishes monographs, edited volumes, lecture notes and professional books on topics relevant to navigation - quickly, up to date and with a high quality. A special focus of the series is the technologies of the Global Navigation Satellite Systems (GNSSs), as well as the latest progress made in the existing systems (GPS, BDS, Galileo, GLONASS, etc.). To help readers keep abreast of the latest advances in the field, the key topics in NST include but are not limited to:

- Satellite Navigation Signal Systems
- GNSS Navigation Applications
- Position Determination
- Navigational instrument
- Atomic Clock Technique and Time-Frequency System
- X-ray pulsar-based navigation and timing
- Test and Evaluation
- User Terminal Technology
- Navigation in Space
- New theories and technologies of navigation
- Policies and Standards

This book series is indexed in **SCOPUS** and **EI Compendex** databases.

Bofeng Li · Zhetao Zhang · Weikai Miao

# GNSS Real-Time Kinematic Positioning

## Theory and Applications

 Springer

Bofeng Li 
College of Surveying and Geo-Infomatics
Tongji University
Shanghai, China

Zhetao Zhang
College of Surveying and Geo-Infomatics
Tongji University
Shanghai, China

Weikai Miao
College of Surveying and Geo-Infomatics
Tongji University
Shanghai, China

ISSN 2522-0454          ISSN 2522-0462  (electronic)
Navigation: Science and Technology
ISBN 978-981-96-9115-9          ISBN 978-981-96-9116-6  (eBook)
https://doi.org/10.1007/978-981-96-9116-6

This work was supported by National Natural Science Funds of China (42225401, 42430109).

**Competing Interests** The authors have no competing interests to declare that are relevant to the content of this manuscript.

# Contents

# Chapter 1
# Overview

## 1.1 GNSS Development

The emergence of the Global Navigation Satellite System (GNSS) has revolutionized human navigation and positioning. Over time, several satellite navigation systems have been developed, including global, regional, and augmented systems.

### *1.1.1 GPS*

The Global Positioning System (GPS) is the first satellite navigation system to be officially operated on a global scale. It remains the most mature and widely used system to date. The United States Department of Defense initiated the research and development of GPS in the 1970s and completed its basic construction in the 1990s. By 1995, the United States Department of Defense officially declared GPS operational. In its early stages, GPS was designed with two types of pseudo-random codes: the precision code (P-code) and the coarse acquisition code (C/A code). The P-code was restricted to military and special users from the United States and its allies, while the C/A code was made available to civilian users worldwide, providing standard positioning services. Currently, the GPS modernization program is an ongoing, multibillion-dollar initiative aimed at enhancing the system features and overall performance. The upgrades include new civilian and military GPS signals [1, 2].

GPS consists of three main segments: the space segment, the control segment, and the user segment. The space segment comprises a constellation of satellites that transmit radio signals to users. The United States is committed to maintaining at least 24 operational GPS satellites under most circumstances. Table 1.1 provides the nominal GPS constellation parameters. Specifically, the system includes 24 satellites evenly distributed across six orbital planes, each inclined at 55°, with four satellites per plane [3]. In recent years, the total number of satellites in the constellation has

© The Author(s) 2025

B. Li et al., *GNSS Real-Time Kinematic Positioning*, Navigation: Science and Technology 17, https://doi.org/10.1007/978-981-96-9116-6_1

**Table 1.1** Nominal GPS constellation parameters

| Parameter | Value |
| --- | --- |
| Number of operational satellites | $t = 24$ |
| Number of orbital planes | $p = 6$ |
| Number of satellites in a plane | $t/p = 4$ |
| Orbit type | Near circular |
| Eccentricity | $e < 0.02$ |
| Inclination | $i = 55°$ |
| Nominal orbital altitude | $h = 20{,}180$ km |
| Period of revolution | $T = 11\,\text{h}\,58\,\text{m}$ |
| Longitude of ascending node between planes | $\Delta\Omega = 60°$ |
| Ground track repeat cycle | $2\ \text{orbit}/1^{\text{d}}_{\text{sid}}$ |

increased to 31. Of these, more than 24 are placed in expandable slots within the baseline 24-satellite constellation [4]. Surplus satellites (beyond the 27th operational satellite) are typically positioned near satellites that are expected to require replacement soonest [4]. The control segment consists of a global network of ground facilities responsible for tracking GPS satellites, monitoring their transmissions, performing analyses, and sending commands and data to the constellation. The current Operational Control Segment (OCS) includes a Master Control Station (MCS), an alternate MCS, 11 command and control antennas, and 16 monitoring sites. The user segment primarily receives navigation signals transmitted by the satellites. It recovers the carrier signal frequency and synchronizes with the satellite clock. Additionally, it demodulates satellite ephemeris data, satellite clock correction parameters, and other relevant information from the navigation message. Using this data, users can determine navigation parameters such as geographic longitude, latitude, altitude, speed, and precise time.

Signal is an essential part of all satellite navigation systems, for GPS, the standard frequency of electromagnetic waves emitted by GPS satellites is $f_0 = 10.23$MHz. Currently, there are both traditional and new signals on L1, L2, and L5, where the frequencies are $f_1 = 1575.42$MHz, $f_2 = 1227.6$MHz, and $f_2 = 1176.45$MHz, respectively.

### *1.1.2  BDS*

China's BeiDou Navigation Satellite System (BDS) is another GNSS, and it has progressed through three major stages since the 1990s. The initial BeiDou Demonstration Navigation System (BDS-1) was established in 2003. It originally consisted of two geostationary orbit (GEO) satellites, with an additional GEO satellite serving as a backup. Initially, BDS-1 was considered a regional positioning system rather than a full navigation system, as it only provided user positioning information. In

2012, the BeiDou Regional Navigation Satellite System (BDS-2) was launched. It featured a constellation of 14 satellites: 5 GEO satellites, 5 inclined geosynchronous orbit (IGSO) satellites, and 4 medium Earth orbit (MEO) satellites [5]. BeiDou Global Navigation Satellite System-3 (BDS-3), which was developed following the stable service provided by BDS-2, began to offer global services in 2020. It mainly covers the Asia-Pacific region with triple-frequency signals and provides global navigation services [6].

Similar to other GNSS, the BDS-3 architecture is composed of three main parts: the space segment, the Ground Control Segment (GCS), and user terminals. The space segment consists of 3 GEO satellites, 3 IGSO satellites, and 24 MEO satellites, with additional backup satellites in orbit. The GEO satellites are positioned at an altitude of 35,786 km and located at longitudes of 80°, 110.5°, and 140°, respectively. The IGSO satellites also have an altitude of 35,786 km, with an orbit inclination angle of 55°. The MEO satellites have an altitude of 21,528 km and an orbit inclination angle of 55°.

The GCS includes the MCS, which incorporates the OCS, Monitor Stations (MSs), and Uplink Stations (ULSs). The primary tasks of the MCS are to collect tracking data from monitoring stations, process this data to determine satellite orbits and clock biases, and generate the satellite navigation messages. The MCS of BDS-3 also supports user position determination via the Radio Determination Satellite Service (RDSS) mode and provides short message communication services. The MSs are strategically distributed across mainland China. They provide code and phase observations to the MCS for satellite orbit determination and for generating wide-area differential products. The ULSs transmit the generated navigation messages and wide-area differential corrections to the satellites for broadcast to users [7].

BDS uses the China Geodetic Coordinate System 2000 (CGCS2000), which is a geocentric geodetic coordinate system. CGCS2000 is realized through the China Terrestrial Reference Frame (CTRF). The definition of this coordinate system follows the criteria established in the 1996 conventions of the International Earth Rotation and Reference Systems Service (IERS). The relevant parameters of the CGCS2000 coordinate system are listed in Table 1.2. The BDS time system is based on BeiDou Time (BDT), which is a continuous navigation time scale without leap seconds and uses the SI second as its basic unit. BDT is typically expressed in terms of BeiDou week number (WN) and seconds of week (SoWs), with values ranging from 0 to 604,799. The reference epoch of BDT is 00:00:00 on January 1, 2006, at which point both WN and SoW are 0 [8].

**Table 1.2** Fundamental parameters of the CGCS2000 system

| Parameter | Value |
| --- | --- |
| Semimajor axis | $a = 6378137.0$ m |
| Flattening | $f = 1/298.257222101$ |
| Gravitational coefficient (incl. atmosphere) | $GM_\otimes = 398600.4418 \times 10^9$ m$^3$/s$^2$ |
| Angular velocity | $\omega_\otimes = 7.292115 \times 10^{-5}$ rad/s |

In addition to the traditional positioning, navigation, and timing (PNT) functions, the BDS user terminal supports six additional featured services. These services include: Global Short Message Communication Service (GSMCS), and MEO Satellite-Based Search and Rescue (MEOSAR) service. These two services are based on MEO satellite features. Regional services offered include: Regional Short Message Communication Service (RSMCS), RDSS, BDS Satellite-Based Augmented Service (BDSBAS), and satellite-based precise point positioning (PPP) service via B2b signal (B2b-PPP). These regional services are based on GEO satellite features [9].

It is also worth noting that the BDS-2 can provide three public service signals, i.e., B1I, B2I and B3I, where the center frequencies of B1, B2, and B3 bands are 1561.098, 1207.140, and 1268.520 MHz, respectively [6]. Whereas in BDS-3, six public service signals B1I, B1C, B2a, B2b, B2a+b, and B3I are provided. Among them, the center frequencies of B1I, B1C, B2a, B2b, B2a+b, and B3I are 1561.098, 1575.420, 1176.450, 1207.140, 1191.795, and 1268.520 MHz, respectively [6].

The Satellite Navigation Interface Control Document (ICD) defines the signal interface relationship between the satellite navigation system and the user. It is an essential technical document for developing manufacturing specifications and chips. The BDS ICD establishes and standardizes the communication interface protocol for the radio link between the space segment and the user segment. It serves as the standard document that user terminals must follow to receive, capture, track, demodulate, and decode BDS navigation satellite signals. While BDS has developed rapidly, it offers unique functions, particularly with BDS-3, which meets the design index requirements for orbit determination accuracy, satellite clock accuracy, signal-in-space accuracy, and PNT service performance [10]. However, its development still faces significant challenges, which can be categorized into four key areas: international competition, lack of national policy, limited initiative in service concepts, and gaps in certain technologies [11].

### *1.1.3  Galileo*

The enormous potential benefits of satellite navigation have led the European Space Agency (ESA) and the European Commission (EC) to collaborate on the development and deployment of the European Navigation Satellite System, named after the Italian astronomer Galileo. The Galileo system is a strategic initiative that not only supports security, defense, and military applications but also plays a significant role in the aerospace sector, offering substantial social and economic benefits. Construction of Galileo began in 2005, and by 2016, it achieved the capability of providing regional independent services following continuous development. In 2017, Galileo entered the operational phase, with the Full Operational Capability (FOC) slated for 2020. This includes the system daily operations, constellation maintenance, and the operation of the ground segment [12]. To meet the corresponding PNT requirements, the Galileo ground segment consists of the Ground Control Segment and the Ground Mission

**Table 1.3**  Nominal Galileo constellation parameters

| Parameter | Value |
| --- | --- |
| Reference constellation type | Walker 24/3/1 + 6 in-orbit spares |
| Semimajor axis | 29,600.318 km |
| Inclination | 56° |
| Period | 14 h 04 m 42 s |
| Ground track repeat cycle | 10 sidereal days/17 orbits |

Segment (GMS). The core facilities of both segments are located at two Galileo Control Centers (GCC) in Oberpfaffenhofen, Germany, and Fucino, Italy [13].

The Galileo constellation consists of 30 satellites, including 24 operational satellites and 6 spare satellites in orbit. These satellites are evenly distributed across 3 orbital planes, with each plane containing 8 operational satellites and 2 spare satellites. The detailed nominal parameters of the Galileo constellation are listed in Table 1.3.

In addition to its global navigation and positioning functions, Galileo also offers other capabilities, such as global search and rescue (SAR). As a fourth service, the Galileo satellite system supports the international satellite search and rescue system Cospas-Sarsat, which was established by the United States, Russia, Canada, and France. The satellites are equipped with transponders to relay distress signals from emergency beacons to rescue coordination centers, which then initiate rescue operations. At the same time, the system is designed to notify users through emergency beacons that their distress signal has been detected and that help is on the way. Each Galileo satellite transmits navigation signals (L-band) across three frequencies. The Galileo system offers three different location services: Open Service: This service is available to all users and offers free access to satellite signals on the E1-B/C, E5a-I/Q, and E5b-I/Q frequencies. Authorized Services (Government Services): This is a publicly regulated service available on the restricted E1-A and E6-A frequencies. Commercial Services: This service utilizes the E6-B/C frequencies (navigation signals on a third frequency, with optional encryption) and is designed to provide future value-added services. Thus, there are five frequencies used by Galileo. The center frequencies of E1, E5a, E5b, E5, and E6 are 1561.098, 1575.420, 1176.450, 1207.140, 1191.795, and 1268.750 MHz, respectively [13, 14].

### *1.1.4  GLONASS*

In the last century, the GLObal Navigation Satellite System (GLONASS) was developed as a second-generation satellite navigation and positioning system by the former Soviet Union and is now managed and maintained by Russia. Similar to the GPS, GLONASS also follows the principle of space-based trilateration, providing users anywhere on Earth and in near-Earth space with continuous and accurate three-dimensional coordinates, speed, and time information. From the launch of the first GLONASS satellite in October 1982 to December 1995, a total of 73 GLONASS satellites were launched. Ultimately, a constellation of 24 operational satellites was established, which was officially completed in 2012 [15].

Similar to GPS, GLONASS consists of three segments: the space segment, the ground segment, and the user segment. GLONASS uses a Walker-type constellation structure. Specifically, the orbital inclination of GLONASS satellites is approximately 10° higher than that of other GNSS satellites. This design provides improved observation conditions for the Russian region. GLONASS users worldwide also benefit from its excellent sky coverage, particularly in polar regions, where more GLONASS satellites are visible than from other systems. The plane positions of each GLONASS satellite are distributed across three orbital planes [16]. The parameters of the GLONASS constellation are provided in Table 1.4.

The ground segment is a crucial component of the GLONASS system. Following the disintegration of the Soviet Union, it is now primarily managed by the Russian Space Agency (RSA), and as a result, this segment is largely confined to Russia. Its core components include a System Control Center (SCC) responsible for planning and coordinating all elements of the ground segment, and central clocks (CCs) that synchronize with Coordinated Universal Time (UTC). Additionally, telemetry, tracking, and command stations (TT&C), along with uplink stations, are used to receive status information from GLONASS satellites, transmit control commands, and determine satellite orbits. The ground segment also includes one-way monitoring stations that collect one-way pseudo-range and carrier-phase measurements [17].

**Table 1.4** Nominal GLONASS constellation parameters

| Parameter | Value |
| --- | --- |
| Number of operational satellites | $t = 24$ |
| Number of orbital planes | $p = 3$ |
| Number of satellites in a plane | $t/p = 8$ |
| Phasing parameter | $f = 1$ |
| Orbit type | Near circular |
| Eccentricity | $e < 0.01$ |
| Inclination | $i = 64° \pm 0.3°$ |
| Nominal altitude | $h = 19100$ km |
| Period of revolution | $T = 11h\ 15m\ 44s\ \pm 5s$ |
| Longitude of ascending node between planes | $\Delta\Omega = 120°$ |
| Argument of latitude difference | $\Delta u = 45°$ |
| Latitude shift between planes | $\Delta uf/n = 45°$ |
| Ground track repeat cycle | 17 orbits/8 d |

Unlike the previous satellite systems, GLONASS is based on the Frequency Division Multiple Access (FDMA) technology. That is, the frequencies of signals transmitted by different satellites are slightly different, and different signals use different signal channels. The GLONASS provides two types of services. The first one is the public service for unencrypted signals. It usually includes two frequency signals L1 and L2 ($f_1 = 1.6$GHz and $f_2 = 1.2$GHz). In recent years, the system has added a third frequency signal L3 ($f_3 = 1202.025$MHz). The unencrypted signals are available to users worldwide. The second one is the services for authorized users. Specific users are currently served using encrypted signals in two frequency bands (L1 and L2) [18].

### *1.1.5 Other Systems*

Currently, several regional or augmented navigation satellite systems are developing rapidly, including the Quasi-Zenith Satellite System (QZSS), Indian Regional Navigation Satellite System (IRNSS), and various Satellite-Based Augmentation Systems (SBAS).

QZSS is a space-based navigation augmentation system developed and built by Japan. It represents the first step in Japan's construction of an autonomous regional navigation satellite system. Initially, the system was planned to be developed in two stages. The first stage involved the construction of a QZSS consisting of 3 satellites, and the second stage involved the addition of 4 more quasi-zenith satellites, along with 3 GEO satellites, creating a regional navigation satellite system with a total of 7 satellites. The first QZSS satellite was successfully launched on September 11, 2010, and has been in operation, significantly enhancing Japan's satellite navigation services [19]. The satellite uses a geostationary communication satellite platform independently developed by Japan, with a mass of 4,100 kg and a design life of 10 years.

The QZSS signal includes L1, L2, L5, and LEX frequencies. Once fully deployed, QZSS will greatly improve the visibility of Japan's satellite navigation signals, especially addressing the urban canyon effect, increasing the availability of navigation signals, and meeting Japan's growing demand for satellite navigation services [20–22].

The IRNSS, also known as Navigation with Indian Constellation (NavIC), is designed to improve positioning accuracy for users to better than 20 m in its main service area. It consists of 7 satellites located in GEO and IGSO orbits, along with a GCS and user segment, covering India and surrounding areas within a 1500 km radius. This system provides improved positioning accuracy [23]. The IRNSS satellites are based on the same platform used for India's Kalpana-1 weather satellite. The payload includes two solid-state power amplifiers, clock management and control units, frequency generation and modulation units, navigation processors, signal generators, and atomic clocks. India has completed the launch of all seven satellites needed for the navigation system, and the system is gradually taking shape. The development of IRNSS positions India to become the fourth country in the world with autonomous satellite navigation capabilities, meeting its military satellite navigation needs. However, due to the limitations of the constellation configuration and the number of satellites, its positioning accuracy and service range are still less comprehensive than GPS, GLONASS, or BDS.

The IRNSS uses S-band (2492.08 MHz) and L-band (L5, 1176.45 MHz) frequencies. Additionally, the GPS-Aided GEO Augmented Navigation (GAGAN) system, developed jointly by the Indian Space Research Organization and the Aviation Authority of India, uses a space segment consisting of 3 GEO satellites over the Indian Ocean. The C-band is mainly used for measurement and control, while the L-band broadcasts navigation information, compatible with GPS. The system covers the entire Indian subcontinent, providing GPS signals and differential corrections to improve GPS positioning accuracy and reliability, particularly for aviation applications in Indian airports.

To further enhance GNSS positioning accuracy and integrity, especially to correct or suppress errors such as ephemeris errors, satellite clock errors, and ionospheric delays, SBAS has been developed and is actively used. The principle of SBAS involves using GEO satellites as communication satellites to forward positioning enhancement information to users, while also broadcasting navigation signals to improve user positioning accuracy. SBAS works by employing a large number of distributed monitoring stations with precisely known locations to continuously observe navigation satellites, calculate correction data (including orbit errors, satellite clock errors, and ionospheric delays), and assess integrity. This information is then transmitted to GEO satellites, which forward it to user terminals. These terminals use the correction data to adjust their positioning and use the navigation signals from GEO satellites to improve accuracy and integrity for users.

Typically, SBAS consists of three parts: the space segment (GEO satellites), the ground segment (monitoring stations, main control stations, and injection stations), and the user segment (devices that receive SBAS signals). Several SBAS systems are already in use, including BDSBAS, GAGAN, the U.S. Wide Area Augmentation System (WAAS), the European Geostationary Navigation Overlay Service (EGNOS), Japan's Multi-Functional Satellite Augmentation System (MSAS), and Russia's System for Differential Corrections and Monitoring (SDCM). Notably, the GEO satellites of BDS-3 provide SBAS services to users in China and surrounding regions, following International Civil Aviation Organization (ICAO) standards, with the goal of achieving APV-I and CAT-I precision approaches [9, 24]. Currently, BDSBAS uses two signals, SBAS-B1C and SBAS-B2a, to offer single-frequency and dual-frequency services, meeting the high accuracy requirements of BDS [25–28].

## 1.2 Techniques for Precise Positioning

Positioning techniques have evolved rapidly over the past few decades, with significant improvements in accuracy, reliability, and efficiency. Today, precise positioning technologies show a trend of diversification, ranging from code-based to phase-based methods, single-station to multi-station setups, post-processing to real-time applications, and undifferenced to differenced approaches.

Single point positioning (SPP) is the first and still widely used positioning method within the GNSS community. Among all positioning modes, SPP is the simplest to implement, although its accuracy is relatively low. SPP uses code observations and broadcast navigation data calculated by the global reference network. Due to the limited precision of orbits and clocks and the relatively simple error processing, the accuracy of SPP is typically at the meter level.

PPP is another positioning method that operates on a global scale and offers absolute positioning. Unlike SPP, PPP uses phase observations in addition to code observations. It also incorporates more precise data, such as precise orbits and clock corrections. When systematic errors, such as ionospheric and tropospheric delays, are properly addressed, PPP can achieve accuracy at the centimeter or even millimeter level. For errors that can be modeled accurately, such as phase center offset (PCO), phase center variation (PCV), phase windup, relativistic effects, solid Earth tides, ocean loading, and Earth rotation, models are applied. Significant errors are handled by adding parameters. Moreover, if uncalibrated phase delays (UPDs) or observable-specific biases (OSBs) are estimated in advance, PPP with ambiguity resolution (PPP-AR) can be applied, significantly reducing convergence time.

In contrast to global-based absolute positioning modes, real-time differenced positioning (RTD) and real-time kinematic positioning (RTK) are becoming increasingly popular. These positioning technologies are typically used on a regional scale, covering areas with radii of a few hundred kilometers or less. Positioning is relative to reference stations. RTD is a code-based technique, while RTK is phase-based. If there is only one nearby reference station, typically with a baseline of about 20 km or less, the techniques are called single-baseline RTD (SRTD) and single-baseline RTK (SRTK), respectively. These are essentially local-scale methods, with the main advantage being that, after double differencing, clock errors and hardware delays at both the receiver and satellite ends are eliminated, and ionospheric and tropospheric delays can often be neglected.

When multiple reference stations are used, network RTD (also known as wide-area DGNSS) or network RTK (NRTK) comes into play. These methods can extend coverage up to hundreds of kilometers. Atmospheric corrections are determined by the network and transmitted to users. In NRTK, ambiguity resolution is crucial, as it is necessary to fix the ambiguities of very long baselines in real time. Ambiguity resolution is typically achieved by using observation data from multiple reference stations and their known coordinates. With accurately known reference station coordinates, prior information or other methods can be employed to correct observational errors. As a result, real-time decimeter-level accuracy can be achieved with RTD, and centimeter-level accuracy can be attained with RTK.

A new positioning technique, called PPP-RTK, has emerged. PPP-RTK can be considered a hybrid of PPP and RTK. Based on traditional PPP, PPP-RTK uses UPDs or OSBs to resolve ambiguities in PPP (i.e., PPP-AR). After resolving ambiguities, ionospheric and tropospheric delays can be estimated using a global or regional network, similar to NRTK. By applying atmospheric correction data at the user end, PPP-RTK can achieve centimeter-level accuracy in a short time.

## 1.3   RTK Benefits

As mentioned earlier, RTK, especially SRTK, is the most popular method in the current GNSS community. It has been widely applied in various fields, such as deformation monitoring and autonomous driving. Compared to other high-precision positioning methods, RTK offers several unique advantages, as illustrated in Fig. 1.1.

First, RTK has an exceptionally fast convergence time. Specifically, for short baselines, centimeter-level accuracy can be achieved within just a few seconds to a minute, with convergence typically taking less than one minute. In some cases, integer ambiguities can be correctly resolved within a single epoch, allowing for instantaneous high-precision positioning. In contrast, PPP and PPP-AR require several minutes to tens of minutes to converge, making them unsuitable for applications demanding real-time results.

Second, RTK is cost-effective in terms of infrastructure. Generally, it requires only a nearby reference station, keeping deployment costs low. In comparison, PPP-RTK

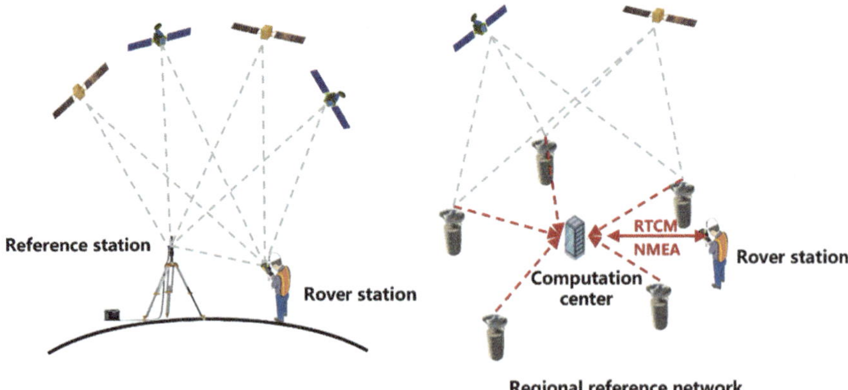


**Fig. 1.1**   Illustration of RTK and NRTK

**Table 1.5** Advantages and disadvantages of all current mainstream high-precision methods

| Methods | Fast covergence time (< 1 min) | No network required and low cost | Concise parameters and convenient algorithm | Broad coverage area (>30 km) |
|---|---|---|---|---|
| SRTK | $\checkmark$ | $\checkmark$ | $\checkmark$ | $\times$ |
| NRTK | $\checkmark$ | $\times$ | $\times$ | $\checkmark$ |
| PPP | $\times$ | $\checkmark$ | $\times$ | $\checkmark$ |
| PPP-AR | $\times$ | $\checkmark$ | $\times$ | $\checkmark$ |
| PPP-RTK | $\checkmark$ | $\times$ | $\times$ | $\checkmark$ |

typically relies on a regional reference network, increasing construction costs. Additionally, NRTK and PPP-RTK require a robust communication network and high-performance servers, whereas SRTK operates without such dependencies, further reducing implementation costs.

From a technical perspective, RTK is the most convenient high-precision positioning method due to its concise parameterization. In RTK, receiver and satellite clock errors, along with corresponding hardware delays, are effectively eliminated through the process of double differencing, provided the baseline length is not excessively long. Additionally, ionospheric and tropospheric delays are significantly mitigated and can even be ignored in short-baseline scenarios. Furthermore, the phase ambiguities exhibit integer properties, making them relatively easy to resolve using appropriate methods. In contrast, other high-precision positioning methods involve more complex parameterization. For example, PPP requires a greater number of parameters, and its ambiguities are typically float solutions unless PPP-AR or PPP-RTK is applied. Similarly, NRTK and PPP-RTK require additional calculations to determine atmospheric corrections, often through interpolation.

The primary limitation of RTK is its restricted coverage area. As the baseline length increases, residual atmospheric delays can degrade the float solution, making ambiguity resolution less reliable. However, this technical challenge can be largely mitigated through NRTK or by employing additional correction techniques. In fact, addressing this issue is one of the key motivations behind writing this monograph. Table 1.5 summarizes the advantages ($\checkmark$) and disadvantages ($\times$) of all current mainstream high-precision positioning methods.

## 1.4  Structure of This Monograph

This monograph fully introduces the principles, methods, and applications of RTK for the first time. At first, the brief introduction of currently existent GNSS is given. Then the typical error sources in RTK are discussed and analyzed, and the estimation methods widely used in RTK are given in detail. As the most important error, the ambiguity is studied systematically in this monograph. At second, several crucial issues are studied comprehensively, including cycle slip, stochastic model, unmodeled error, and quality control. Last but not least, almost all important RTK modes are presented. Specifically, they are the long-range RTK (LRTK), Extra-wide-lane RTK (ERTK), RTK with BDS short-message communication (SMC-RTK), antenna-array aided RTK (ARTK), cost-effective RTK (CRTK), and state-space-representation-based RTK (SSR-RTK). Highlights of each chapter are summarized as follows.

This chapter provides an overview of GNSS. It begins with an introduction to the major satellite navigation systems, focusing on global navigation systems such as GPS, BDS, Galileo, and GLONASS, as well as regional or augmented navigation systems like QZSS, IRNSS, and SBAS. Each system has its own unique structure, functions, and signal specifications, contributing to the global satellite navigation landscape. Next, the chapter traces the evolution of positioning technology, from SPP to PPP and RTK, leading to their derivative technologies, such as NRTK and PPP-RTK. Their respective operating principles are also discussed. Finally, the chapter highlights the unique advantages of RTK and provides a comparative analysis of the strengths and weaknesses of mainstream high-precision positioning methods.

Chapter 2 discusses the GNSS error sources in RTK. This chapter mainly introduces some errors that affect the RTK accuracy. The GNSS observation equations are given firstly, where the errors are all given. The errors can be mainly divided into three parts: the satellite-related errors (satellite ephemeris error, satellite clock offset, satellite hardware delay, phase windup, PCO, and PCV of the satellite antenna), the path-related errors (ionospheric error and tropospheric error) and the receiver-related errors (receiver clock offset, receiver hardware delay, multipath effect and receiver noise, phase windup, PCO, and PCV of the receiver antenna). Due to the characteristics of RTK, orbit error and clock offset, ionospheric delay, and tropospheric delay are discussed in detail. Finally, since other errors can be basically eliminated or ignored, multipath effect cannot be mitigated. This chapter will dedicate relatively substantial coverage to multipath effects.

Chapter 3 systematically studies the estimation methods in RTK. It first establishes the mathematical foundation through least squares adjustment, with rigorous analysis of its statistical characteristics and geometric interpretation. The discussion then progresses to sequential adjustment techniques, where the approaches are formulated for time-independent scenarios, while a parameter estimation framework is developed for time-dependent cases. The core focus resides in Kalman filter theory, which not only elucidates the conventional Kalman filter derivation but also introduces an enhanced window-recursive estimation algorithm incorporating sliding window mechanisms. The chapter objectively assesses the dynamic adaptability, computational efficiency, and precision stability of these methods, ultimately establishing their complementary relationships in modern navigation system implementations.

Chapter 4 offers an in-depth examination of integer ambiguity resolution in GNSS positioning, a key process for achieving high-precision solutions. It begins by presenting the mixed-integer GNSS model, which serves as the foundational framework for all integer ambiguity resolution methods. The chapter then explores various strategies for integer estimation, addressing how unmodeled errors affect ambiguity resolution. Following this, the chapter outlines methods for evaluating and validating integer solutions. It also highlights the practical benefits of partial ambiguity resolution techniques, particularly in real-world GNSS applications. In response to the growing adoption of multi-frequency and multi-GNSS systems, the chapter further discusses methods for resolving multi-frequency ambiguities, providing valuable theoretical insights for GNSS users.

Chapter 5 establishes a comprehensive methodological framework for cycle slip detection and repair. Commencing with multi-frequency signal processing architecture, the chapter develops a geometry-based ionosphere-weighted estimator that innovatively integrates single-differenced ionospheric biases for effective cycle slip and data gap repair, validated by extensive experiments. Progressing to single-frequency scenarios, the analysis introduces a dual-domain detection paradigm combining positional polynomial fitting in coordinate domain with partial cycle slip resolution in ambiguity domain. The results demonstrate significant improvements in accuracy and reliability, ensuring continuous high-precision positioning across various conditions.

Chapter 6 constructs a rigorous theoretical framework for advanced stochastic modeling in RTK. The discussion first introduces a variance and covariance component estimation method, where an efficient approach is also given. This technique meticulously quantifies measurement noise, ensuring that least squares adjustments yield unbiased estimates with minimal variance. The chapter argues for approaches that address the unique constellation and signal characteristics of BDS. Comprehensive experimental validations confirm that these tailored models significantly improve positioning reliability under diverse applications.

Chapter 7 establishes a systematic approach to addressing unmodeled errors in GNSS observations, recognizing their inevitable presence due to their complex spatiotemporal characteristics. Building on prior findings that suggest the existence of such errors, the chapter focuses on both their detection and mitigation. Commencing with an in-depth analysis of error detection, advanced methodologies to identify unmodeled disturbances within GNSS measurements is introduced firstly. Following this, compensation strategies are explored, presenting innovative techniques to minimize their impact on positioning accuracy. Through a combination of theoretical insights, experimental validation, and practical applications, the chapter provides a comprehensive framework for understanding, quantifying, and addressing unmodeled errors, ultimately enhancing the positioning performance.

Chapter 8 studies the quality control methods for RTK, introducing robust estimation and the detection, identification, adaptation method for outlier management. Outliers in GNSS data necessitate specialized processing to mitigate their biasing effects on least-squares estimators. Two principal outlier detection frameworks are outlined, categorized by based on whether outliers follow a non-stochastic (mean shift model) or stochastic (variance inflation model). It also emphasizes the importance of realistic stochastic models in statistical reliability testing, which can minimize false alarms and enhance detection accuracy. Proper modeling of physical correlations, such as those related to satellite elevation and observation time, is shown to significantly improve the reliability of GNSS positioning tests.

Chapter 9 presents the LRTK: long-range RTK. It explores the capability of long-range single-baseline RTK with multi-frequency multi-constellation observations in high-precision positioning from both theoretical and practical aspects. Regarding the big city with Shanghai-like area, Tongji real-time kinematic (TJRTK) is able to provide centimeter-level positioning service in Shanghai based on multi-frequency and multi-constellation LRTK instead of NRTK. The costs of the LRTK infrastructure maintenance needed by NRTK will be dramatically reduced by TJRTK.

Chapter 10 presents the ERTK: extra-wide-lane RTK. It dedicates to fully exploit the RTK capability of virtual extra-wide-lane (EWL) signals over long baselines, which is referred to as ERTK. Ionosphere-ignored and ionosphere-float models which are two ERTK models are formulated. And then the ionosphere-smoothed ERTK model is introduced. In addition, the ERTK equivalence of using any two EWL observations is proven, and the condition of selecting either ionosphere-ignored or ionosphere-float model is discussed. Through the experiment and analysis, we find some useful remarks. The ERTK is promising and can be applied in many applications.

Chapter 11 presents the SMC-RTK: RTK with short-message communication. This chapter introduces the SMC-RTK method, which can realize high-precision positioning. And the SMC-RTK technique overcomes the problem of communication at sea by sending corrections through the BDS short message service based on an efficient encoding and broadcasting strategy. Moreover, SMC-RTK reduces the dependence on reference stations by using only a single reference station. The service radius of the single reference station is extended to 300 km by applying an asynchronous, time-differenced, precise ephemerides-aided and ionosphere-weighted positioning model.

Chapter 12 presents the ARTK: antenna-array aided RTK. It explores the potential benefits of antenna-array aided PPP to the long-range RTK, which is referred to as array-aided RTK. We formulate the platform array model and show how its data can be reduced. Then, three different ionosphere-weighted differential array models are described, and closed-form formulae for their ambiguity variance matrices are presented. These matrices determine the success rates for estimating the integer ambiguities. Finally, the ARTK model for platforms is outlined, where -static and kinematic experiments are also presented.

Chapter 13 presents the CRTK: cost-effective RTK. It addresses the challenges and factors influencing ambiguity resolution in smartphone-based RTK. The chapter begins by formulating the methods for estimating smartphone observation precisions, followed by a detailed explanation of the experimental setup and datasets used throughout the study. It then analyzes how smartphone brands, operating systems, and antenna attitudes impact the ambiguity integer property, data quality, and positioning performance. Two kinematic experiments are presented to demonstrate the ambiguity resolution and positioning performance for different smartphone models with both embedded and external antennas.

Chapter 14 presents the SSR-RTK: RTK with SSR corrections. This chapter proposes a novel SSR-RTK method achieving fast ambiguity resolution through PPP with B2b and supplementary SSR corrections. A full-rank undifferenced and uncombined PPP-B2b model is formulated. After analyzing PPP-B2b product characteristics, satellite-specific phase biases and atmospheric corrections from a single reference station are integrated to augment positioning. This enables single-station SSR-RTK within PPP-B2b infrastructure. Kinematic experiments validate the positioning performance, with discussions on ambiguity resolution and atmospheric augmentation methods.

# References


1. Parkinson BW, Enge PK, Spilker JJ (1996) Global positioning system: theory and applications, vol I. AIAA, Florida
2. Misra P, Enge P (2006) Global positioning system: signals. Measure Perf, pp 466–490
3. Green CGB, Massatt PD, Rhodus NW (1989) The GPS 21 primary satellite constellation. Navigation: J Inst Navigat 36:9–24
4. DOD (2008) Global positioning system standard positioning service performance standard, 4th edn
5. CSNO (2012) Report on the development of BeiDou Navigation Satellite System, Version 2.1
6. CSNO (2021) BeiDou Navigation Satellite System Open Service Performance Standard, Version 3.0
7. CSNO (2020a) BeiDou Navigation Satellite System Signal in Space Interface Control Document Open Service Signal B2b, Version 1.0
8. Yang Y, Lu M, Han C (2016) Some notes on interoperability of GNSS. Acta Geodaetica et Cartographica Sinica 45:253
9. Yang Y, Liu L, Li J, Yang Y, Zhang T, Mao Y, Sun B, Ren X (2021) Featured services and performance of BDS-3. Sci Bull 66:2135–2143
10. Yang Y, Mao Y, Sun B (2020) Basic performance and future developments of BeiDou global navigation satellite system. Satellite Navigat 1:1
11. Yang Y (2010) Progress, contribution and challenges of compass/Beidou Satellite Navigation System. Acta Geodaetica et Cartographica Sinica 1–6
12. GSA (2020) Galileo High Accuracy Service (HAS) Info Note
13. OS SIS ICD (2015) European GNSS (Galileo) Open Service Signal in Space Interface Control Document
14. OS SIS ICD (2021) European GNSS (Galileo) Open Service Signal-in-Space Interface Control Document
15. Fatkulin R, Kossenko V, Storozhev S, Zvonar V, Chebotarev V (2012) GLONASS space segment: satellite constellation, GLONASS-M and GLONASS-K spacecraft, main features. Proceedings of the ION GNSS 2012, pp 3912–3930
16. Walker JG (1984) Satellite constellations. J Br Interplanet Soc 37:559–572
17. Sadovnikov MA, Shargorodskiy VD (2014) Stages of development of stations, networks and SLR usage methods for global space geodetic and navigation systems in Russia. Proceedings of the 19th ILRS workshop, pp 1–23
18. Russian Institute of Space Device Engineering (2008) Global Navigation Satellite System GLONASS—interface control document
19. Murai Y (2014) Project overview of the quasi-zenith satellite system. Proceedings of the ION GNSS+ 2014, pp 8–12
20. CAO (2016a) Quasi-Zenith Satellite System Interface Specification—Satellite Positioning, Navigation and Timing Service, IS-QZSS-PNT-001
21. CAO (2016b) Quasi-Zenith Satellite System Interface Specification—Centimeter Level Augmentation Service, IS-QZSSL6-001

22. CAO (2016c) Quasi-Zenith Satellite System Interface Specification—Positioning Technology Verification Service, IS-QZSS-TV-001
23. Ganeshan AS, Rathnakara SC, Gupta R, Jain AK (2005) Indian Regional Navigation Satellite System (IRNSS) concept. J Spacecr Technol 15:19–23
24. ICAO (2006) International Standards and Recommended Practices (SARPS), Annex 10-Aeronautical Telecommunications
25. CSNO (2020b) BeiDou navigation satellite system signal in space interface control document Satellite-Based Augmentation System service signal BDSBAS-B1C, Version 1.0
26. CSNO (2022) Interface specification for signal in space of BeiDou satellite-based augmentation system—dual-frequency augmentation service signal BDSBAS-B2a
27. EUROCAE (2019) ED-259—Minimum Operational Performance Standards for Galileo—Global Positioning System—Satellite-Based Augmentation System Airborne Equipment
28. RTCA (2016) Minimum operational performance standards for global positioning system/satellite-based augmentation system airborne equipment, RTCA DO229E

# Chapter 2
# GNSS Error Sources in RTK

## 2.1 GNSS Observation Equations

To generate pseudorange observations, a Global Navigation Satellite System (GNSS) receiver measures the apparent signal travel time from the navigation satellite to the user. The receiver delay lock loop (DLL) generates a replica of the signal code based on its internal frequency source and aligns it with the received signal. The required time shift represents the apparent transit time, modulo the code chip length. This shift is then combined with the number of complete code chips, complete code repeats, and additional information from the satellite navigation data to determine the unambiguous apparent signal travel time. Multiplying this by the speed of light yields the pseudorange.

In addition to pseudorange measurements, the receiver also measures the signal carrier phase using its phase lock loop (PLL). The receiver generates a replica of the carrier signal, aligns it with the incoming carrier from the satellite, and measures the fractional phase shift between the two signals. When the range between the user and the satellite changes by more than one wavelength cycle, the receiver counts the full cycles, providing a continuous measurement. Due to the short wavelength of the carrier phase, approximately 19–25 cm, depending on the frequency, carrier-phase measurements are significantly more precise than pseudorange measurements.

Typically, the undifferenced and uncombined (UDUC) GNSS observation equations for receiver $r$ and satellite $s$ on frequency $j$ at epoch $k$ are expressed as follows [1, 2]

$$
\begin{aligned}
P_{r,j}^{s}(k) = {}& \varrho_{r}^{s}(k, k - \delta) + cdt_{r}(k) - cdt^{s}(k - \delta) + D_{r,j}(k) \\
& - d_{j}^{s}(k - \delta) + \iota_{r,j}^{s}(k) + \tau_{r}^{s}(k) + M_{r,j}^{s}(k) + \varepsilon_{r,j}^{s}(k)
\end{aligned}
\tag{2.1}
$$

$$
\begin{aligned}
\Phi_{r,j}^{s}(k) = {}& \varrho_{r}^{s}(k, k - \delta) + cdt_{r}(k) - cdt^{s}(k - \delta) + B_{r,j}(k) \\
& - b_{j}^{s}(k - \delta) + \lambda_{j} a_{r,j}^{s} - \iota_{r,j}^{s}(k) + \tau_{r}^{s}(k) + m_{r,j}^{s}(k) + \epsilon_{r,j}^{s}(k)
\end{aligned}
\tag{2.2}
$$

© The Author(s) 2025

B. Li et al., *GNSS Real-Time Kinematic Positioning*, Navigation: Science and Technology 17, https://doi.org/10.1007/978-981-96-9116-6_2

The notations in (2.1) and (2.2) are as follows.

$P_{r,j}^s$ denotes the code/pseudorange observation (m),
$\Phi_{r,j}^s$ denotes the phase/carrier phase observation (m),
$\varrho_r^s$ denotes the satellite-to-receiver range (m),
$\delta$ denotes the signal travel time (s),
$c$ denotes the speed of light in a vacuum (m/s),
$dt_r$ denotes the receiver clock offset (s),
$dt^s$ denotes the satellite clock offset (s),
$D_{r,j}$ denotes the receiver code hardware delay (m),
$d_j^s$ denotes the satellite code hardware delay (m),
$B_{r,j}$ denotes the receiver phase hardware delay (m),
$b_j^s$ denotes the satellite phase hardware delay (m),
$\iota_{r,j}^s$ denotes the ionospheric delay (m),
$\tau_r^s$ denotes the tropospheric delay (m),
$\lambda_j$ denotes the wavelength (m/cycle),
$a_{r,j}^s$ denotes the phase ambiguity (cycle),
$M_{r,j}^s$ denotes the code multipath effect (m),
$m_{r,j}^s$ denotes the phase multipath effect (m),
$\varepsilon_{r,j}^s$ denotes the code noise (m),
$\epsilon_{r,j}^s$ denotes the phase noise (m).

Unlike the constellations that transmit signals based on the Code Division Multiple Access (CDMA), the frequency in constellation based on the Frequency Division Multiple Access (FDMA) like GLONASS is different per channel, hence there exist inter-frequency code bias and inter-frequency phase bias, which are not shown in Eqs. (2.1) and (2.2). The other error terms such as phase center offset (PCO) and phase center variation (PCV), phase windup, solid earth tide, ocean tide loading, pole tide, relativistic effect, and earth rotation are assumed to be corrected in advance. It is worth noting that there may exist unmodeled errors in GNSS code and phase observations mainly due to the complicated mechanism and limited knowledge on them, which will be discussed in the following chapters.

According to Eqs. (2.1) and (2.2), there are various error sources that may contaminate the GNSS observations. They can be divided into four parts, the satellie-related errors, the signal propagation errors, the receiver-related errors, and other errors. Specifically, the satellite-related errors refer to the errors introduced during the orbit and clock determination and the signal production of the satellites, which mainly include the ephemeris error, the satellite clock offset, the satellite hardware delay, and the phase windup, PCO, and PCV of the satellite antenna. The signal propagation errors refer to the errors introduced during the propagation of satellite signals, which mainly include the ionospheric delay and the tropospheric delay. The receiver-related errors refer to the errors introduced during the signal reception, demodulation and interpretation of the receiver, which mainly include the receiver clock offset, the receiver hardware delay, the multipath effect, and the phase windup, PCO, and PCV of the receiver antenna. The other errors refer to various additional factors that may affect GNSS observations, such as solid earth tide, ocean tide loading, pole tide, relativistic effect, earth rotation, and other geophysical or environmental influences. Table 2.1 and Fig. 2.1 are the error table and error diagram of GNSS observations, respectively.

Considering two receivers observing the same satellites at the same nominal times, three types of differences can be computed from these observations. The

**Table 2.1**  Various error sources that may contaminate the GNSS observations

| Error source | Main error classification |
| --- | --- |
| Satellite | Ephemeris error |
| | Satellite clock offset |
| | Satellite hardware delay |
| | Phase windup, PCO, and PCV of the satellite antenna |
| Signal propagation | Ionospheric delay |
| | Tropospheric delay |
| Receiver | Receiver clock offset |
| | Rceiver hardware delay |
| | Multipath effect and receiver noise |
| | Phase windup, PCO, and PCV of the receiver antenna |
| Other | Solid earth tide |
| | Ocean tide loading |
| | Pole tide |
| | Relativistic effect |
| | Earth rotation |

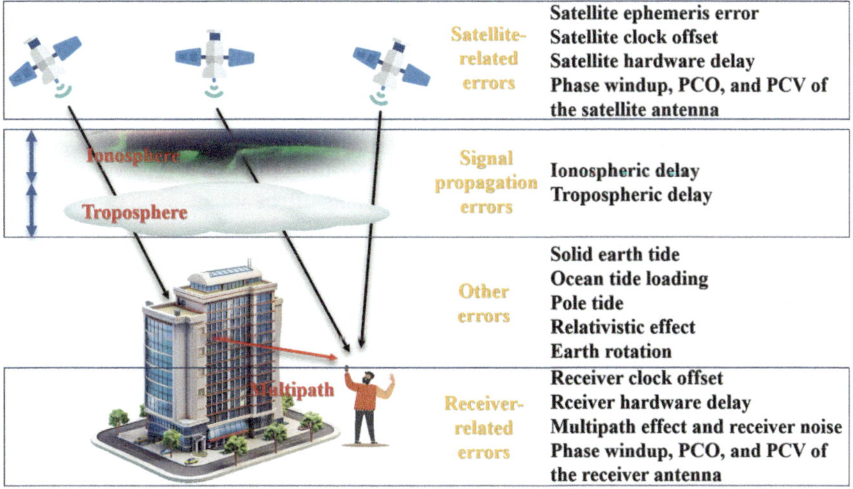


**Fig. 2.1**  The various GNSS error sources

first is the between-receiver difference, obtained by differencing the observations of two stations tracking the same satellites. Another is the between-satellite difference, which results from differencing observations from the same station but across different satellites. The third type, known as the between-time difference, is the difference between observations from the same station tracking the same satellite at different epochs.

A double difference can be formed when two receivers observe two satellites simultaneously, or at least near simultaneously. One can either difference two between-receiver differences or two between-satellite differences. In real-time kinematic (RTK) positioning, thanks to the between-receiver and between-satellite double differenced (DD) operator, the DD GNSS observation equations for CDMA constellations read [1, 2]

$$P_{rg,j}^{sq}(k) = \varrho_{rg}^{sq}(k, k - \delta) + \iota_{rg,j}^{sq}(k) + \tau_{rg}^{sq}(k) + M_{rg,j}^{sq}(k) + \varepsilon_{rg,j}^{sq}(k) \tag{2.3}$$

$$\Phi_{rg,j}^{sq}(k) = \varrho_{rg}^{sq}(k, k - \delta) + \lambda_j a_{rg,j}^{sq} - \iota_{rg,j}^{sq}(k) + \tau_{rg}^{sq}(k) + m_{rg,j}^{sq}(k) + \epsilon_{rg,j}^{sq}(k) \tag{2.4}$$

The new notations in (2.3) and (2.4) are as follows.

$s$ denotes the reference satellite,
$q$ denotes the common satellite,
$r$ denotes the base station,
$g$ denotes the rover station.

Compared with the UDUC GNSS observation equations, the receiver and satellite clock offsets, the receiver and satellite code hardware delays, and the receiver and satellite phase hardware delays can all be eliminated in RTK. However, in practical RTK applications, perfect simultaneity is often not achievable due to slight time offsets between observations at the base and rover stations. These asynchronous measurements arise from factors such as processing delays. To mitigate this issue, interpolation techniques or high-rate synchronized data logging can be employed, ensuring that the time offsets remain within acceptable limits for precise positioning. In addition, some error sources such as ionospheric and tropospheric delays are mitigated to a great extent. Therefore, in RTK, the orbit error and clock offset, ionospheric delay, tropospheric delay, and multipath effect are the main error sources which will hinder the precision and reliability of RTK. To have a insight into the effects of these errors on RTK, we try to introduce them in detail in the following texts.

## 2.2  Orbit Error and Clock Offset

The orbits of GNSS satellites are theoretically well-known with high precision; however, discrepancies often exist between the real satellite orbits provided by satellite ephemerides and their calculated counterparts. According to the theory of artificial satellite orbits, if a satellite orbit is precisely known, its position and velocity in space can be determined. Conversely, if the position and velocity of a satellite are known, its orbit can be determined.

Since satellite positions are determined through continuous tracking and monitoring by ground-based systems, orbit errors primarily stem from inaccuracies in satellite ephemerides. As these orbit errors affect receiver-to-satellite distance measurements, the satellite range can be expanded accordingly

$$\rho_r^s = ||X_r - X^s|| + \mathrm{los}_r^s \cdot \mathrm{rec}_r + \mathrm{orb}_r^s \tag{2.5}$$

The new notations in (2.5) are as follows.

$X_r$ denotes the given coordinates of the receiver $r$.

$X^s$ denotes the given coordinates of the satellite $s$.

$\mathrm{los}_r^s$ denotes the line-of-sight (LOS) vector.

$\mathrm{rec}_r$ denotes the receiver coordinate error.

$\mathrm{orb}_r^s$ denotes the satellite orbit error introduced by the satellite ephemeris error.

$|| \cdot ||$ denotes the operator of calculating the quadratic norm.

GNSS satellite ephemerides contain correction terms for satellite orbits and clock errors. There are two main types of ephemerides. The first is the broadcast ephemeris, which is predicted and provides meter-level orbit accuracy to users. It is directly modulated onto the satellite signal and is accessible worldwide. The second is the precise ephemeris, which includes ultra-rapid, rapid, and final products. These precise products can be obtained from International GNSS Service (IGS) Analysis Centers (ACs) as of 2025, including Natural Resources Canada (EMR), Wuhan University (WHU), Geodetic Observatory Pecny (GOP), the Space Geodesy Team of CNES (GRG), the European Space Agency (ESA), GeoForschungsZentrum (GFZ), Geospacial Information Authority of Japan and Japan Aerospace Exploration Agency (JGX), the Center for Orbit Determination in Europe (CODE), the Jet Propulsion Laboratory (JPL), the Massachusetts Institute of Technology (MIT), the National Geodetic Survey (NGS), the Scripps Institution of Oceanography (SIO), and the U.S. Naval Observatory (USNO). Additionally, the Tongji BeiDou Analysis Center (TJBAC), established by the authors, provides precise orbits with centimeter-level accuracy and clock offsets with 0.1-ns-level precision. Although broadcast ephemerides offer lower precision, their direct modulation onto satellite signals ensures global availability.

As shown in Fig. 2.2, the satellite position calculated from the ephmerides has a discrepancy to the real satellite position. The discreancy is denoted as $\vec{\vartheta}$, then the projection of $\vec{\vartheta}$ on the LOS from the satellite $s$ to the receiver $r$ is the orbit error

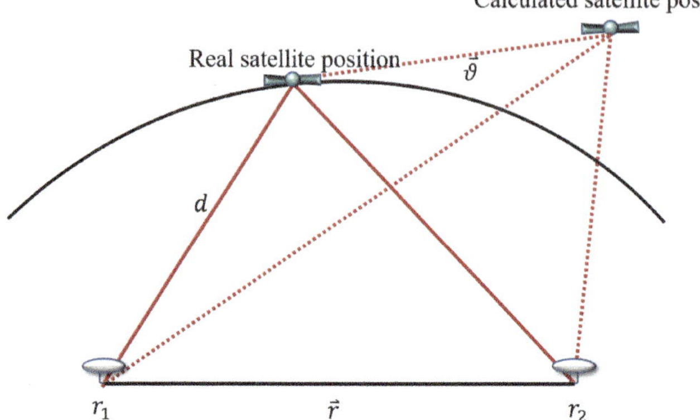


**Fig. 2.2** Orbit error of a baseline

$orb_r^s$. For a baseline in the RTK application, the observations from two receivers are subtracted. Hence, the orbit errors of two receivers are combined for the single differenced (SD) observations. The between-receiver SD orbit error can be approximated as below

$$\text{orb}_{r_1 r_2}^s = \frac{\vec{r}}{D} \| \vec{\vartheta} \| \tag{2.6}$$

where $\vec{r}$ is the vector from the receiver $r_1$ to $r_2$ and $D$ is the approximated distance from the satellite to the receiver. If we take $D = 20{,}000$ km, $\| \vec{r} \| = 20$ km and $\| \vec{\vartheta} \| = 1$ m, the SD orbit error of the satellite $s$ is about 1 mm, which is ignorable for RTK. However, when the baseline extends to more than 100 km, the orbit error can be centimeter-level. In such case, the precise ephemerides are recommended to minimize the effects of the orbit error.

In RTK, it is theoretically assumed that the observations from the rover and base stations are perfectly synchronized, meaning they share identical observation timestamps. However, in practice, real-time synchronization is rarely achieved due to inevitable time delays or asynchronicity, commonly referred to as the age of differential. RTK operating under these conditions is known as asynchronous RTK. Because the rover and base stations record observations at different epochs, positioning errors in asynchronous RTK differ from those in traditional RTK, with satellite clock offsets being a primary source of error. Furthermore, in asynchronous RTK, single differencing between stations effectively mitigates the impact of satellite orbit errors, ionospheric delays, and tropospheric delays. However, satellite clock offsets cannot be eliminated and will introduce additional positioning errors.

## 2.3 Ionospheric Delay

The ionosphere, located at altitudes between approximately 50 and 1000 km above the Earth surface, is a region where neutral gas molecules are ionized under the influence of solar ultraviolet rays, X-rays, $\gamma$-rays, and high-energy particles. This ionization process results in a high concentration of free electrons and positive ions, forming an ionized layer. The electron density in the ionosphere depends on the intensity of solar radiation and atmospheric density. Moreover, it is influenced by factors such as altitude, solar and celestial radiation intensity, seasonal variations, and geographical location [3].

In atmospheric physics, a medium is considered dispersive if the propagation speed of electromagnetic waves within it depends on their frequency. This dispersion phenomenon arises due to the interaction between the medium internal electric field and the external electric field of the incident wave. Like other electromagnetic signals, GNSS signals are affected by the dispersive properties of the ionosphere. As they travel through this region, signal paths experience bending (although this curvature has a negligible impact on ranging results and is generally ignored), and propagation speeds are altered, leading to measurement errors known as ionospheric delays.

In GNSS measurement, the code pseudorange measurement is related to the group velocity, and the carrier phase measurement is related to the phase velocity. When the electromagnetic wave passes through the ionosphere, the propagation path error of code pseudorange measurement $\Delta\rho_P$ and carrier phase measurement $\Delta\rho_\phi$ caused by the change of the refractive index can generally be expressed as [4, 5]

$$\Delta\rho_P = \frac{40.3}{f^2} \int_0^S N_e dS + \frac{2.2566 \cdot 10^{12}}{f^3} \int_0^s \int N_e B \cos\theta dS$$
$$- \frac{2437.4}{f^4} \int_0^s \int N_e^2 dS \tag{2.7}$$

$$\Delta\rho_\phi = -\frac{40.3}{f^2} \int_0^S N_e dS - \frac{1.1283 \cdot 10^{12}}{f^3} \int_0^s \int N_e B \cos\theta dS$$
$$- \frac{812.47}{f^4} \int_0^s \int N_e^2 dS \tag{2.8}$$

where $f$ denotes the frequency value; $S$ and $N_e$ denote the propagation path and electron density, respectively; $B$ denotes magnetic field strength of the geomagnetic field; $\theta$ denotes angle between the geomagnetic field direction and the electromagnetic wave propagation path. Actually, $\int_0^S N_e dS$ is the total electron content (TEC) of the electromagnetic wave on its propagation path. For the same ionosphere, the TEC in the direction from a station to each satellite is different. The smaller the satellite elevation, the longer the propagation path of the satellite signal in the ionosphere, and the larger the TEC value. That is, when the propagation direction of the electromagnetic wave deviates from the zenith direction, the TEC will increase obviously. There is a minimum value in the TEC in all directions of the station, that is, the TEC in the zenith direction, which is called vertical total electron content (VTEC). VTEC has nothing to do with the satellite elevation, and can reflect the overall characteristics of the ionosphere above the station, hence the concept of VTEC is widely used. In GNSS positioning and navigation, the ranging difference caused by the ionospheric delay can reach up to 50 m in the zenith direction, and exceed 150 m when the elevation is 5°. Therefore, the ionospheric delay must be carefully corrected in GNSS applications including RTK.

There are three main approaches to mitigate the ionospheric delay. The first approach is to utilize the dual-frequency correction method. As aforementioned, the ionospheric delays have dispersion characteristics, hence the ionospheric delays between frequencies $i$ and $j$ can be expressed as follows [6]

$$\iota_{r,j}^{s} = \frac{f_i^2}{f_j^2} \iota_{r,i}^{s} = \frac{\lambda_j^2}{\lambda_i^2} \iota_{r,i}^{s} \tag{2.9}$$

This method leverages the ionospheric dispersion properties to establish a dual-frequency ionospheric correction model, commonly known as the ionosphere-free (IF) model. High-precision satellite positioning typically uses the IF model to mitigate the impact of ionospheric delay, achieving effectiveness of no less than 95%. It is important to note that different dual-frequency combinations yield varying correction effects on the ionospheric impact. Moreover, because higher-order ionospheric delays are neglected, a residual error, up to the centimeter level, remains even after dual-frequency correction. This residual error becomes significantly larger if the observations are made at noon when sunspot activity peaks.

The second approach involves using the DD operator to mitigate or even eliminate ionospheric delays. When the baseline between two stations is relatively short (usually less than 30 km) and the atmospheric conditions along the satellite-to-station propagation paths are similar, the systematic error introduced by the atmosphere, including the ionosphere, can be largely canceled out by differencing the observations. Typically, the residual ionospheric delay between these observations does not exceed $10^{-6}$ times the baseline length. However, for very long baselines, ionospheric delays become significant and must be explicitly estimated in the mathematical model.

The third approach is model correction, which can be divided into two main types. The first type is the empirical ionospheric model, usually developed on a global scale. Such models rely on mathematical formulas to describe the spatiotemporal variations of parameters such as electron density, ion density, electron temperature, ion temperature, ion composition, and TEC in the ionosphere. By fitting these formulas to extensive observational data gathered over long periods from ionospheric monitoring stations around the world, an empirical ionospheric delay correction model can be established. Commonly used global empirical models include the Klobuchar model, the International Reference Ionosphere (IRI), the Global Ionospheric Map (GIM), the BeiDou Global Ionospheric Model (BDGIM), and NeQuick. The second type is based on measured observations. GNSS data consist of decimeter-level pseudorange measurements and millimeter-level carrier phase measurements, both affected by the ionosphere. However, because the phase observations include unknown ambiguities, the absolute ionospheric delay cannot be directly determined. Instead, the phase-smoothed pseudorange (PSP) method is typically used to indirectly calculate the VTEC at the ionospheric pierce point (IPP). Since VTEC varies over time and space, directly introducing numerous parameters for its estimation can increase computational complexity and instability. To address this, VTEC is often modeled as a function of time and space, with the function parameters being solved for rather than VTEC itself. This approach reduces the number of parameters in the observation equation and enhances computational efficiency. Common mathematical models used to describe ionospheric VTEC include polynomial functions, spherical harmonic functions, trigonometric series, and multifaceted function models.

In RTK, methods such as the IF model are commonly used to reduce the impact of first-order ionospheric delay. However, the interference from second-order and higher-order ionospheric delays is often overlooked. These higher-order delays tend to manifest as more complex, nonlinear disturbances, which are especially pronounced at high latitudes or under extreme weather conditions. The impact of higher-order ionospheric delays is typically accounted for using the final GIM products provided by the IGS Analysis Center CODE. However, there is limited research on using TEC derived from GNSS observations themselves for correction, and even less discussion on the alignment between geomagnetic models and TEC. Additionally, phenomena such as scintillation, magnetic storms, and plasma bubbles can occur in the ionospheric environment, causing severe fluctuations in the signal and further degrading positioning accuracy.

## 2.4 Tropospheric Delay

The tropospheric delay in GNSS positioning and navigation refers to the signal delay caused when the electromagnetic wave passes through the non-ionized neutral atmosphere, including the troposphere and stratosphere, at altitudes below 50 km. Nearly 99% of the mass of the entire atmosphere is concentrated in this layer. The troposphere is in direct contact with the ground and receives radiant energy from it. Since more than 80% of the neutral atmospheric delay occurs in the troposphere, the signal delay in the neutral atmosphere is collectively referred to as the tropospheric delay [7]. The density of the atmosphere in the troposphere is higher than that in the ionosphere. Similar to the ionosphere, electromagnetic waves bend and delay their propagation paths as they travel through the troposphere, distorting distance measurements.

Because the troposphere is neutral, it can be considered non-dispersive for electromagnetic wave frequencies below 30 GHz. Thus, the propagation velocity of electromagnetic waves in this neutral atmosphere is independent of frequency. Unlike ionospheric delay, both code pseudorange and carrier phase measurements are equally affected by the neutral atmosphere. For GNSS measurements, the dual-frequency observation method cannot be used to eliminate tropospheric delay. Instead, the tropospheric delay can only be estimated by integrating the atmospheric refraction coefficient along the entire signal propagation path.

The refractive index of the troposphere is closely related to atmospheric pressure, temperature, and humidity. Due to the strong convective effect of the atmosphere in this layer, and the complex changes of atmospheric pressure, temperature, humidity, and other factors, it is still difficult to accurately model the tropospheric refractive index and its changes. In general, the refraction index of tropospheric delay $N_T$ reads [8, 9]

$$N_T = N_h + N_w \tag{2.10}$$

where $N_h$ and $N_w$ denote the hydrostatic and wet components, respectively. According to (2.10), it shows that the observed delay caused by tropospheric delay can be divided into two parts: hydrostatic delay and wet delay. When the electromagnetic wave propagates along the zenith direction of the ground observation station, the zenith tropospheric delay (ZTD) $\Delta\rho^Z$ reads [10–12]

$$\Delta\rho^Z = \int_0^S N_T dS = \Delta\rho_h^Z + \Delta\rho_w^Z \tag{2.11}$$

where $\Delta\rho_h^Z$ and $\Delta\rho_w^Z$ denote the ZTD caused by the hydrostatic delay and wet delay, respectively.

In GNSS applications, a mapping function is typically used to project the $\Delta\rho^Z$ ZTD onto the signal propagation path at any given satellite elevation. The variation range of the tropospheric delay in the zenith direction at sea level is approximately 2.30–2.60 m. At a satellite elevation of $3°$, the tropospheric delay can reach 50 m. The contribution of the wet component of the atmosphere is usually much smaller than that of the hydrostatic component, which accounts for approximately 90% of the total tropospheric delay. While water vapor is primarily concentrated within 2 km above the ground, its changes over time and space are complex and irregular, making it difficult to accurately describe or estimate the influence of the hydrostatic component. Given that the tropospheric delay is frequency-independent and the complex variations of the troposphere, especially the wet delay, it is a significant factor limiting future multi-frequency and multi-constellation GNSS precise positioning, including RTK.

Three main approaches exist to mitigate tropospheric delays. The first is to use external data to directly estimate tropospheric delays. To calculate the tropospheric delay along the signal path, it is necessary to know the atmospheric refractive index at every point along the path. This requires information about meteorological elements, such as temperature, air pressure, and water vapor pressure, at various locations along the propagation path. Available data sources include microwave radiometer observation data, radiosonde data, and numerical weather model data. However, it is challenging to measure the meteorological elements along the signal path, and typically only the meteorological data at ground stations are available.

The second approach involves using the DD operator to mitigate or eliminate tropospheric delays. Like ionospheric delays, if the baseline between two stations is relatively short (usually less than 30 km) and the tropospheric conditions along the propagation paths are similar, the systematic error caused by the troposphere can be minimized through the difference between the observations. However, for long baselines, tropospheric delays cannot be ignored and need to be estimated in the mathematical model.

The third approach is model correction, which includes two main types. The first type is the tropospheric empirical model. This model uses ground station meteorological data (such as temperature, air pressure, and water vapor pressure) and location information (e.g., latitude, longitude, height) to calculate the tropospheric delay along the signal path. The tropospheric empirical model consists of two parts: the ZTD model and the mapping function. Common ZTD models include the Saastamoinen, Hopfield, and New Brunswick 3 (i.e., UNB3) models. The widely used mapping functions include the Ifadis, Chao, Neill, Davis, Herring, Black, and Vienna Mapping Function 1 (VMF1)/Vienna Mapping Function 3 (VMF3). To account for the asymmetry of the atmosphere, gradient mapping functions can also be applied. The second type of model uses measured observations. ZTD estimators are often modeled as functions of time and space, similar to the modeling of VTEC.

However, in RTK, the traditional tropospheric delay model primarily addresses the dry component of the tropospheric delay, while the wet component is more difficult to handle. The wet delay is mainly caused by water vapor, which is difficult to model. Water vapor varies with climate, meteorological conditions, and geographical location, resulting in strong temporal and spatial variability. In the case of long baselines or large height differences, residual tropospheric delay can become quite significant and can be corrected using precise products and refined parameter estimation. However, introducing tropospheric parameters alongside coordinates can result in a serious ill-conditioned model due to its strong correlation with the height parameter. As a result, precise solutions require sufficient observation accumulation [13]. Additionally, tropospheric delays can sometimes lead to abnormal phenomena such as turbulence, bubbles, and cyclones, which further affect positioning accuracy.

## 2.5  Multipath Effect

Theoretically, what the receiver should receive is only the signal directly from the satellite, but because the signal tends to be reflected, diffracted, and even occluded near the station and generate an indirect signal, the signal received by the receiver not only contains the direct signal, indirect signals are also included, which can be called multipath effect. As usual, the multipath effects include traditional multipath, diffraction, and even none-line-of-sight (NLOS) reception [14]. In GNSS carrier phase observations, there always exist unmodeled errors mainly due to their spatiotemporal complexity [15–17]. Unlike the other types of unmodeled errors, the multipath cannot be effectively mitigated by the DD technique. Therefore, the multipath is one of the major concerns for high-precision GNSS applications.

Multipath effects will directly affect the accuracy of pseudorange and phase observations. The influence of multipath on pseudorange observations is usually between 10 and 20 m, and even up to 100 m in severe cases [18, 19]. In addition, when multipath is severe, it will also cause signal loss of lock. For the phase observations, the multipath is usually between a few millimeters and a few centimeters. For instance, multipath caused by reflection can reach a quarter cycles of the wavelength at most, while the multipath caused by diffraction can be as large as one cycle of wavelength or even more. Therefore, according to the needs of positioning accuracy, it is necessary to pay attention to whether this error can be ignored in practical applications. However, the NLOS reception has a larger range of variation, and the amplitude can reach hundreds of meters, and there is no upper limit in theory.

Apart from the NLOS reception, the multipath falls into two categories: the reflective multipath and the diffractive multipath. In general, the multipath signal $S_m$ can be formulated as [20–22]

$$S_m = A_d \cos \varphi + A_i \cos(\varphi + \Delta\varphi) \tag{2.12}$$

where $A_d$ and $A_i$ denote the amplitudes of direct and indirect signals, respectively; $\varphi$ denotes the phase of the direct signal, and $\Delta\varphi$ denotes the phase shift delayed by the indirect signal. It leads to the relations as follows

$$\Delta\varphi_m = \arctan\left(\frac{A_i \sin \Delta\varphi}{A_d + A_i \cos \Delta\varphi}\right) \tag{2.13}$$

$$A_m = \sqrt{A_d^2 + A_i^2 + 2A_d A_i \cos \Delta\varphi} \tag{2.14}$$

where $\Delta\varphi_m$ and $A_m$ denote the phase shift and amplitude influenced by the multipath (i.e., the composite signal).

Based on (2.13) and (2.14), it can be found that the multipath is influenced by the phase shift, which can be determined in the case of a horizontal reflector

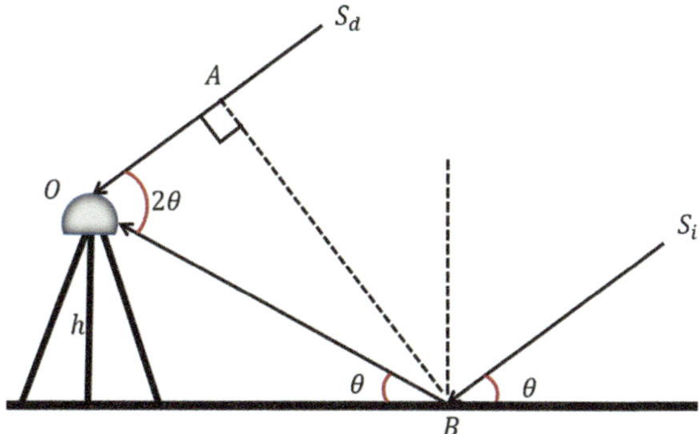

**Fig. 2.3** Schematic diagram of typical multipath



$$S = \mathrm{BO} - \mathrm{AO} = \mathrm{BO}(1 - \cos 2\theta) = \frac{h}{\sin\theta}(1 - \cos 2\theta) = 2h\sin\theta \qquad (2.15)$$

$$\Delta\varphi = \frac{2\pi}{\lambda}s = \frac{2\pi}{\lambda}2h\sin\theta \qquad (2.16)$$

where $S$ and $\lambda$ denote the path delay and wavelength, respectively; $h$ denotes the vertical distance between the antenna phase center and ground, and $\theta$ denotes the elevation. Obviously, the phase shift is a function of wavelength and receiver-satellite geometry, which can be shown in Fig. 2.3.

Since the multipath is difficult to eliminate, the multipath evaluation is essenatial and widely used since one can mitigate or even corret these errors more efficiently while ensuring accuracy in the meantime. Four multipath assessment methods that are especially suitable for the low-cost receivers are comprehensively deduced and assessed. First, two traditional methods are given, i.e., the geometry-free (GF) and IF method, the geometry-based (GB) and ionospheric-corrected (IC) method. Second, two easy-to-implement methods, including the geometry-fixed (GFix) and IC method, the GF and IC method are deduced. Specifically, the first method is the GF and IF method, which are the most widely used methods to assessing multipath effects. This method requires at least two frequencies of phase observations, which can be expressed as [1]

$$\overline{M}^s_{r,i} = P^s_{r,i} - \frac{f_i^2 + f_j^2}{f_i^2 - f_j^2}\Phi^s_{r,i} + \frac{2f_j^2}{f_i^2 - f_j^2}\Phi^s_{r,j} + \frac{f_i^2 + f_j^2}{f_i^2 - f_j^2}\lambda_i a_i$$

$$- \frac{2f_j^2}{f_i^2 - f_j^2}\lambda_i a_i - D_{r,i} + d^s_j + \varepsilon_{\overline{M}^s_{i,i}} \qquad (2.17)$$

where $s$ and $r$ denote the satellite and receiver, $i$ and $j$ the frequencies, $P$ and $\Phi$ the code and phase observations, $\lambda$ and $a$ the wavelength and ambiguity, and $D/d$ and $\varepsilon$ the code hardware delay and observation noise. $\overline{M}$ denotes the equivalent code multipath consisting of the ambiguities of two frequencies, and receiver and satellite code hardware delays. Without cycle slips, the above error terms can be regarded as a constant during some period. Hence, the code multipath can be obtained

$$M_{r,i}^{s} = \overline{M}_{r,i}^{s} - \frac{1}{n}\sum_{k=1}^{n}\left[\overline{M}_{r,i}^{s}(k)\right] \tag{2.18}$$

where $n$ denotes the epoch number. The GF and IF method is widely used due to its high reliability. However, there are mainly two limitations. First, only the peak-to-peak behaviors of the code multipath can be estimated. Second, it can only work when there are two or more frequencies for a certain constellation.

The second method is the GB and IC method based on the relative or standalone mode. In the precise relative mode that is more suitable for low-cost devices, the satellite and receiver clock errors and hardware delays can all be eliminated after double differencing. If coupled with a relatively short baseline, the residual atmospheric delays can also be ignored. Hence, the DD code and phase observation equations are derived as [1]

$$P_{rg,i}^{sq} = \varrho_{rg}^{sq} + M_{rg,i}^{sq} + \varepsilon_{rg,i}^{sq} \tag{2.19}$$

$$\Phi_{rg,i}^{sq} = \varrho_{rg}^{sq} + \lambda_i a_{rg,i}^{sq} + m_{rg,i}^{sq} + {}_{rg,i}^{sq} \tag{2.20}$$

where $\varrho$ denotes the satellite-to-receiver range, $M$ and $m$ the code and phase multipath effects, $\epsilon$ the phase observation noise, and $\langle\cdot\rangle_{qr,i}^{ks} = \langle\cdot\rangle_{r,i}^{s} - \langle\cdot\rangle_{r,i}^{k} - \langle\cdot\rangle_{q,i}^{s} + \langle\cdot\rangle_{q,i}^{k}$, with the reference satellite $q$, and the master receiver $g$. To apply the GB model, the satellite-to-receiver range needs to be converted to coordinate components $x$, $y$ and $z$. Also, the multipath effects and observation noise need to be combined. Thus, the DD multipath can be assessed according to the DD residuals.

Although the GB and IC method is precise enough, it has several limitations. The first one is that a nearby base station is often needed that limits the availability of the method. Also, the ambiguities should be fixed precisely, requiring the non-linear model to be used, which will lead to the potential problems with changes in the number of satellites and selection of the reference satellite. The other main limitation is that the multipath effects from the master receiver and reference satellite are usually included.

The third method is GFix and IC method, in which the observation need to be preprocessed. Specifically, the satellite clock errors should be corrected according to the broadcast or precise ephemeris. After the atmospheric delays are corrected, the code and phase observation equations can be deduced as

$$P_{r,i}^{s} = \varrho_{r}^{s} + dt_r + D_{r,j} + M_{r,i}^{s} + \varepsilon_{r,i}^{s} \tag{2.20}$$

$$\Phi_{r,i}^{s} = \varrho_{r}^{s} + \lambda_i a_{r,i}^{s} + dt_r + B_{r,j} - b_j^{s} + d_j^{s} + m_{r,i}^{s} + \epsilon_{r,i}^{s} \tag{2.21}$$

where $dt_r$ and $B/b$ denote the receiver clock error and phase hardware delay, respectively. Since the receiver clock error and hardware delay need to be eliminated, the between-satellite single differencing is formed

$$P_{r,i}^{sq} = \varrho_r^{sq} + M_{r,i}^{sq} + \varepsilon_{r,i}^{sq} \tag{2.22}$$

$$\Phi_{r,i}^{sq} = \varrho_r^{sq} + \lambda_i a_{r,i}^{sq} - b^{sq,i} + d^{sq,i} + m_{r,i}^{sq} + \epsilon_{r,i}^{sq} \tag{2.23}$$

Based on the precise coordinates of the test station and satellites used, the GFix model is applied. When estimating the coordinates of the satellite, the broadcast or precise ephemeris can be used according to the demands of the users. Hence, the code multipath can be estimated as

$$E\left(M_{r,i}^{sq}\right) = P_{r,i}^{sq} - \varrho_r^{sq} \tag{2.24}$$

where "$E(\cdot)$" denotes the expectation operator. For the phase multipath, the bias term $b = \lambda_i a_{r,i}^{sq} - b^{sq,i} + d^{sq,i}$ can be treated as a constant when there are no cycle slips. The bias term can be removed by averaging over a certain period. Accordingly, the phase multipath effects are estimated as

$$E\left(m_{r,i}^{sq}\right) = \Phi_{r,i}^{sq} - \varrho_r^{sq} - b \tag{2.25}$$

with $b = \frac{1}{n}\sum_{t=1}^{n}\left[\Phi_{r,i}^{sq}(t) - \varrho_r^{sq}(t)\right]$.

The biggest advantage of the GFix and IC method is that it can still work even if there is only one observable satellite in addition to the reference satellite. However, the accuracy is highly dependent on the precision of the coordinates of the test station and used satellites. As usual, the accuracies of orbit and satellite clock are approximately 100 cm and 5 ns when using the broadcast ephemeris. For the precise ephemeris, these values can reach approximately 2.5 cm and 75 ps. Hence it is better to apply the precise ephemeris. Besides, the multipath effects of the reference satellite are also included.

The last method is GF and IC method, and its preprocessing is like that of the GFix and IC method. After preprocessing, the time-differenced operator is used, thus the hardware delays and the phase ambiguities can be regarded as eliminated in case of no cycle slips. The corresponding code and phase observations are deduced as

$$\Delta P_{r,i}^{s} = \Delta\varrho_r^{s} + \Delta dt_r + \Delta M_{r,i}^{s} + \Delta\varepsilon_{r,i}^{s} \tag{2.26}$$

$$\Delta\Phi_{r,i}^{s} = \Delta\varrho_r^{s} + \Delta dt_r + \Delta m_{r,i}^{s} + \Delta\epsilon_{r,i}^{s} \tag{2.27}$$

where "$\Delta$" denotes the time-differenced operator. Since the above observation models are rank deficient, the parameters need to be combined as

$$\Delta P_{r,i}^{s} = \Delta \tilde{\varrho}_{r}^{s} + \Delta M_{r,i}^{s} + \Delta \varepsilon_{r,i}^{s} \tag{2.28}$$

$$\Delta \Phi_{r,i}^{s} = \Delta \tilde{\varrho}_{r}^{s} + \Delta \tilde{\epsilon}_{r,i}^{s} \tag{2.29}$$

with $\Delta \tilde{\varrho}_{r}^{s} = \Delta \varrho_{r}^{s} + \Delta dt_{r}$, $\Delta \tilde{\epsilon}_{r,i}^{s} = \Delta m_{r,i}^{s} + \Delta \epsilon_{r,i}^{s}$. Hence, the time-differenced code multipath effects can be estimated as

$$\mathrm{E}\left(\Delta M_{r,i}^{s}\right) = \Delta P_{r,i}^{s} - \Delta \Phi_{r,i}^{s} \tag{2.30}$$

This method is also convenient and can work under any conditions, but the method also has several limitations. First, only the time-differenced code multipath effects can be depicted, where the undifferenced multipath is missing. Second, the phase multipath effects cannot be obtained.

In real-world applications, there are three main approaches to mitigate multipath effects. The first approach is selecting an ideal observation environment. The simplest way to suppress multipath is to place the station in a low-multipath environment. For instance, it is beneficial to choose an open area that avoids signal reflectors. Users should avoid urban canyon areas with many high-rise buildings. When selecting a station location, opt for areas with rough terrain, such as bushes or grass, and avoid places with high reflection coefficients, such as water, snow, or glass walls. However, this approach is often impractical in real GNSS applications, as ideal environments are rarely available.

The second approach involves using advanced receiver and antenna hardware. If the satellite signal employs right-handed circularly polarized (RHCP) electromagnetic waves, the reflected signal will become left-handed circularly polarized (LHCP). RHCP antennas help attenuate LHCP signals. If possible, a choke ring can be installed beneath the antenna to suppress multipath signals at lower elevations. Additionally, improvements in signal processing methods within the receiver can enhance performance in mitigating multipath effects. Techniques such as narrow correlation, multipath estimation, and multipath elimination using DLL can all improve receiver performance.

The third approach is applying appropriate data processing methods. Given the time-space complexity and unpredictability of multipath changes, data processing is currently the most widely used mitigation technique. One common method is sidereal filtering (SF), which leverages the temporal repeatability of satellite constellations. SF can be implemented either in the coordinate domain or the observation domain. Multipath tends to exhibit periodic repeatability in a static environment, due to the temporal repeatability of GNSS satellites. Another method that can be adopted is the multipath hemispherical map (MHM). The fundamental concept of MHM is that multipath effects exhibit spatial repeatability as long as the station surroundings remain unchanged or relatively stable. However, the original MHM method cannot precisely correct multipath errors. Recent studies have focused on refining MHM for better accuracy. Additionally, multipath can be processed using techniques such as carrier-to-noise power density ratio (C/N0) analysis, wavelet analysis, ray-tracing, and support vector machine. With the advent of multiple frequencies, multipath errors can also be parameterized and reduced in the observation domain [23].

# References


1. Leick A, Rapoport L, Tatarnikov D (2015) GPS satellite surveying. Wiley, New Jersey
2. Teunissen PJG, Montenbruck O (2017) Springer handbook of global navigation satellite systems. Springer, Cham
3. Webb DF, Howard RA (1994) The solar cycle variation of coronal mass ejections and the solar wind mass flux. J Geophys Res: Space Phys 99:4201–4220
4. Datta-Barua S, Walter T, Blanch J, Enge P (2008) Bounding higher-order ionosphere errors for the dual-frequency GPS user. Radio Sci 43:1–15
5. Petrie EJ, Hernández-Pajares M, Spalla P, Moore P, King MA (2011) A review of higher order ionospheric refraction effects on dual frequency GPS. Surv Geophys 32:197–253
6. Hoque MM, Jakowski N (2008) Estimate of higher order ionospheric errors in GNSS positioning. Radio Sci 43:1–15
7. Mendes V de B (1998) Modeling the neutral-atmosphere propagation delay in radiometric space techniques. UNB geodesy and geomatics engineering technical report
8. Thayer GD (1974) An improved equation for the radio refractive index of air. Radio Sci 9:803–807
9. Askne J, Nordius H (1987) Estimation of tropospheric delay for microwaves from surface weather data. Radio Sci 22:379–386
10. Hopfield HS (1969) Two-quartic tropospheric refractivity profile for correcting satellite data. J Geophys Res 1896–1977:4487–4499
11. Saastamoinen J (1972) Atmospheric correction for the troposphere and stratosphere in radio ranging satellites. Proceedings of the AGU, pp 247–251
12. Davis JL, Herring TA, Shapiro II, Rogers AEE, Elgered G (1985) Geodesy by radio interferometry: effects of atmospheric modeling errors on estimates of baseline length. Radio Sci 20:1593–1607
13. Li B, Feng Y, Shen Y, Wang C (2010) Geometry-specified troposphere decorrelation for subcentimeter real-time kinematic solutions over long baselines. J Geophys Res: Solid Earth 115:B11404
14. Zhang Z, Dong Y, Wen Y, Luo Y (2023) Modeling, refinement and evaluation of multipath mitigation based on the hemispherical map in BDS2/BDS3 relative precise positioning. Measurement 213:112722
15. Zhang Z, Li B, Shen Y (2017) Comparison and analysis of unmodelled errors in GPS and BeiDou signals. Geodesy Geodyn 8:41–48
16. Zhang Z, Li B, Shen Y (2018) Efficient approximation for a fully populated variance-covariance matrix in RTK positioning. J Surv Eng 144:04018005
17. Li B, Zhang Z, Shen Y, Yang L (2018) A procedure for the significance testing of unmodeled errors in GNSS observations. J Geod 92:1171–1186
18. Braasch MS (1997) Autocorrelation sidelobe considerations in the characterization of multipath errors. IEEE Trans Aerosp Electron Syst 33:290–295
19. Zhang Z, Yuan H, Li B, He X, Gao S (2021) Feasibility of easy-to-implement methods to analyze systematic errors of multipath, differential code bias, and inter-system bias for low-cost receivers. GPS Solut 25:116
20. Bilich A, Larson KM (2007) Mapping the GPS multipath environment using the signal-to-noise ratio (SNR). Radio Sci 42:1–16
21. Bilich A, Larson KM, Axelrad P (2008) Modeling GPS phase multipath with SNR: case study from the Salar de Uyuni, Boliva. J Geophys Res: Solid Earth 113:B04401

22. Rost C, Wanninger L (2010) Carrier phase multipath corrections based on GNSS signal quality measurements to improve CORS observations. Proceedings of the IEEE/ION position, location and navigation symposium, pp 1162–1167
23. Zhang Z (2021) Code and phase multipath mitigation by using the observation-domain parameterization and its application in five-frequency GNSS ambiguity resolution. GPS Solut 25:144

# Chapter 3
# Estimation Methods in RTK

## 3.1 Least Squares Adjustment

Least squares (LS) adjustment is a mathematical optimization technique that estimates unknown parameters by minimizing the sum of squares of errors. This method has a wide range of applications in statistics, regression analysis, and curve fitting. In Global Navigation Satellite System (GNSS) real-time kinematic positioning (RTK), without loss of generality, the multi-GNSS functional model can be simplified as follows [1–3]

$$y = Ax + e \tag{3.1}$$

where $y$ denotes the observation vector; $A$ denotes the design matrix of the unknown parameters; $x$ denotes the vector of unknown parameters including coordinate components, ambiguities, and others, while $e$ denotes the observation noise. Then the corresponding stochastic model reads

$$D = \sigma^2 Q \tag{3.2}$$

where $\sigma$ denotes the variance factor, and $Q$ denotes the cofactor matrix. Let the estimate of $x$ as $\hat{x}$, then

$$v = A\hat{x} - y \tag{3.3}$$

where $v$ denotes the vector of LS adjustment residuals. According to the law of covariance propagation, the variance-covariance matrix of the observation vector $l$ can be expressed as $D$. LS, that is, the estimation $\hat{x}$, is required to minimize the quadratic form $\varphi(\hat{x})$ as follows

$$\varphi(\hat{x}) = v^{\mathrm{T}} D^{-1} v = \min \tag{3.4}$$



B. Li et al., *GNSS Real-Time Kinematic Positioning*, Navigation: Science and Technology 17, https://doi.org/10.1007/978-981-96-9116-6_3

Since $\hat{x}$ is an independent parameter, to find the extremum of $\varphi(\hat{x})$ with respect to $\hat{x}$, and set its first derivative to zero, we get

$$\frac{\partial \varphi(\hat{x})}{\partial \hat{x}} = 2v^{\mathrm{T}} D^{-1} \frac{\partial v}{\partial \hat{x}} = 2v^{\mathrm{T}} D^{-1} A \tag{3.5}$$

Then (3.5) can be reformulated as follows

$$AD^{-1}v = AD^{-1}(A\hat{x} - y) = 0 \tag{3.6}$$

$$AD^{-1}A\hat{x} = AD^{-1}y \tag{3.7}$$

The estimated unknown parameters $\hat{x}$ can be derived as follows

$$\hat{x} = \left(A^{\mathrm{T}} D^{-1} A\right)^{-1} A^{\mathrm{T}} D^{-1} y \tag{3.8}$$

The corresponding variance-covariance matrix $D_{\hat{x}}$ can be derived as follows

$$D_{\hat{x}} = \left(A^{\mathrm{T}} D^{-1} A\right)^{-1} \tag{3.9}$$

Since the measurement errors are random, we have

$$\mathrm{E}(e) = 0 \tag{3.10}$$

Therefore

$$\mathrm{E}(l) = Ax \tag{3.11}$$

$$\mathrm{E}(\hat{x}) = \left(A^{\mathrm{T}} D^{-1} A\right)^{-1} A^{\mathrm{T}} D^{-1} \mathrm{E}(y) = x \tag{3.12}$$

Based on (3.12), the $\hat{x}$ is unbiased. Therefore, the LS adjustment is an unbiased estimator. Essentially, the LS adjustment is derived from the deterministic principles of orthogonality and minimum distance. That is, no probabilistic considerations are involved. The LS estimators do not inherently possess optimal properties in a probabilistic sense. They are merely unbiased estimators, independent of the choice of the weight matrix, and their variance can be minimized by selecting the weight matrix as the inverse of the covariance matrix of the observations. Moreover, they are linear estimators if they are based on the linear functions of the observations. As is well known, estimators that are linear, unbiased and have minimum variance are called best linear unbiased estimators (BLUEs), where "best" in this case refers to having minimum variance. The LS estimators are BLUEs when the weight matrix equals the inverse of the covariance matrix of the observations [4–6].

In RTK, the float solution can be obtained using (3.7) and (3.8). Then, the real-valued ambiguities must be resolved to integer values. After validating the integer ambiguity solution, the baseline is updated to obtain the final estimated coordinate parameters, as discussed in Chap. 4.

## 3.2  Sequential Adjustment

The mathematical model consisting of (3.1) and (3.2) applies to a single-epoch case. If multiple epochs are considered and the observations between epochs are independent, the mathematical model can be expressed as follows

$$Y = BX + E \tag{3.13}$$

$$D_{YY} = \begin{bmatrix} D_{y_1 y_1} & \cdots & 0 \\ \vdots & \ddots & \vdots \\ 0 & \cdots & D_{y_k y_k} \end{bmatrix} \tag{3.14}$$

with $Y = \left[ y_1^{\mathrm{T}}, \ldots, y_K^{\mathrm{T}} \right]^{\mathrm{T}}$, $B = \mathrm{blkdiag}(B_1, \ldots, B_k)$, $X = \left[ x_1^{\mathrm{T}}, \ldots, x_k^{\mathrm{T}} \right]^{\mathrm{T}}$, $E = \left[ e_1^{\mathrm{T}}, \ldots, e_k^{\mathrm{T}} \right]^{\mathrm{T}}$. According to the (3.14), we can find that the variance-covariance matrix $D_{YY}$ consisting of the variance-covariance matrices of each epoch $D_{y_k y_k}$ is a block diagonal matrix (i.e., $D_{y_i y_j} = 0(i \neq j)$). Therefore, the kinematic solutions (i.e., $x_1 \neq \cdots \neq x_k$) of epoch $k$ can be deduced as follows

$$\hat{x}_k = \left( B_k^{\mathrm{T}} D_{y_k y_k}^{-1} B_k \right)^{-1} B_k^{\mathrm{T}} D_{y_k y_k}^{-1} y_k \tag{3.15}$$

Based on (3.15), the corresponding variance-covariance matrix $D_{\hat{x}_k}$ can be derived as follows

$$D_{\hat{x}_k} = \left( B_k^{\mathrm{T}} D_{y_k y_k}^{-1} B_k \right)^{-1} \tag{3.16}$$

If the parameters to be estimated remain unchanged during this period (i.e., $x_1 = \cdots = x_k$), it is a static solution at this time. Then one can use the overall solution or the superposition solution of the normal equations. For the convenience of calculation in real applications, the sequential adjustment can be adopted. The sequential adjustment processes measurement data in time or space order, and each time a set of measurement data is processed, the results of the previous processing are used to update the parameter estimation. This section introduces it from the perspective of time.

The prior expectation and variance of $x$ are

$$\mathrm{E}(x) = \hat{x}_0, \quad \mathrm{D}(x) = D_0 \tag{3.17}$$

And the prior expectation and variance of $E$ are

$$\mathrm{E}(e_i) = 0, \quad \mathrm{D}(e_i) = D_{e_i} \quad (i = 1, 2, \ldots, k) \tag{3.18}$$

When filtering based on the generalized LS principle, the variable $x$ can be regarded as nonrandomized, and its prior expectation $\hat{x}$ can be regarded as a virtual observation with variance of $D_{\hat{x}_0}$. Then, by the method of indirect adjustment, the error equation can be written as follows

$$\begin{cases} V_{x_0} = \hat{x} + \hat{x}_0 \\ V_1 = B_1\hat{x} + y_1 \\ V_2 = B_2\hat{x} + y_2 \\ \qquad \cdots \\ V_k = B_k\hat{x} + y_k \end{cases} \tag{3.19}$$

Based on the above model, the $\hat{x}_1$ and $D_{\hat{x}_1}$ obtained from the first adjustment of sequential filtering with the method of sequential indirect adjustment.

$$\begin{cases} \hat{x}_1 = \left(B_1^{\mathrm{T}}D_{e_1}^{-1}B_1 + D_{\hat{x}_0}^{-1}\right)^{-1}\left(B_1^{\mathrm{T}}D_{e_1}^{-1}y_1 + D_{\hat{x}_0}^{-1}\hat{x}_0\right) \\ D_{\hat{x}_1} = \left(B_1^{\mathrm{T}}D_{e_1}^{-1}B_1 + D_{\hat{x}_0}^{-1}\right)^{-1} \end{cases} \tag{3.20}$$

By applying the matrix inversion formula, we obtain the following result

$$\begin{cases} \hat{x}_1 = \hat{x}_0 + D_{\hat{x}_0}B_1^{\mathrm{T}}\left(B_1D_{\hat{x}_0}B_1^{\mathrm{T}} + D_{e_1}^{-1}\right)^{-1}\left(y_1 + B_1\hat{x}_0\right) \\ D_{\hat{x}_1} = D_{\hat{x}_0} - D_{\hat{x}_0}B_1^{\mathrm{T}}\left(B_1D_{\hat{x}_0}B_1^{\mathrm{T}} + D_{e_1}^{-1}\right)^{-1}B_1D_{\hat{x}_0} \end{cases} \tag{3.21}$$

After obtaining $\hat{x}_{k-1}$ and $D_{\hat{x}_{k-1}}$ from the $(k-1)$-th adjustment, the result of the $k$-th adjustment is

$$\begin{cases} \hat{x}_k = \left(B_k^{\mathrm{T}}D_{e_k}^{-1}B_k + D_{\hat{x}_{k-1}}^{-1}\right)^{-1}\left(B_k^{\mathrm{T}}D_{e_k}^{-1}y_k + D_{\hat{x}_{k-1}}^{-1}\hat{x}_{k-1}\right) \\ D_{\hat{x}_k} = \left(B_k^{\mathrm{T}}D_{e_k}^{-1}B_k + D_{\hat{x}_{k-1}}^{-1}\right)^{-1} \end{cases} \tag{3.22}$$

Similarly, we obtain the following result

$$\begin{cases} \hat{x}_k = \hat{x}_{k-1} + D_{\hat{x}_{k-1}}B_k^{\mathrm{T}}\left(B_kD_{\hat{x}_{k-1}}B_k^{\mathrm{T}} + D_{e_k}^{-1}\right)^{-1}\left(y_k + B_k\hat{x}_{k-1}\right) \\ D_{\hat{x}_k} = D_{\hat{x}_{k-1}} - D_{\hat{x}_{k-1}}B_k^{\mathrm{T}}\left(B_kD_{\hat{x}_{k-1}}B_k^{\mathrm{T}} + D_{e_k}^{-1}\right)^{-1}B_kD_{\hat{x}_{k-1}} \end{cases} \tag{3.23}$$

In case of white noise, (3.23) is the recursive formula for static sequential filtering. If there is also an estimation signal $x'$, the prior expectation and prior variance be set as follows

$$\mathrm{E}(x') = \hat{x}'_0, \mathrm{D}(x') = D'_0 \tag{3.24}$$

The covariance of $x'$ between $x$ and $e_i$ is

$$\begin{cases} \mathrm{cov}(x', x) = C_0 \\ \mathrm{cov}(x', e_i) = 0 \quad (i = 1, 2, \ldots, k) \end{cases} \tag{3.25}$$

The observation equation of $y_k$ can be written as follows

$$y_k = \begin{bmatrix} B_k & 0 \end{bmatrix} \begin{bmatrix} x_k \\ x_k' \end{bmatrix} + e_k \tag{3.26}$$

Then, we obtain the following result

$$\begin{bmatrix} \hat{x}_k \\ \hat{x}_k' \end{bmatrix} = \begin{bmatrix} \hat{x}_{k-1} \\ \hat{x}_{k-1}' \end{bmatrix} + \begin{bmatrix} D_{\hat{x}_{k-1}} & C_{k-1}^{\mathrm{T}} \\ C_{k-1} & D_{\hat{x}_{k-1}}' \end{bmatrix} \begin{bmatrix} B_k^{\mathrm{T}} \\ 0 \end{bmatrix} \left\{ \begin{bmatrix} B_k & 0 \end{bmatrix} \begin{bmatrix} D_{\hat{x}_{k-1}} & C_{k-1}^{\mathrm{T}} \\ C_{k-1} & D_{\hat{x}_{k-1}}' \end{bmatrix} \begin{bmatrix} B_k^{\mathrm{T}} \\ 0 \end{bmatrix} + D_{e_k} \right\}^{-1}$$
$$\left( y_k - B_k \hat{x}_{k-1} \right) \tag{3.27}$$

where $D_{\hat{x}_{k-1}}'$ denotes the variance of the $k-1$-th estimation $\hat{x}_{k-1}'$ of $x$, and $C_{k-1}$ denotes the covariance between $\hat{x}_{k-1}'$ and $\hat{x}_k'$. Expanding (3.27), we can obtain

$$\hat{x}_k' = \hat{x}_{k-1}' + C_{k-1} B_k^{\mathrm{T}} \left( B_k D_{\hat{x}_{k-1}} B_k^{\mathrm{T}} + D_{e_k} \right)^{-1} \left( y_k - B_k \hat{x}_{k-1} \right) \tag{3.28}$$

It can also be obtained according to (3.23)

$$\begin{bmatrix} D_{\hat{x}_k} & C_k^{\mathrm{T}} \\ C_k & D_{\hat{x}_k}' \end{bmatrix} = \begin{bmatrix} D_{\hat{x}_{k-1}} & C_{k-1}^{\mathrm{T}} \\ C_{k-1} & D_{\hat{x}_{k-1}}' \end{bmatrix}$$
$$- \begin{bmatrix} D_{\hat{x}_{k-1}} B_k^{\mathrm{T}} \\ C_{k-1} B_k^{\mathrm{T}} \end{bmatrix} \left( B_k D_{\hat{x}_{k-1}} B_k^{\mathrm{T}} + D_{e_k} \right)^{-1} \begin{bmatrix} B_k D_{\hat{x}_{k-1}} & B_k C_{k-1}^{\mathrm{T}} \end{bmatrix} \tag{3.29}$$

Therefore

$$D_k' = D_{k-1}' + C_{k-1} B_k^{\mathrm{T}} \left( B_k D_{\hat{x}_{k-1}} B_k^{\mathrm{T}} + D_{e_k} \right)^{-1} B_k C_{k-1}^{\mathrm{T}} \tag{3.30}$$

with

$$C_k = C_{k-1} - C_{k-1} B_k^{\mathrm{T}} \left( B_k D_{\hat{x}_{k-1}} B_k^{\mathrm{T}} + D_{e_k} \right)^{-1} B_k D_{\hat{x}_{k-1}} \tag{3.31}$$

Then, (3.31) can be rewritten as follows

$$C_k = C_{k-1} + C_{k-1} D_{\hat{x}_{k-1}}^{-1} \left( D_{\hat{x}_k} - D_{\hat{x}_{k-1}} \right) \tag{3.32}$$

Thus, we have

$$C_k = C_{k-1} D_{\hat{x}_{k-1}}^{-1} D_{\hat{x}_k} \tag{3.33}$$

Also, it can be obtained from (3.33)

$$C_k = C_0 D_{\hat{x}_0}^{-1} D_{\hat{x}_k} \tag{3.34}$$

In the case of white noise, the sequential extrapolation formulas are the above (3.28) and (3.31).

On the other hand, residual systematic errors exist in GNSS observations; hence, the GNSS observations are time-correlated. In this case, the sequential adjustment mentioned above cannot be applied. Instead, a sequential adjustment method that accounts for time correlations in the observations must be adopted. First, we study the time-correlated observation model without between-epoch common parameters. For instance, this applies to GNSS pseudorange or carrier phase measurements with fixed ambiguities used for kinematic positioning. The corresponding observation equation and its variance-covariance matrix are given as follows

$$Y = BX + E \tag{3.35}$$

$$D_{YY} = \begin{bmatrix} D_{y_1 y_1} & \cdots & D_{y_1 y_k} \\ \vdots & \ddots & \vdots \\ D_{y_k y_1} & \cdots & D_{y_k y_k} \end{bmatrix} \tag{3.36}$$

Since the between-epoch observations are dependent, i.e., $D_{y_i y_j} \neq \mathbf{0}$, the methods of decorrelation transformation, differential transformation, and maximum a posteriori (MAP) estimation are applied in this section.

The first method is the decorrelation transformation. The $\mathbf{LDL}^{\mathrm{T}}$ decomposition method is applied to transform the variance-covariance matrix $D_{YY}$, and set the decomposition form as $D_{YY} = UDU^{\mathrm{T}}$, where $U$ is a unit lower triangular matrix. The corresponding recursive formula reads [7].

$$D_j = D_{jj} - \sum_{k=1}^{j-1} U_{jk} D_k U_{jk}^{\mathrm{T}}, \quad U_{ij} = \left( D_{ij} - \sum_{k=1}^{j-1} U_{ik} D_k U_{jk}^{\mathrm{T}} \right) D_j^{-1}, \quad 1 \leq j < i \tag{3.37}$$

where $U_{ij}$ denotes the $i$-th row and the $j$-th column sub-matrix of the unit lower triangular matrix, and $D = \mathrm{blkdiag}(D_1, \ldots, D_k)$ is the block diagonal matrix. We set $LD_{YY}L^{\mathrm{T}} = D$, where the $L_{ij}$ is as follows

$$L_{ij} = - \sum_{k=j+1}^{i} L_{ik} U_{kj}, \quad 1 \leq j < i \tag{3.38}$$

The basic idea of the decorrelation transformation is to obtain a new set of independent observations by transforming the time-correlated observations. Specifically, by multiplying the matrix $\boldsymbol{L}$, the new observation model is obtained

$$\overline{Y} = AX + \overline{E} \tag{3.39}$$

with $\overline{Y} = LY = \left[\overline{y}_1^{\mathrm{T}}, \ldots, \overline{y}_K^{\mathrm{T}}\right]^{\mathrm{T}}$, $\overline{y}_i = y_i + \sum_{k=1}^{i-1} L_{ik} y_k$, $\overline{E} = LE = \left[\overline{e}_1^{\mathrm{T}}, \ldots, \overline{e}_K^{\mathrm{T}}\right]^{\mathrm{T}}$, $\overline{e}_i = e_i + \sum_{k=1}^{i-1} L_{ik} e_k$, $A = LB = \left[A_1^{\mathrm{T}}, \ldots, A_K^{\mathrm{T}}\right]^{\mathrm{T}}$, $A_i = \left[L_{i1}B_1, \ldots, L_{i,i-1}B_{i-1}, B_i, 0, \ldots\right]$. The variance-covariance matrix of the transformed observation vector $\overline{y}$ reads

$$D_{\overline{y}\overline{y}} = L D_{yy} L^{\mathrm{T}} = D \tag{3.40}$$

where $D_{\overline{y}_i\overline{y}_i} = D_i$. Since $D$ is a block diagonal matrix, it is obvious that the observations are independent. Set $\overline{x}_{i-1} = \left[x_1^{\mathrm{T}}, \ldots, x_{i-1}^{\mathrm{T}}\right]^{\mathrm{T}}$, and the observation equation and the variance-covariance matrix of all the epochs read

$$\begin{bmatrix} l_i \\ \overline{y}_i \end{bmatrix} = \begin{bmatrix} \overline{A}_i & 0 \\ E_i & B_i \end{bmatrix} \begin{bmatrix} \overline{x}_{i-1} \\ x_i \end{bmatrix} + \begin{bmatrix} \epsilon_i \\ \overline{e}_i \end{bmatrix}, \quad \mathrm{cov}\begin{bmatrix} \epsilon_i \\ \overline{e}_i \end{bmatrix} = \begin{bmatrix} D_{l_i l_i} & 0 \\ 0 & D_{\overline{y}_i\overline{y}_i} \end{bmatrix} \tag{3.41}$$

with $l_i = \left[\overline{y}_1^{\mathrm{T}}, \ldots, \overline{y}_{i-1}^{\mathrm{T}}\right]^{\mathrm{T}}$, $\overline{A}_i = \begin{bmatrix} B_1 & & & \\ L_{21}B_1 & B_2 & & \\ \vdots & \vdots & \ddots & \\ L_{i-1,1}B_1 & \cdots & L_{i-1,i-2}B_{i-2} & B_{i-1} \end{bmatrix}$, $E_i = \left[L_{i1}B_1, \ldots, L_{i,i-1}B_{i-1}\right]$, $\epsilon_i = \left[\overline{e}_1^{\mathrm{T}}, \ldots, \overline{e}_{i-1}^{\mathrm{T}}\right]^{\mathrm{T}}$, $D_{l_i l_i} = \mathrm{blkdiag}(D_1, \ldots, D_{i-1})$. It is worth noting that all the $i$ parameters are included in (3.41) at this time. The corresponding LS normal equation is as follows

$$\begin{bmatrix} \overline{A}_i^{\mathrm{T}} D_{l_i l_i}^{-1} \overline{A}_i + E_i^{\mathrm{T}} D_{\overline{y}_i\overline{y}_i}^{-1} E_i & E_i^{\mathrm{T}} D_{\overline{y}_i\overline{y}_i}^{-1} B_i \\ B_i^{\mathrm{T}} D_{\overline{y}_i\overline{y}_i}^{-1} E_i & B_i^{\mathrm{T}} D_{\overline{y}_i\overline{y}_i}^{-1} B_i \end{bmatrix} \begin{bmatrix} \hat{\overline{x}}_{i-1} \\ \hat{x}_i \end{bmatrix}$$
$$= \begin{bmatrix} \overline{A}_i^{\mathrm{T}} D_{l_i l_i}^{-1} l_i + E_i^{\mathrm{T}} D_{\overline{y}_i\overline{y}_i}^{-1} \overline{y}_i \\ B_i^{\mathrm{T}} D_{\overline{y}_i\overline{y}_i}^{-1} \overline{y}_i \end{bmatrix} \tag{3.42}$$

with $D_{\hat{\overline{x}}_{i-1}\hat{\overline{x}}_{i-1}}^{-1} = \overline{A}_i^{\mathrm{T}} D_{l_i l_i}^{-1} A_i, \overline{A}_i^{\mathrm{T}} D_{l_i l_i}^{-1} l_i = D_{\hat{\overline{x}}_{i-1}\hat{\overline{x}}_{i-1}}^{-1} \hat{\overline{x}}_{i-1}$. Then $\overline{x}_{i-1}$ is eliminated by using the normal equation reduction, and the LS estimates of the $i$-th epoch can be derived as follows

$$\hat{x}_i = \left(B_i^{\mathrm{T}} D_{\overline{y}_i|\overline{x}_{i-1}}^{-1} B_i\right)^{-1} B_i^{\mathrm{T}} D_{\overline{y}_i|\overline{x}_{i-1}}^{-1} \left(\overline{y}_i - E_i \hat{\overline{x}}_{i-1}\right), \quad D_{\hat{x}_i\hat{x}_i} = \left(B_i^{\mathrm{T}} D_{\overline{y}_i|\overline{x}_{i-1}}^{-1} B_i\right)^{-1} \tag{3.43}$$

with $D_{\overline{y}_i|\overline{x}_{i-1}} = D_{\overline{y}_i\overline{y}_i} + E_i D_{\hat{\overline{x}}_{i-1}\hat{\overline{x}}_{i-1}} E_i^{\mathrm{T}}$.

Then the second method is the differential transformation. The (3.35) is separated by the first $i$ epochs and the $i$-th epoch, and the equation of the first $i$ epochs is as follows

$$l_i = C_i \bar{x}_{i-1} + e_{l_i} \tag{3.44}$$

with $C_i = \mathrm{blkdiag}(B_1, \ldots, B_{i-1})$, $e_{l_i} = \left[e_1^{\mathrm{T}}, \ldots, e_{i-1}^{\mathrm{T}}\right]^{\mathrm{T}}$. The variance-covariance matrix of (3.36) reads

$$\mathrm{cov}\left(\begin{bmatrix} \epsilon_{l_i} \\ e_i \end{bmatrix}\right) = \begin{bmatrix} D_{l_i l_i} & D_{l_i y_i} \\ D_{y_i l_i} & D_{y_i y_i} \end{bmatrix} \tag{3.45}$$

with $D_{y_i l_i} = D_{l_i y_i}^{\mathrm{T}} = \left[D_{y_i y_1}, \ldots, D_{y_i y_{i-1}}\right]$.

The observation vector of the $i$-th epoch is transformed as $\bar{y}_i = y_i - G^{\mathrm{T}} l_i$, then the new observation equation is obtained

$$\bar{y}_i = B_i x_i - G^{\mathrm{T}} C_i \bar{x}_{i-1} + \bar{e}_i \tag{3.46}$$

with $\bar{e}_i = e_i - G^{\mathrm{T}} e_{l_i}$. According to the covariance propagation law, the new variance-covariance matrix reads

$$\mathrm{cov}\left(\begin{bmatrix} e_{l_i} \\ \bar{e}_i \end{bmatrix}\right) = \begin{bmatrix} D_{l_i l_i} & D_{l_i y_i} - D_{l_i l_i} G \\ D_{y_i l_i} - G^{\mathrm{T}} D_{l_i l_i} & D_{y_i y_i} + G^{\mathrm{T}} D_{l_i l_i} G - G^{\mathrm{T}} D_{l_i y_i} - D_{y_i l_i} G \end{bmatrix} \tag{3.47}$$

Apparently, if $G = D_{l_i l_i}^{-1} D_{l_i y_i}$, the observation vectors $\bar{y}_i$ and $l_i$ are independent. At this time, the new observation equation and the variance-covariance matrix can be written as follows

$$\begin{bmatrix} l_i \\ \bar{y}_i \end{bmatrix} = \begin{bmatrix} C_i & 0 \\ -G^{\mathrm{T}} C_i & B_i \end{bmatrix} \begin{bmatrix} \bar{x}_{i-1} \\ x_i \end{bmatrix} + \begin{bmatrix} e_{l_i} \\ \bar{e}_i \end{bmatrix}, \quad \mathrm{cov}\begin{bmatrix} \epsilon_i \\ \bar{e}_i \end{bmatrix} = \begin{bmatrix} D_{l_i l_i} & 0 \\ 0 & D_{\bar{y}_i \bar{y}_i} \end{bmatrix} \tag{3.48}$$

with $D_{\bar{y}_i \bar{y}_i} = D_{y_i y_i} - D_{y_i l_i} D_{l_i l_i}^{-1} D_{l_i y_i}$. The corresponding LS normal equation is as follows

$$\begin{bmatrix} D_{\hat{\bar{x}}_{i-1} \hat{\bar{x}}_{i-1}}^{-1} + C_i^{\mathrm{T}} G D_{\bar{y}_i \bar{y}_i}^{-1} G^{\mathrm{T}} C_i & -C_i^{\mathrm{T}} G D_{\bar{y}_i \bar{y}_i}^{-1} B_i \\ -B_i^{\mathrm{T}} D_{\bar{y}_i \bar{y}_i}^{-1} G^{\mathrm{T}} C_i & B_i^{\mathrm{T}} D_{\bar{y}_i \bar{y}_i}^{-1} B_i \end{bmatrix} \begin{bmatrix} \hat{\bar{x}}_{i-1} \\ \hat{x}_i \end{bmatrix}$$
$$= \begin{bmatrix} D_{\hat{\bar{x}}_{i-1} \hat{\bar{x}}_{i-1}}^{-1} \hat{\bar{x}}_{i-1} - C_i^{\mathrm{T}} G D_{\bar{y}_i \bar{y}_i}^{-1} \bar{y}_i \\ B_i^{\mathrm{T}} D_{\bar{y}_i \bar{y}_i}^{-1} \bar{y}_i \end{bmatrix} \tag{3.49}$$

with $D_{\hat{\bar{x}}_{i-1} \hat{\bar{x}}_{i-1}}^{-1} = C_i^{\mathrm{T}} D_{l_i l_i}^{-1} C_i$, $C_i^{\mathrm{T}} D_{l_i l_i}^{-1} l_i = D_{\hat{\bar{x}}_{i-1} \hat{\bar{x}}_{i-1}}^{-1} \hat{\bar{x}}_{i-1}$. The $\bar{x}_{i-1}$ is eliminated by using the normal equation reduction, and the LS estimates of the $i$-th epoch is derived as follows

$$\hat{x}_i = \left(B_i^{\mathrm{T}} D_{\bar{y}_i|\bar{x}_{i-1}}^{-1} B_i\right)^{-1} B_i^{\mathrm{T}} D_{\bar{y}_i|\bar{x}_{i-1}}^{-1} \left(\bar{y}_i + G^{\mathrm{T}} C_i \hat{\bar{x}}_{i-1}\right), \quad D_{\hat{x}_i \hat{x}_i} = \left(B_i^{\mathrm{T}} D_{\bar{y}_i|\bar{x}_{i-1}}^{-1} B_i\right)^{-1}$$

$$(3.50)$$

with $D_{\bar{y}_i|\bar{x}_{i-1}} = D_{\bar{y}_i \bar{y}_i} + G^{\mathrm{T}} C_i D_{\hat{\bar{x}}_{i-1} \hat{\bar{x}}_{i-1}} C_i^{\mathrm{T}} G$.

Compared with the decorrelation transformation, it can be found that we have $G^{\mathrm{T}} C_i = -E_i$ and $D_{y_i y_i} - D_{y_i l_i} D_{l_i l_i}^{-1} D_{l_i y_i} = D_{ii}$. That is, the estimations of the decorrelation transformation and the differential transformation are equivalent. However, the differential transformation requires that the observation type and the dimension of the observations between epochs are the same, whereas the decorrelation transformation is not.

Finally, we can use the MAP estimation. The essence of MAP estimation theory is to modify the prior $\bar{x}$ with the observation $y$, and the modification is determined by the observation variances and the correlations between the observations and the parameters [8]. First, the standard form of MAP estimation is presented, where the prior statistical information of variables $\underline{x}$ and $\underline{y}$ are as follows

$$\mathrm{E}\left(\begin{bmatrix} \underline{x} \\ \underline{y} \end{bmatrix}\right) = \begin{bmatrix} \bar{x} \\ \bar{y} \end{bmatrix}, \quad \mathrm{D}\left(\begin{bmatrix} \underline{x} \\ \underline{y} \end{bmatrix}\right) = \begin{bmatrix} D_{xx} & D_{xy} \\ D_{yx} & D_{yy} \end{bmatrix} \tag{3.51}$$

where E denotes the expectation operator. According to the MAP estimation theory, when the sample (real observation) $y$ of the random variable $\underline{y}$ exists, the estimation $\hat{x}$ corresponding to the random variable $\underline{x}$ can be derived [8]

$$\begin{cases} \hat{x} = \mathrm{E}(\underline{x}) + D_{xy} D_{yy}^{-1} \left(y - \mathrm{E}(\underline{y})\right) = \bar{x} + D_{xy} D_{yy}^{-1} (y - \bar{y}) \\ D_{\hat{x}\hat{x}} = D_{xx} + D_{xy} D_{yy}^{-1} D_{yx} \end{cases} \tag{3.52}$$

where the underlined variables $\underline{x}$ and $\underline{y}$ denote the random variables, and the underlined variable $y$ denote the sample of $y$.

Then the MAP estimation is applied to derive kinematic solutions for time-correlated observations, and the block observation equation is used via the differential transformation. As a result, we have $y_i \to \bar{x}$, $l_i \to \bar{y}$, $\hat{l}_i \to y$, and obtain the new observation of the $i$-th epoch

$$\tilde{y}_i = y_i - D_{y_i l_i} D_{l_i l_i}^{-1} \left(l_i - \hat{l}_i\right) = y_i - G^{\mathrm{T}} \hat{e}_{l_i} \tag{3.53}$$

$$D_{\tilde{y}_i \tilde{y}_i} = D_{y_i y_i} + G^{\mathrm{T}} D_{\hat{l}_i \hat{l}_i} G - D_{y_i \hat{l}_i} G - G^{\mathrm{T}} D_{\hat{l}_i y_i} \tag{3.54}$$

Since the observation vectors $l_i$ and $y_i$ are correlated, the observation vector $y_i$ should be updated. At the same time, the updated observation vector $\tilde{y}_i$ actually considers the time correlations, so the updated observation $\tilde{y}_i$ can obtain the equivalent parameter estimation of the $i$-th epoch compared with the overall parameter estimation

$$\hat{x}_i = \left(B_i^{\mathrm{T}} D_{\tilde{y}_i \tilde{y}_i}^{-1} B_i\right)^{-1} B_i^{\mathrm{T}} D_{\tilde{y}_i \tilde{y}_i}^{-1} \tilde{y}_i, \quad D_{\hat{x}_i \hat{x}_i} = \left(B_i^{\mathrm{T}} D_{\tilde{y}_i \tilde{y}_i}^{-1} B_i\right)^{-1} \tag{3.55}$$

Since it has been proved that the decorrelation transformation is equivalent to the differential transformation, only the equivalence between the kinematic solution (3.55) and (3.50) is discussed. Since $D_{\hat{e}_{l_i} \hat{e}_{l_i}} = D_{l_i l_i} - C_i D_{\hat{\bar{x}}_{i-1} \hat{\bar{x}}_{i-1}} C_i^T$, $D_{y_i \hat{e}_{l_i}} = D_{y_i l_i} - G^{\mathrm{T}} C_i D_{\hat{\bar{x}}_{i-1} \hat{\bar{x}}_{i-1}} C_i^T$, we can substitute them into (3.54)

$$D_{\tilde{y}_i \tilde{y}_i} = D_{y_i y_i} - D_{y_i l_i} D_{l_i l_i}^{-1} D_{l_i y_i} + G^{\mathrm{T}} C_i D_{\hat{\bar{x}}_{i-1} \hat{\bar{x}}_{i-1}} C_i^T G \tag{3.56}$$

Considering $D_{\bar{y}_i \bar{y}_i} = D_{y_i y_i} - D_{y_i l_i} D_{l_i l_i}^{-1} D_{l_i y_i}$ and $Q_{\bar{y}_i | \bar{x}_{i-1}} = D_{\bar{y}_i \bar{y}_i} + G^{\mathrm{T}} C_i D_{\hat{\bar{x}}_{i-1} \hat{\bar{x}}_{i-1}} C_i^T G$, we have

$$D_{\tilde{y}_i \tilde{y}_i} = D_{\bar{y}_i | \bar{x}_{i-1}} \tag{3.57}$$

In addition, since $\tilde{y}_i = y_i - G^{\mathrm{T}}\left(l_i - C_i \hat{\bar{x}}_{i-1}\right)$ and $\bar{y}_i = y_i - G^{\mathrm{T}} l_i$, then

$$\tilde{y}_i = \bar{y}_i + G^{\mathrm{T}} C_i \hat{\bar{x}}_{i-1} \tag{3.58}$$

In the end, we can obtain (3.50) by substituting (3.57) and (3.58) into (3.55), thus proving that the kinematic solutions of the MAP estimation and differential transformation are equivalent. Similarly, the MAP estimation does not require that the observation type and the dimension of the between-epoch observations are the same, whereas the differential transform is not.

On the other hand, the time-correlated observation model with between-epoch common parameters is presented. The traditional multi-epoch adjustment method often leads to low efficiency. Therefore, in order to ensure that the solutions remain consistent and, at the same time, maximize the efficiency of the solution, we will explore the kinematic data processing method of the time-correlated observation model with between-epoch common parameters. Suppose the observations of consecutive $K$ epochs have time correlations. At this point, there are parameter $X$ and between-epoch common parameters $\xi$. Then the observation equation of $K$ epochs is defined as follows

$$Y = BX + C\xi + E \tag{3.59}$$

where $C = \left[C_1^{\mathrm{T}}, \ldots, C_K^{\mathrm{T}}\right]^{\mathrm{T}}$, and the variance-covariance matrix is defined as (3.19). Since the equivalence of decorrelation transformation, differential transformation and MAP estimation methods has been proved in the previous section, this monograph uses the decorrelation transformation method to derive as an example.

By multiplying the matrix $L$, the new observation model is obtained that is similar to (3.39)

$$\overline{Y} = AX + \overline{C}\xi + \overline{E} \tag{3.60}$$

with $\overline{C} = \left[\overline{C}_1^{\mathrm{T}}, \ldots, \overline{C}_K^{\mathrm{T}}\right]^{\mathrm{T}}$, $\overline{C}_i = C_i + \sum_{k=1}^{i-1} L_{ik} C_k$. Apparently, the observations are independent of each other after the transformation, and the transformed $i$-th observation equation contains the parameters of all the $i$ epochs. In order to reduce the normal equation, the $i$-th observation equation and its variance-covariance matrix are reorganized

$$\begin{bmatrix} l_i \\ \overline{y}_i \end{bmatrix} = \begin{bmatrix} \overline{A}_i & F_i & 0 \\ E_i & \overline{C}_i & B_i \end{bmatrix} \begin{bmatrix} \overline{x}_{i-1} \\ \xi \\ x_i \end{bmatrix} + \begin{bmatrix} \epsilon_i \\ \overline{e}_i \end{bmatrix}, \quad \mathrm{D}\left(\begin{bmatrix} \epsilon_i \\ \overline{e}_i \end{bmatrix}\right) = \begin{bmatrix} Q_{l_i l_i} & 0 \\ 0 & Q_{\overline{y}_i \overline{y}_i} \end{bmatrix} \tag{3.61}$$

where $F_i = \left[\overline{C}_1^{\mathrm{T}}, \ldots, \overline{C}_{i-1}^{\mathrm{T}}\right]^{\mathrm{T}}$, and the other variables are defined as (3.40). Obviously, $\zeta_{i-1} = \left[\overline{x}_{i-1}^{\mathrm{T}}, \xi_{i-1}^{\mathrm{T}}\right]^{\mathrm{T}}$, and $\xi_{i-1}$ denote the estimates of the parameter $\xi$ of the previous $i$-1 epoch observations. Let the parameters obtained from the previous $i$-1 epoch observations be evaluated as follows

$$\hat{\xi}_{i-1} = D_{\hat{\xi}_{i-1}} \begin{bmatrix} \overline{A}_i^{\mathrm{T}} D_{l_i l_i}^{-1} l_i \\ F_i^{\mathrm{T}} D_{l_i l_i}^{-1} l_i \end{bmatrix}, \quad Q_{\hat{\xi}_{i-1}} = \begin{bmatrix} \overline{A}_i^{\mathrm{T}} D_{l_i l_i}^{-1} \overline{A}_i & \overline{A}_i^{\mathrm{T}} D_{l_i l_i}^{-1} F_i \\ F_i^{\mathrm{T}} D_{l_i l_i}^{-1} \overline{A}_i & F_i^{\mathrm{T}} D_{l_i l_i}^{-1} F_i \end{bmatrix}^{-1} \tag{3.62}$$

Then the second equation of (3.61) and (3.62) is fused by the LS criterion, and the normal equation is obtained

$$\begin{bmatrix} D_{\hat{\xi}_{i-1}}^{-1} + H_i^{\mathrm{T}} D_{\overline{y}_i \overline{y}_i}^{-1} H_i & H_i^{\mathrm{T}} D_{\overline{y}_i \overline{y}_i}^{-1} B_i \\ B_i^{\mathrm{T}} D_{\overline{y}_i \overline{y}_i}^{-1} H_i & B_i^{\mathrm{T}} D_{\overline{y}_i \overline{y}_i}^{-1} B_i \end{bmatrix} \begin{bmatrix} \delta_{i-1} \\ \hat{x}_i \end{bmatrix} = \begin{bmatrix} D_{\hat{\xi}_{i-1}}^{-1} \hat{\xi}_{i-1} + H_i^{\mathrm{T}} D_{\overline{y}_i \overline{y}_i}^{-1} \overline{y}_i \\ B_i^{\mathrm{T}} D_{\overline{y}_i \overline{y}_i}^{-1} \overline{y}_i \end{bmatrix} \tag{3.63}$$

with $H_i = \left[E_i \ \overline{C}_i\right]$, $\delta_{i-1} = \left[\tilde{\overline{x}}_{i-1}^{\mathrm{T}}, \hat{\xi}^{\mathrm{T}}\right]^{\mathrm{T}}$. It is worth noting that the $\tilde{\overline{x}}_{i-1}$ denotes the estimate of $\overline{x}_{i-1}$ according to all the $i$ epochs, which is different from the $\hat{\overline{x}}_{i-1}$. By using the normal equation reduction, we derive the LS estimates of the $i$-th epoch

$$\hat{x}_i = \left(B_i^{\mathrm{T}} D_{\overline{y}_i | \zeta_{i-1}}^{-1} B_i\right)^{-1} B_i^{\mathrm{T}} D_{\overline{y}_i | \zeta_{i-1}}^{-1} \left(\overline{y}_i - H_i \hat{\zeta}_{i-1}\right), \quad D_{\hat{x}_i \hat{x}_i} = \left(B_i^{\mathrm{T}} D_{\overline{y}_i | \zeta_{i-1}}^{-1} B_i\right)^{-1} \tag{3.64}$$

with $D_{\overline{y}_i | \zeta_{i-1}} = D_{\overline{y}_i \overline{y}_i} + H_i D_{\hat{\zeta}_{i-1} \hat{\zeta}_{i-1}} H_i^{\mathrm{T}}$.

Although the above theoretical formulas are rigorous and accurate, they involve a large number of matrix operations, making the calculations complicated and inefficient. A compromise approach is to use a variance-covariance matrix that is as simple as possible yet effectively describes the time correlations, balancing calculation efficiency with the actual time-dependent characteristics. Assuming that the variance-covariance matrix of each observation is the same, the time correlation coefficient is estimated using the autocorrelation function (ACF). If the time correlation coefficient between adjacent epochs is $\rho$, the original variance-covariance matrix is as follows

$$D_{yy} = \begin{bmatrix} 1 & \rho & \rho^2 & \cdots & \rho^{K-1} \\ \rho & \ddots & \ddots & \vdots & \vdots \\ \rho^2 & \rho & 1 & \ddots & \rho^2 \\ \vdots & \vdots & \ddots & \ddots & \rho \\ \rho^{K-1} & \cdots & \rho^2 & \rho & 1 \end{bmatrix} \otimes Q = R \otimes Q \tag{3.65}$$

where the symbol $\otimes$ denotes the Kronecker product. The above variance-covariance matrix can also be called partial continuation mode [9]. The corresponding decompo-

sition $LRL^{\mathrm{T}} = D$, with $L = \begin{bmatrix} 1 & & & \\ -\rho & 1 & & \\ & \ddots & \ddots & \\ & & -\rho & 1 \end{bmatrix}$, $D = \mathrm{diag}\big([1, 1-\rho^2, \ldots, 1-\rho^2]\big)$.

Taking the observation Eq. (3.35) as an example, the decorrelation transformation method is applied for derivation. Specifically, by left multiplying matrix for observation equation $L \otimes I$, the variables in the observation Eq. (3.39) corresponding to the decorrelation transformation change to $\bar{y}_i = y_i - \rho y_{i-1}$, $\bar{e}_i = e_i - \rho e_{i-1}$, $A_i = [\ldots, 0, -\rho B_{i-1}, B_i, 0, \ldots]$. The variance-covariance matrix of the observations changes to $D_{yy} = D \otimes Q$ with $D_{y_i y_i} = (1 - \rho^2)Q$. According to $E_i = [0, \ldots 0, -\rho B_{i-1}]$, substitute the corresponding variables into (3.42), then

$$\hat{x}_i = \left(B_i^{\mathrm{T}} D_{\bar{y}_i|\hat{x}_{i-1}}^{-1} B_i\right)^{-1} B_i^{\mathrm{T}} D_{\bar{y}_i|\hat{x}_{i-1}}^{-1} \left(\bar{y}_i + \rho B_{i-1}\hat{x}_{i-1}\right), \quad D_{\hat{x}_i,\hat{x}_i} = \left(B_i^{\mathrm{T}} D_{\bar{y}_i|\hat{x}_{i-1}}^{-1} B_i\right)^{-1}$$
$$\tag{3.66}$$

with $D_{\bar{y}_i|\hat{x}_{i-1}} = (1 - \rho^2)Q + \rho^2 B_{i-1} D_{\hat{x}_{i-1}\hat{x}_{i-1}} B_{i-1}^{\mathrm{T}}$. Obviously, the ACF-based degradation form is extremely simple. Specifically, $E_i$, $G^{\mathrm{T}} C_i$ and $\tilde{y}_i$ can be simplified by $\rho$ by using ACF. Besides, the $\hat{\bar{x}}_{i-1}$ is reduced to $\hat{x}_{i-1}$, where the relationship between epochs are simplified. Therefore, the reduced form can not only meet the actual time-dependent error characteristics, but also significantly improve the computational efficiency.

## 3.3   Kalman Filter

In real-world applications of navigation and positioning, the state vector can be estimated using both the predicted state and the system observations. In this context, the Kalman filter can be employed, as it utilizes a series of observations and the system dynamics model to estimate the state vector. To achieve high-precision results, the Kalman filter is widely used in RTK. The conceptual basis of the Kalman filter is a weighting method, where the optimal estimator is obtained by correcting the weight of the estimated value and the observed value. Furthermore, the Kalman filter is suitable for linear Gaussian systems, where linearity refers to superposition and homogeneity.

Actually, internal or external constraints are often available. If these constraints can be precisely determined, equality constraints, such as the state equation, can be applied. The general principles of the Kalman filter primarily include unconstrained extremum and constrained extremum. Based on these two principles, the solutions of the Kalman filter can be classified into two categories. According to the dynamic model and observation equations, different filter solutions can be obtained by using various estimation criteria. The Kalman filter is a minimum mean square error estimation method, which is equivalent to the LS estimate.

To solve the LS solution of the state parameters, the dynamic system equation and the observation equation are rewritten in the form of error equations, that is

$$\begin{cases} v_{\bar{x}_k} = \hat{x}_k - \bar{x}_k \\ v_k = A_k \hat{x}_k - l_k \end{cases} \tag{3.67}$$

where $v_k$ and $v_{\bar{x}_k}$ denote the matrix of residual of $l_k$ and $\bar{x}_k$, respectively; $\bar{x}_k$ denotes the predicted parameters

$$\bar{x}_k = \boldsymbol{\Phi}_{k,k-1} \hat{x}_{k-1} \tag{3.68}$$

where $\boldsymbol{\Phi}_{k,k-1}$ denotes the state transition matrix of adjacent epochs $k$ and $k-1$. Then the variance matrix of $\bar{x}_k$ can be derived as follows

$$\sum_{\bar{x}_k} = \boldsymbol{\Phi}_{k,k-1} \sum_{\bar{x}_{k-1}} \boldsymbol{\Phi}_{k,k-1}^{\mathrm{T}} + \sum_{w_k} \tag{3.69}$$

where $w_k$ denotes the state noise vector.

According to the LS algorithm, the following objective function can be constructed as follows

$$\Omega(k) = v_k^{\mathrm{T}} P_k v_k + v_{\bar{x}_k}^{\mathrm{T}} P_{\bar{x}_k} v_{\bar{x}_k} = \min \tag{3.70}$$

where $P_k = \sum_k^{-1}$ and $P_{\bar{x}_k} = \sum_{\bar{x}_k}^{-1}$ denotes the weight matrices of $l_k$ and $\bar{x}_k$, respectively. Considering (3.67), and take the derivative of (3.70) and let it to be zero, we have:

$$\frac{\mathrm{d}\Omega(k)}{\mathrm{d}\hat{x}_k} = 2v_k^{\mathrm{T}} P_k A_k + 2v_{\bar{x}_k}^{\mathrm{T}} P_{\bar{x}_k} = 0 \tag{3.71}$$

Then, we can obtain (3.72) by substituting (3.71) into (3.67) as follows

$$\left(A_k^{\mathrm{T}} P_k A_k + P_{\bar{x}_k}\right) \hat{x}_k = \left(A_k^{\mathrm{T}} P_k l_k + P_{\bar{x}_k} \bar{x}_k\right) \tag{3.72}$$

Therefore, the LS solution of the state vector of $x_k$ can be derived as follows

$$\hat{x}_k = \left(A_k^{\mathrm{T}} P_k A_k + P_{\bar{x}_k}\right)^{-1}\left(A_k^{\mathrm{T}} P_k l_k + P_{\bar{x}_k}\bar{x}_k\right) \tag{3.73}$$

Based on (3.73), the corresponding variance-covariance matrix $\hat{x}_k$ can be derived as follows

$$\sum_{\hat{x}_k} = \left(A_k^{\mathrm{T}} P_k A_k + P_{\bar{x}_k}\right)^{-1}\hat{\sigma}_0^2 \tag{3.74}$$

where $\hat{\sigma}_0^2$ denotes variance factor. Considering $P_{\hat{x}_k} = A_k^{\mathrm{T}} P_k A_k + P_{\bar{x}_k}$, we have

$$\sum_{\hat{x}_k} = P_{\hat{x}_k}^{-1}\hat{\sigma}_0^2 \tag{3.75}$$

where $P_{\hat{x}_k}$ denotes the weight matrix of $\hat{x}_k$. Through the identical transformation of matrices, we have

$$\hat{x}_k = \bar{x}_k + K_k(l_k - A_k\bar{x}_k) \tag{3.76}$$

where $K_k = \sum_{\bar{x}_k} A_k^{\mathrm{T}}\left(A_k \sum_{\bar{x}_k} A_k^{\mathrm{T}} + \sum_k\right)^{-1}$ denotes the gain matrix. According to the law of covariance propagation, the variance-covariance matrix of $\hat{x}_k$ in (3.76) can be derived as follows

$$\sum_{\hat{x}_k} = \sum_{\bar{x}_k} A_k^{\mathrm{T}}\left(A_k \sum_{\bar{x}_k} A_k^{\mathrm{T}} + \sum_k\right)^{-1} A_k P_k \tag{3.77}$$

$$\sum_{\hat{x}_k} = (I_k - K_k A_k)\sum_{\bar{x}_k} \tag{3.78}$$

It also can be obtained by the identical transformation of matrices of (3.74).

The Kalman filter solution can also be obtained by the method of constrained extremum, and the objective function can be constructed as follows

$$\Omega(k) = v_k^{\mathrm{T}} P_k v_k + v_{\bar{x}_k}^{\mathrm{T}} P_{\bar{x}_k} v_{\bar{x}_k} - 2\lambda^{\mathrm{T}}(A_k\hat{x}_k - l_k - v_k) = \min \tag{3.79}$$

where $\lambda$ denotes Lagrange multiplier vector. Take the derivative of $\Omega(k)$ in (3.79) with respect to $v_k$ and $\hat{x}_k$, respectively, and let them be zero

$$\frac{\mathrm{d}\Omega(k)}{\mathrm{d}v_k} = 2v_k^{\mathrm{T}} P_k + 2\lambda^{\mathrm{T}} = 0 \tag{3.80}$$

$$\frac{\mathrm{d}\Omega(k)}{\mathrm{d}\hat{x}_k} = 2v_{\bar{x}_k}^{\mathrm{T}} P_{\bar{x}_k} - 2\lambda^{\mathrm{T}} A_k = 0 \tag{3.81}$$

According to (3.80)

$$v_k = -P_k^{-1}\lambda = -\sum_k \lambda \tag{3.82}$$

According to (3.81)

$$v_{\overline{x}_k} = P_{\overline{x}_k}^{-1} A_k^{\mathrm{T}} \lambda \tag{3.83}$$

After substituting (3.82) and (3.83) into (3.67), we have

$$-\sum_k \lambda = A_k \hat{x}_k - l_k \tag{3.84}$$

$$\sum_{\overline{x}_k} A_k^{\mathrm{T}} \lambda = \hat{x}_k - \overline{x}_k \tag{3.85}$$

Specifically, by left multiplying the matrix $A_k$, (3.85) can be rewritten as follows

$$A_k \sum_{\overline{x}_k} A_k^{\mathrm{T}} \lambda = A_k \hat{x}_k - A_k \overline{x}_k \tag{3.86}$$

After subtracting (3.84) from (3.86), $\lambda$ can be derived as follows

$$\lambda = \left( A_k \sum_{\overline{x}_k} A_k^{\mathrm{T}} + \sum_k \right)^{-1} (l_k - A_k \overline{x}_k) \tag{3.87}$$

We can substitute (3.86) into (3.85), the estimated unknown parameter $\hat{x}_k$ can be derived as follows

$$\hat{x}_k = \overline{x}_k + \sum_{\overline{x}_k} A_k^{\mathrm{T}} \left( A_k \sum_{\overline{x}_k} A_k^{\mathrm{T}} + \sum_k \right)^{-1} (l_k - A_k \overline{x}_k) \tag{3.88}$$

Obviously, (3.88) is equivalent to (3.77). In addition, (3.73), (3.77) and (3.88) can also be rewritten as follows

$$\hat{x}_k = (I_k - K_k A_k)\overline{x}_k + K_k l_k \tag{3.89}$$

Since $\overline{x}_k$ and $l_k$ are uncorrelated, according to the law of covariance propagation, the variance-covariance matrix of $\hat{x}_k$ is

$$\sum_{\overline{x}_k} = (I_k - K_k A_k) \sum_{\overline{x}_k} \left( I_k - A_k^{\mathrm{T}} K_k^{\mathrm{T}} \right) + K_k \sum_k K_k^{\mathrm{T}} \tag{3.90}$$

From a computational perspective, the covariance matrix for estimating the state parameter vector obtained by (3.90) or (3.73) exhibits higher numerical stability than that of (3.78), because the covariance matrix in (3.78) may be negative or non-positive definite, whereas (3.73) and (3.90) can ensure that the matrix is non-negative definite. By substituting (3.87) into (3.82), the observation residual vector $v_k$ can be derived as follows

$$v_k = -\sum_k \left( A_k \sum_{\bar{x}_k} A_k^{\mathrm{T}} + \sum_k \right)^{-1} (l_k - A_k \bar{x}_k) \tag{3.91}$$

Obviously, the solutions of the unconstrained extremum and the constrained extremum are strictly equivalent. In addition, (3.77) can also be written as follows

$$\hat{x}_k = \boldsymbol{\Phi}_{k,k-1} \hat{x}_{k-1} - K_k \bar{v}_k \tag{3.92}$$

$$\bar{v}_k = A_k \bar{x}_k - l_k \tag{3.93}$$

The above equation indicates that the state parameter vector estimated by the Kalman filter is equivalent to the predicted state vector at time $t_k$ plus a correction vector $-K_k \bar{v}_k$, which is the product of the gain matrix and the innovation vector.

However, the above dynamic model predicts the current state based on only the previous single epoch. In principle, the states of the most recent multiple epochs can provide more information for a more reliable prediction of the current position, thus improving the solution. The window-recursive approach (WRA) is introduced. Assuming that the vehicle motion is stable over a short period (a short time window), the state of the current epoch can be predicted more reliably using multiple recent epochs rather than just one. Suppose that the time-window length contains $n$ epochs, and then, the dynamic model of current epoch $k$ is expressed as follows

$$x_k = \boldsymbol{\Phi}_{(k,k-n:k-1)} x_{(k-n:k-1)} + w_k \tag{3.94}$$

where $x_k$ denotes the state vector to be estimated; $\boldsymbol{\Phi}_{(k,k-n:k-1)}$ denotes the transition matrix that transfers the state information of the previous $n$ epochs into that of the current one, $x_{(k-n:k-1)} = \left( x_{k-n}^{\mathrm{T}}, x_{k-n+1}^{\mathrm{T}}, \ldots, x_{k-1}^{\mathrm{T}} \right)$ denotes the vector consists of the stacked state vectors from the epoch $(k-n)$ to the epoch $(k-1)$ and the corresponding covariance matrix is $D_{\hat{x}_{(k-n:k-1)}}$; $w_k \sim N\left(0, D_{w_k}\right)$ denotes the process noise with zero mean normal distribution. The predicted state $\hat{x}_k$ and its covariance matrix $D_{\bar{x}_k}$ read [10, 11]

$$\bar{x}_k = \boldsymbol{\Phi}_{(k,k-n:k-1)} \hat{x}_{(k-n:k-1)} \tag{3.95}$$

$$D_{\bar{x}_k} = \boldsymbol{\Phi}_{(k,k-n:k-1)} D_{\hat{x}_{(k-n:k-1)}} \boldsymbol{\Phi}_{(k,k-n:k-1)}^{\mathrm{T}} + D_{w_k} \tag{3.96}$$

The linearized GNSS observation model at epoch $k$ reads

$$l_k = H_k x_k + \varepsilon_k \tag{3.97}$$

where $H_k$ denotes the design matrix connecting the state vector $x_k$ with the observation vector $l_k$; $\varepsilon_k \sim N(0, D_k)$ denotes the observation noise with also zero-mean normal distribution. Moreover, the noises $w_k$ of dynamic model and the noise $\varepsilon_k$ of observations are assumed noncorrelated. The Bayesian risk function is established in the sense of generalized LS principle

$$\min_{x_k}: v_k^T D_k^{-1} v_k + \bar{v}_{\bar{x}_k}^T D_{\bar{x}_k}^{-1} \bar{v}_{\bar{x}_k} \tag{3.98}$$

The Kalman filter type of sequential solution for (3.98) is formulated as follows

$$\hat{x}_k = \bar{x}_k + K_k(l_k - H_k \bar{x}_k) \tag{3.99}$$

$$D_{\hat{x}_k} = (I - K_k H_k) D_{\bar{x}_k} \tag{3.100}$$

$$K_k = D_{\bar{x}_k} H_k^T \left( H_k D_{\bar{x}_k} H_k^T + D_k \right)^{-1} \tag{3.101}$$

where $\hat{x}_k$ and $D_{\hat{x}_k}$ denote the estimated state and its covariance; $K_k$ denotes the so-called gain matrix and $I$ denotes the identity matrix which has the same dimension with state vector. The correlation between $\hat{x}_k$ and $\hat{x}_{(k-n:k-1)}$ must be rigorously handled when the time-window moves forward. Inserting (3.95) into (3.99) and applying the error propagation law yields the covariance matrix $D_{\hat{x}_k \hat{x}_{(k-n:k-1)}}$ between $\hat{x}_k$ and $\hat{x}_{(k-n:k-1)}$ as follows

$$D_{\hat{x}_k \hat{x}_{(k-n:k-1)}} = (I - K_k H_k) \Phi_{(k,k-n:k-1)} D_{\hat{x}_{(k-n:k-1)}} \tag{3.102}$$

where $D_{\hat{x}_{(k-n:k-1)}}$ denotes the covariance matrix of states $\hat{x}_{(k-n:k-1)}$ of the first $n$ epochs in the window. Then, the covariance matrix of states $\hat{x}_{(k-n:k)}$ of all $(n+1)$ epochs in the window is symbolized as follows

$$D_{\hat{x}_{(k-n:k)}} = \begin{bmatrix} D_{\hat{x}_{(k-n:k-1)}} & D_{\hat{x}_k \hat{x}_{(k-n:k-1)}}^T \\ D_{\hat{x}_k \hat{x}_{(k-n:k-1)}} & D_{\hat{x}_k} \end{bmatrix} \tag{3.103}$$

with the window moving forward by one epoch, it is easy to analogously derive the filtering solution and its covariance matrix $\hat{x}_{(k-n+1:k+1)}$ and $D_{\hat{x}_{(k-n+1:k+1)}}$.

It points out that the computational burden of the WRA does not strongly depend on the window length, because the only inversion computation in (3.101) is needed and it is not related to the window length. As a special case when the window length $n = 1$, the WRA reduces to the Kalman filter. The main difference between the WRA and the conventional Kalman filter is that the more information from multiple historical epochs is employed than using only one epoch. The key is how to construct a transition matrix to transform the information of these multiple historical epochs into the current epoch, namely how to predict the state of current epoch using the states of multiple historical epochs. The current dynamic characteristics of vehicle motion can be adequately identified by its motion trajectory in the most recent epochs of a window. Therefore, the dynamic model construction can be realized by fitting the vehicle trajectory. In this study, polynomial fitting is used to model the vehicle trajectory. Two important factors are involved in the polynomial fitting with states of multiple epochs in a time-window, which are the window length and the polynomial order. Let the window length be $n$, the polynomial order $m(m \leq n)$ and the between-epoch sampling interval $\delta t$, then the polynomial model of trajectory can be established as follows

$$
\begin{cases}
\hat{x}_{k-n} = a_1 + \eta_1 \\
\hat{x}_{k-n+1} = a_1 + a_2 \delta t + a_3 \delta t^2 + \cdots + a_m \delta t^{m-1} + \eta_2 \\
\cdots \\
\hat{x}_{k-1} = a_1 + a_2((n-1)\delta t) + a_3((n-1)\delta t)^2 + \cdots + a_m((n-1)\delta t)^{m-1} + \eta_n
\end{cases}
\tag{3.104}
$$

where $a_i = \begin{bmatrix} a_{iX} & a_{iY} & a_{iZ} \end{bmatrix}^{\mathrm{T}}$ is the column parameter vector to be estimated; $\eta_i$ is the normal distributed noise vector, $i \in \{1, 2, \ldots, m\}$. The compact matrix form of (3.104) is

$$
\hat{x}_{(k-n:k+1)} = Ma + \eta
\tag{3.105}
$$

with

$$
M = \begin{bmatrix}
1 & 0 & \cdots & 0 \\
1 & \delta t & \cdots & \delta t^{m-1} \\
\cdots & \cdots & \cdots & \cdots \\
1 & (n-1)\delta t & \cdots & ((n-1)\delta t)^{m-1}
\end{bmatrix} \otimes I_3;
$$

$$
a = \begin{bmatrix} a_1 \\ a_2 \\ \cdots \\ a_m \end{bmatrix}; \eta = \begin{bmatrix} \eta_1 \\ \eta_2 \\ \cdots \\ \eta_n \end{bmatrix}.
$$
Treating $\hat{x}_{(k-n:k-1)}$ as the measurement vector, the LS estimate of $a$ is

$$
\hat{a} = \left( M^{\mathrm{T}} Q_{\hat{x}_{(k-n:k-1)}}^{-1} M \right)^{-1} M^{\mathrm{T}} D_{\hat{x}_{(k-n:k-1)}}^{-1} \hat{x}_{(k-n:k-1)}
\tag{3.106}
$$

Then the state vector at epoch $k$ is predicted as follows

$$
x_k = u\hat{a} + w_k
\tag{3.107}
$$

with $\boldsymbol{u} = \left(1, n\delta t, \ldots, (n\delta t)^{m-1}\right) \otimes \boldsymbol{I}_3$. Inserting (3.106) into (3.107) yields

$$\boldsymbol{x}_k = \boldsymbol{u}\left(\boldsymbol{M}^T \boldsymbol{D}_{\hat{\boldsymbol{x}}_{(k-n:k-1)}}^{-1} \boldsymbol{M}\right)^{-1} \boldsymbol{M}^T \boldsymbol{D}_{\hat{\boldsymbol{x}}_{(k-n:k-1)}}^{-1} \hat{\boldsymbol{x}}_{(k-n:k-1)} + \boldsymbol{w}_k \tag{3.108}$$

The corresponding transition matrix becomes

$$\boldsymbol{\Phi}_{(k,k-n:k-1)} = \boldsymbol{u}\left(\boldsymbol{M}^{\mathrm{T}} \boldsymbol{D}_{\hat{\boldsymbol{x}}_{(k-n:k-1)}}^{-1} \boldsymbol{M}\right)^{-1} \boldsymbol{M}^{\mathrm{T}} \boldsymbol{D}_{\hat{\boldsymbol{x}}_{(k-n:k-1)}}^{-1} \tag{3.109}$$

It should be noticed that when $n = m$, Eq. (3.16) reduces to $\boldsymbol{\Phi}_{(k,k-n:k-1)} = \boldsymbol{u}\boldsymbol{M}^{-1}$, which indicates that the transition matrix is independent with the covariance matrix of states in the time-window. In this case, one can alternatively use the Newton's forward differential extrapolation model instead of (3.16) for the higher computation efficiency [10, 11].

# References


1. Leick A, Rapoport L, Tatarnikov D (2015) GPS satellite surveying. Wiley, New Jersey
2. Teunissen PJG, Kleusberg A (1998) GPS for Geodesy. Springer, Berlin
3. Koch KR (1999) Parameter estimation and hypothesis testing in linear models. Springer, Berlin
4. Teunissen PJG (2010) Integer least-squares theory for the GNSS compass. J Geod 84:433–447
5. Teunissen PJG (2024) Adjustment theory: an introduction. TU Delft OPEN Publishing, Delft
6. Teunissen PJG (2007) Best prediction in linear models with mixed integer/real unknowns: theory and application. J Geod 81:759–780
7. Ashcraft C, Grimes RG, Lewis JG (1998) Accurate symmetric indefinite linear equation solvers. SIAM J Matrix Anal Appl 20:513–561
8. Li B (2014) Theory and method of parameter estimation in mixed integer GNSS model. Surveying and Mapping Press, Beijing
9. Guo J, Ou J, Chao R (2005) Partial continuation model and its application in mitigating systematic errors of double-differenced GPS measurements. Prog Nat Sci 15:246–251
10. Zhou Z, Shen Y, Li B (2010) A windowing-recursive approach for GPS real-time kinematic positioning. GPS Solut 14:365–373
11. Zhou Z, Li B (2015) GNSS windowing navigation with adaptively constructed dynamic model. GPS Solut 19:37–48

# Chapter 4
# Integer Ambiguity Resolution

## 4.1 Introduction

The process of resolving carrier-phase integer ambiguities in Global Navigation Satellite System (GNSS) is known as carrier-phase ambiguity resolution. It involves determining the carrier-phase ambiguities as integer values. This process is crucial for fast and precise GNSS parameter estimation and is widely applied across various GNSS models used in fields such as navigation, surveying, geodesy, and geophysics. The underlying theory of GNSS carrier-phase ambiguity resolution is based on the concept of integer inference. This chapter focuses on explaining this theory and its practical applications.

Carrier-phase integer ambiguity resolution is essential for achieving fast and accurate GNSS parameter estimation. Once successfully resolved, high-precision carrier-phase data can be treated as precise pseudorange data, enabling accurate positioning and navigation.

GNSS ambiguity resolution is applicable to a wide range of current and future GNSS models, with applications in fields such as surveying, navigation, geodesy, and geophysics. These models can vary significantly in terms of complexity and diversity. For example, they range from simple single-receiver or single-baseline models used for kinematic positioning to more complex multi-baseline models that are used for studying geodynamic phenomena. Some models include the relative geometry between the receiver and satellites, while others may not. Additionally, models can differ based on whether the slave receiver(s) are stationary or moving, or whether differential atmospheric delays (e.g., ionospheric and tropospheric effects) are treated as unknowns.

The structure of this chapter is as follows: The mixed-integer GNSS model is introduced, which forms the foundation for all integer ambiguity resolution methods. Next, the strategies for integer estimation are presented, followed by a discussion on the impact of unmodeled errors on ambiguity resolution and their role in the solution. The methods for integer evaluation and validation are then outlined. This



B. Li et al., *GNSS Real-Time Kinematic Positioning*, Navigation: Science and Technology 17, https://doi.org/10.1007/978-981-96-9116-6_4

chapter also highlights the advantages of partial ambiguity resolution techniques in practical GNSS applications. In response to the evolving trends of multi-frequency and multi-GNSS systems, methods for resolving multi-frequency ambiguities are presented, providing theoretical guidance for GNSS users.

## 4.2  Mixed-Integer Model and Integer Ambiguity Estimation

The GNSS observation model for integer ambiguity resolution can be summarized as a mixed integer model as

$$\mathrm{E}(y) = Aa + Bb, \quad \mathrm{D}(y) = Q_{yy} \tag{4.1}$$

where $y \in \mathbb{R}^m$ is the vector of pseudo-range and carrier-phase observables. $a \in \mathbb{Z}^n$ is the vector of unknown integer ambiguities. $b \in \mathbb{R}^p$ is the vector of real-valued unknown parameters (e.g., baseline vector). $(A,B) \in \mathbb{R}^{m \times (n+p)}$ is the full-rank coefficient matrix. $Q_{yy}$ is the variance-covariance matrix of $y$. $\mathrm{E}(\cdot)$ and $\mathrm{D}(\cdot)$ denote the expectation operator and the dispersion operator, respectively. The objective function of (4.1) solved using the least squares (LS) criterion is expressed as

$$(y - Aa - Bb)^{\mathrm{T}} Q_{yy}^{-1} (y - Aa - Bb) = \min \tag{4.2}$$

Actually, the objective function can be transformed into [1]

$$\|y - A\hat{a} - B\hat{b}\|_{Q_{yy}}^2 + \|\hat{a} - a\|_{Q_{\hat{a}\hat{a}}}^2 + \|\hat{b}(a) - b\|_{Q_{\hat{b}(a)\hat{b}(a)}}^2 = \min \tag{4.3}$$

where $\| \cdot \|_Q^2 = (\cdot)^{\mathrm{T}} Q^{-1} (\cdot)$. The extremum problem in Eq. (4.3) is now decomposed into three criteria, expressed as

$$\|y - A\hat{a} - B\hat{b}\|_{Q_{yy}}^2 = \min, \hat{a} \in \mathbb{R}^n \tag{4.4}$$

$$\|\hat{a} - a\|_{Q_{\hat{a}\hat{a}}}^2 = \min, a \in \mathbb{Z}^n \tag{4.5}$$

$$\|\hat{b}(a) - b\|_{Q_{\hat{b}(a)\hat{b}(a)}}^2 = \min \tag{4.6}$$

The solution to the three extremum problems above is computed in four steps. In fact, this four-step procedure is also called integer estimation [2].

Step 1: Float Solution

Disregarding the integer property of the ambiguities $\boldsymbol{a} \in \mathbb{Z}^n$, the solution to Eq. (4.4) is obtained as

$$\begin{bmatrix} \hat{\boldsymbol{a}} \\ \hat{\boldsymbol{b}} \end{bmatrix}, \quad \begin{bmatrix} \boldsymbol{Q}_{\hat{a}\hat{a}} & \boldsymbol{Q}_{\hat{a}\hat{b}} \\ \boldsymbol{Q}_{\hat{b}\hat{a}} & \boldsymbol{Q}_{\hat{b}\hat{b}} \end{bmatrix} \tag{4.7}$$

where $\hat{\boldsymbol{a}} = (\hat{a}_1, \ldots, \hat{a}_n)^{\mathrm{T}} \in \mathbb{R}^n$ is the float ambiguity solution.

Step 2: Integer Estimation

Apply criterion (4.5) to find the integer solution of the ambiguities. With an admissible integer map $S : \mathbb{R}^n \mapsto \mathbb{Z}^n$, the fixed integer ambiguity vector $\check{\boldsymbol{a}} = (\check{a}_1, \ldots, \check{a}_n)^{\mathrm{T}} \in \mathbb{Z}^n$ is obtained as

$$\check{\boldsymbol{a}} = S(\hat{\boldsymbol{a}}) \tag{4.8}$$

The integer map is admissible when its pull-in-regions $\mathcal{P}_z = \{x \in \mathbb{R}^n | S(x) = z\}, z \in \mathbb{Z}^n$ cover $\mathbb{R}^n$ while being disjoint and integer translational invariant [3]. Some popular choices of mapping function $S$ are available, including integer rounding (IR), integer bootstrapping (IB) and integer least squares (ILS). Of all choices, ILS is proven to be optimal and can be efficiently mechanized in the least-squares ambiguity decorrelation adjustment (LAMBDA) method [2].

Step 3: Integer Evaluation and Validation

An integer evaluation and validation test are devised to determine whether or not the integer solution $\check{\boldsymbol{a}}$ from Step 2 is sufficiently more reliable than any other integer candidate. Several validation tests are currently used in practice, such as the R-ratio test [4], W-ratio test [5], difference test [6] and project test [7]. If the integer validation test is passed, the reliable solution $\check{\boldsymbol{a}}$ is used to update the baseline parameter. Otherwise, the float solution in Step 1 is adopted for users. Regardless of the validation method used, the underlying goal is essentially the same: to assess the distinguishability among the integer candidates. The evaluation determines whether the integer solution is sufficiently distinct and reliable compared to other alternatives. If the integer solution is validated, it is adopted as the optimal solution; otherwise, the float solution from Step 1 is retained for the users.

Step 4: Fixed Solution

The float solution of the baseline parameters and its variance-covariance matrix are updated with the fixed integer solution, written as

$$\check{\boldsymbol{b}} = \hat{\boldsymbol{b}} - \boldsymbol{Q}_{\hat{b}\hat{a}} \boldsymbol{Q}_{\hat{a}\hat{a}}^{-1} (\hat{\boldsymbol{a}} - \check{\boldsymbol{a}}) \tag{4.9}$$

$$\boldsymbol{Q}_{\check{b}\check{b}} = \boldsymbol{Q}_{\hat{b}\hat{b}} - \boldsymbol{Q}_{\hat{b}\hat{a}} \boldsymbol{Q}_{\hat{a}\hat{a}}^{-1} \boldsymbol{Q}_{\hat{a}\hat{b}} \tag{4.10}$$

where $\breve{b}$ is the fixed baseline solution. $Q_{\hat{a}\hat{b}}$ and $Q_{\hat{b}\hat{a}}$ are the cross-covariance of $\hat{a}$ and $\hat{b}$. $Q_{\hat{b}\hat{b}}$ is the variance-covariance matrix of $\hat{b}$. It should be noted that if the success rate in Step 3 is not high enough, the fixed solution $\breve{b}$ is not necessarily better than the float solution $\hat{b}$ [8].

To gain a deeper understanding of the three criteria in Eq. (4.3) and their relationship with Eq. (4.2), we start with Eq. (4.2). Based on the principle that real parameters are differentiable while integer parameters are non-differentiable, we differentiate Eq. (4.2) to obtain

$$-2A^{\mathrm{T}}Q_{yy}^{-1}\left(y - A\hat{a} - B\hat{b}\right) = 0 \tag{4.11}$$

The real parameters are expressed in terms of the observations and the integer parameters as

$$\hat{b} = \left(A^{\mathrm{T}}Q_{yy}^{-1}A\right)^{-1}\left(A^{\mathrm{T}}Q_{yy}^{-1}y - A^{\mathrm{T}}Q_{yy}^{-1}Ba\right) \tag{4.12}$$

Substituting Eq. (4.12) into Eq. (4.2) gives

$$(a - a_0)^{\mathrm{T}}H(a - a_0) + \zeta = \min \tag{4.13}$$

where $\zeta = y^{\mathrm{T}}Q_{yy}^{-1}QQ_{yy}^{-1}y - y^{\mathrm{T}}Q_{yy}^{-1}QQ_{yy}^{-1}AH^{-1}A^{\mathrm{T}}Q_{yy}^{-1}QQ_{yy}^{-1}y$, $Q = Q_{yy} - B(B^{\mathrm{T}}Q_{yy}^{-1}B)^{-1}B^{\mathrm{T}}$, $H = A^{\mathrm{T}}Q_{yy}^{-1}QQ_{yy}^{-1}A$, $a_0 = H^{-1}A^{\mathrm{T}}Q_{yy}^{-1}QQ_{yy}^{-1}y$. It is clear that $\zeta$ is a constant, so Eq. (4.13) is equivalent to

$$(a - a_0)^{\mathrm{T}}H(a - a_0) = \min \tag{4.14}$$

It is easy to prove that $Q_{\hat{a}\hat{a}} = H^{-1}$, $a_0 = \hat{a}$, therefore Eq. (4.14) is equivalent to

$$(a - \hat{a})^{\mathrm{T}}Q_{\hat{a}\hat{a}}^{-1}(a - \hat{a}) = \min \tag{4.15}$$

The above equation is equivalent to Eq. (4.2). Furthermore, since Eq. (4.15) is the same as Eq. (4.5), it shows that solving the mixed-integer model (4.1) using the LS criterion is equivalent to the second criterion (4.5) proposed by Teunissen. The other two criteria (4.4) and (4.6) can be understood as auxiliary computations for solving (4.1). Furthermore, when the mixed-integer model is subject to constraints, such as constraints on both integer parameters $a$ and real parameters $b$, the following conclusion can be drawn: Constraints on the real parameters can improve the quality of the float solutions for the integer parameters, whereas constraints on the integer parameters provide no significant benefit. In other words, constraints on the integer parameters serve only to eliminate incorrect candidates during the integer search process.

### *4.2.1  Strategy of Integer Estimation*

Regardless of the integer estimation method employed, the objective is to solve the minimization problem presented in Eq. (4.5). However, since the ambiguity $a$ is an integer and cannot be directly differentiated, the core approach of integer estimation methods is to search for a set of integer candidate combinations within the infinite integer space, and then assess whether they minimize the objective function by substituting them into Eq. (4.5). The fewer the integer candidate combinations, the higher the efficiency in determining the optimal integer. Consequently, all proposed integer estimation methods aim to reduce the number of integer candidates, thereby facilitating the rapid determination of the optimal integer solution.

Over the past few decades, several integer estimation strategies have been developed to address the discrete integer nature of the ambiguity $a$. These strategies include enumeration, fast ambiguity resolution approach (FARA) [9], ambiguity function method (AFM) [10, 11], least squares search method (LSSM) [12], and the LAMBDA [2], among others. Among them, the LAMBDA method performs particularly well in terms of integer search efficiency and has been widely applied.

In the LAMBDA method, the integer solution to (4.5) is found by means of an efficient search over the ellipsoidal search space defined as

$$(a - \hat{a})^{\mathrm{T}} Q_{\hat{a}\hat{a}}^{-1} (a - \hat{a}) \leq \chi^2 \tag{4.16}$$

The search speed depends on the size $\chi^2$ and the shape of the ellipsoid. The positive constant $\chi^2$ can be predetermined using different strategies [13] and then gradually shrunk during the search [13, 14]. The shape and orientation of the ellipsoid are defined by the variance–covariance matrix $Q_{\hat{a}\hat{a}}$ of the float ambiguity estimates. Since a high correlation among the ambiguities may lead to search halting which in turn makes the search time-consuming, the decorrelated ambiguities are used instead of the original ones in the LAMBDA method. After decorrelation, the original ambiguities are transformed to the decorrelated ones using $z = Z^{\mathrm{T}} a$, and then the search is conducted in the transformed ellipsoid

$$(z - \hat{z})^{\mathrm{T}} Q_{\hat{z}\hat{z}}^{-1} (z - \hat{z}) \leq \chi^2, \forall z \in \mathbb{Z}^n \tag{4.17}$$

where $\hat{z} = Z^{\mathrm{T}} \hat{a}$ and $Q_{\hat{z}\hat{z}} = Z^{\mathrm{T}} Q_{\hat{a}\hat{a}} Z$. Let the triangular factorization of the decorrelated variance-covariance matrix be $Q_{\hat{z}\hat{z}} = L^{\mathrm{T}} D L$, the search over the ellipsoid (4.17) is then based on the evaluation of the scalar intervals

$$\tilde{z}_i - \sigma_{z_{i|I}} \sqrt{\chi^2 - \sum_{j=i+1}^{n} \frac{(z_j - \tilde{z}_j)^2}{d_j}} \leq \hat{z}_i \leq \tilde{z}_i - \sigma_{z_{i|I}} \sqrt{\chi^2 + \sum_{j=i+1}^{n} \frac{(z_j - \tilde{z}_j)^2}{d_j}},$$

$$(i = 1, \ldots, n) \tag{4.18}$$

with

$$\tilde{z}_i = \hat{z}_i - \sum_{j=i+1}^{m} \left(z_j - \tilde{z}_j\right) l_{ji} \tag{4.19}$$

where $L$ is a unit lower triangular matrix and $l_{ji}(j > i)$ is its element of the $j$th row and the $i$th column; D is a diagonal matrix whose $i$th element, $d_i = \sigma^2_{\hat{z}_{i|I}}$, is the conditional variance of the $i$th transformed ambiguity $z_i$ conditioned on the transformed ambiguities $I = \{i + 1, \cdots, n\}$. Based on these bounds, the search is performed. We must emphasize that the LAMBDA method primarily enhances the efficiency of integer ambiguity search and focuses solely on the variance-covariance matrix of the ambiguity $\hat{a}$. However, unmodeled errors in the GNSS model can introduce biases in the ambiguity $\hat{a}$, which presents a limitation when applying the LAMBDA method. In such cases, neglecting the bias in $\hat{a}$ while relying solely on the LAMBDA method is problematic.

### 4.2.2  *Integer Ambiguity Resolution in the Presence of Biases*

In GNSS data processing, the observational errors inherent in the GNSS signals are typically modeled and corrected. The residuals obtained after model correction are subsequently treated as the parameters to be estimated, thereby constituting the widely adopted mixed-integer GNSS linearized model, which is expressed as

$$E(y) = Aa + Bb, D(y) = Q_{yy} \tag{4.20}$$

However, some GNSS observational errors, such as multipath errors or atmospheric biases that are often neglected, are difficult to model. In practice, these errors cannot always be ignored in GNSS observation models, as they may lead to biased estimates of the ambiguity parameters. The mixed-integer GNSS linearized model that accounts for these errors is expressed as

$$E(y) = Aa + Bb + C\nabla, D(y) = Q_{yy} \tag{4.21}$$

where $\nabla \in \mathbb{R}^q$ is other nuisance parameter vector. Their design matrices are $A \in \mathbb{R}^{m \times n}$, $B \in \mathbb{R}^{m \times p}$ and $C \in \mathbb{R}^{m \times q}$ with $[ABC]$ of full column rank. It is emphasized that the nuisance parameter $\nabla$ is set up in (4.21) to compensate the non-ignorable systematic biases. However, if $\nabla \neq 0$, i.e., if the atmospheric biases are not so small to be completely ignorable, biased parameter estimates will be obtained with (4.21).

In principle, reducing the parameters in an adjustment system can improve the model strength but as trade-off introduces biases in the parameter estimates if the systematic effects specified by these parameters cannot be completely ignored. The three-step procedure for solving model (4.21) in the presence of atmospheric biases ($\nabla \neq 0$) will thus lead to a biased solution, as described next.

Step 1: Float solution

The float solution is again obtained by disregarding the integer constraints on the ambiguities,

$$\begin{bmatrix} \hat{a}^b \\ \hat{b}^b \end{bmatrix} \sim N\left( \begin{bmatrix} a + \Delta a \\ b + \Delta b \end{bmatrix}, \begin{bmatrix} Q^b_{\hat{a}\hat{a}} & Q^b_{\hat{a}\hat{b}} \\ Q^b_{\hat{b}\hat{a}} & Q^b_{\hat{b}\hat{b}} \end{bmatrix} \right) \tag{4.22}$$

with the variance-covariance matrix

$$\begin{bmatrix} Q^b_{\hat{a}\hat{a}} & Q^b_{\hat{a}\hat{b}} \\ Q^b_{\hat{b}\hat{a}} & Q^b_{\hat{b}\hat{b}} \end{bmatrix} = \begin{bmatrix} A^{\mathrm{T}}Q_{yy}^{-1}A & A^{\mathrm{T}}Q_{yy}^{-1}B \\ A^{\mathrm{T}}Q_{yy}^{-1}A & B^{\mathrm{T}}Q_{yy}^{-1}B \end{bmatrix}^{-1} \tag{4.23}$$

and the bias vector

$$\begin{bmatrix} \Delta a \\ \Delta b \end{bmatrix} = \begin{bmatrix} Q^b_{\hat{a}\hat{a}} & Q^b_{\hat{a}\hat{b}} \\ Q^b_{\hat{b}\hat{a}} & Q^b_{\hat{b}\hat{b}} \end{bmatrix} \begin{bmatrix} A^{\mathrm{T}}Q_{yy}^{-1}C \\ A^{\mathrm{T}}Q_{yy}^{-1}C \end{bmatrix} \nabla \tag{4.24}$$

where the superscripts "b" are used to denote the biased terms. It can be easily shown that the variance-covariance matrix (4.23) is smaller than the unbiased one and also $Q^b_{\hat{a}\hat{a}} \leq Q_{\hat{a}\hat{a}}$ and $Q^b_{\hat{b}\hat{b}} \leq Q_{\hat{b}\hat{b}}$. The reason is that the number of unknown parameters is reduced.

Step 2: Integer estimation

Similarly, as with the unbiased model, the float ambiguity estimate $\hat{a}^b$ is used to compute its integer counterpart:

$$\check{a}^b = S(\hat{a}^b) \tag{4.25}$$

Step 3: Fixed solution

Once the ambiguities are fixed by applying Steps 1 and 2 described here, one should never disregard the biases for computing the precise baseline solution, since even if $\check{a}^b$ is correct, the bias $\Delta b$ will propagate in the fixed baseline solution. Therefore, the float solution of the baseline parameters and its variance-covariance matrix are updated with the fixed integer solution, written as

$$\check{b} = \hat{b} - Q_{\hat{b}\hat{a}}Q_{\hat{a}\hat{a}}^{-1}(\hat{a} - \check{a}^b) \tag{4.26}$$

$$Q_{\breve{b}\breve{b}} = Q_{\hat{b}\hat{b}} - Q_{\hat{b}\hat{a}}Q_{\hat{a}\hat{a}}^{-1}Q_{\hat{a}\hat{b}} \tag{4.27}$$

This is equivalent to solving the model

$$\mathrm{E}\left(\boldsymbol{y} - \boldsymbol{A}\breve{\boldsymbol{a}}^{b}\right) = \boldsymbol{B}\boldsymbol{b} + \boldsymbol{C}\nabla, \boldsymbol{D}(\boldsymbol{y}) = \boldsymbol{Q}_{yy} \tag{4.28}$$

Note from (4.10) and (4.27) that the precision $Q_{\breve{b}\breve{b}}$ of the fixed baseline solution with $\breve{a}$ is the same as with $\breve{a}^{b}$ if in both cases the uncertainty of the fixed ambiguity solution can be ignored. Therefore, it will be better to use $\breve{a}^{b}$ if the probability that this integer solution is correct is higher than that $\breve{a}$ being correct. However, in practical applications, these unmodeled biases are difficult to handle through explicit modeling, yet they still exist. Consistently neglecting these biases can lead to biased ambiguity estimates. In practice, however, a simplified model, as expressed in Eq. (4.20), is typically used for processing.

### 4.2.3  Integer Evaluation and Validation

The correctness of the final integer solution obtained through integer estimation methods is crucial throughout the ambiguity resolution process. Once an incorrect integer solution is accepted, the remaining real parameter solutions, such as position estimates, may become significantly erroneous. Therefore, when applying the integer solution obtained from integer estimation methods, it is essential to perform both evaluation and validation test. Evaluation test involves assessing the correctness of the integer solution from a probabilistic perspective, providing an internally consistent measure of integer correctness. Validation test, on the other hand, involves evaluating the correctness of the integer solution through mutual comparison among the integer candidates.

A very high positioning performance can only be guaranteed if the estimated integer ambiguities are correct. It is therefore very important to assess the probability of correct integer estimation. This probability is called the success rate and only if it is very close to 1, it is possible to rely on the integer solution without further validation [15]. In that case the integer ambiguity solution can be assumed to be deterministic, and the variance-covariance matrix of the fixed baseline solution is obtained in Eqs. (4.10) and (4.27).

The essence of correct integer estimation was described previously. It is thus important to have means available to evaluate the ambiguity success rate, i.e. the probability of correct integer estimation $P_s$ [15]. This success rate is equal to the probability that $\hat{a}$ resides in the correct pull-in region $\mathcal{P}_a$ with $a$ the true but unknown ambiguity vector

$$P_{s} = P(\breve{a} = a) = P(\hat{a} \in \mathcal{P}_{a}) = \int_{\mathcal{P}_{a}} f_{\hat{a}}(\boldsymbol{x}|\boldsymbol{a})d\boldsymbol{x} \tag{4.29}$$

The probability density function (PDF) of the float ambiguities, $f_{\hat{a}}(x|a)$, is assumed to be the normal PDF with mean $a$:

$$f_{\hat{a}}(x|a) = \frac{1}{\sqrt{det(2\pi Q_{\hat{a}\hat{a}})}} \exp\left\{-\frac{1}{2}(x-a)^{\mathrm{T}} Q_{\hat{a}\hat{a}}^{-1}(x-a)\right\} \tag{4.30}$$

As the pull-in regions of the integer estimators are integer translation invariant, the success rate can also be evaluated as

$$P_s = \int_{\mathcal{P}_0} f_{\hat{a}}(x|0)d\mathbf{x} \tag{4.31}$$

The success rates also depend on the selected integer estimation method, since the pull-in region is different for IR, IB and ILS. In [16] it was proven that

$$P(\check{a}_{\mathrm{IR}} = a) \leq P(\check{a}_{\mathrm{IB}} = a) \leq P(\check{a}_{\mathrm{ILS}} = a) \tag{4.32}$$

The success rate cannot be evaluated exactly in all cases due to the complex integration over the pull-in region. It is of course important to be able to have good approximations of the success rate in case exact evaluation is not feasible. A lower bound is an approximation of the success rate, which is guaranteed to be smaller than or equal to the actual success rate. As such it is particularly useful. However, if the lower bound is not tight, this may result in an unnecessarily high rejection rate as the success rate is deemed too low. An upper bound can be useful as well, especially in combination with a lower bound, since it then tells the user in which range the success rate will be. If the upper bound is below a user-defined threshold, one cannot expect ambiguity resolution to be successful. In addition, for IR and IB it may be useful to have an upper bound which is invariant for the class of admissible ambiguity transformations. Different approximations and bounds were proposed in the literature, an evaluation of some of the bounds was made in [8].

Integer rounding success rates

The $n$-fold integral over the IR pull-in region is difficult to evaluate. Only if the variance-covariance matrix $Q_{\hat{a}\hat{a}}$ is diagonal will the success rate become equal to the $n$-fold product of the univariate success rates. When $Q_{\hat{a}\hat{a}}$ is not diagonal, a lower bound written as

$$P_{s,\mathrm{IR}} = P(\check{a}_{\mathrm{IR}} = a) \geq \prod_{i=1}^{n}\left(2\Phi\left(\frac{1}{2\sigma_{\hat{a}_i}}\right) - 1\right) \tag{4.33}$$

with $\Phi(x) = \frac{1}{\sqrt{2\pi}}\int_{-\infty}^{x}\exp\left\{-\frac{1}{2}t^2\right\}dt$.

According to Eq. (4.32), IB will always result in a success rate higher than or equal to the IR success rate if the same parameterization of the float ambiguities is used. Hence, the IB success rate can be used as an upper bound for IR. In the next subsection it will be shown that the IB success rate can in fact be evaluated exactly.

Integer Bootstrapping success rates

In the case of bootstrapping, the success rate can be evaluated exactly using

$$P_{s,\text{IB}} = P(\check{a}_{\text{IB}} = a) = \prod_{i=1}^{n}\left(2\Phi\left(\frac{1}{2\sigma_{\hat{a}_{i|I}}}\right) - 1\right) \tag{4.34}$$

For bootstrapping we thus have an exact and easy-to-compute formula for the success rate. An upper bound is given by

$$P_{s,\text{IB}} \leq \left(2\Phi\left(\frac{1}{2\text{ADOP}}\right) - 1\right)^{n} \tag{4.35}$$

with $\text{ADOP} = \sqrt{\det(Q_{\hat{a}\hat{a}})}^{1/n}$, representing the ambiguity dilution of precision (ADOP) and expressed in units of cycles. The ADOP is a diagnostic that captures the main characteristics of the ambiguity precision. When the ambiguities are completely decorrelated, the ADOP equals the geometric mean of the standard deviations of the ambiguities, hence, it can be considered as a measure of the average ambiguity precision.

Integer least squares success rates

Due to the complex geometry of the ILS pull-in region, the multivariate integral can only be evaluated by using monte carlo simulation. In addition, several lower and upper bounds of the ILS success rate have been proposed. It was already mentioned that IB may perform close to optimal if applied to decorrelated ambiguities. Therefore, the corresponding IB success rate can be used as a lower bound for the ILS success rate

$$P_{s,\text{ILS}} = P(\check{a}_{\text{ILS}} = a) \geq P_{s,\text{IB}} \tag{4.36}$$

The conditional standard deviations $\sigma_{\hat{z}_{i|I}}$ of the decorrelated ambiguities must be used.

Consequently, the invariant upper bound of the IB success rate may serve as an approximation of the ILS success rate. Furthermore, an upper bound for the ILS success rate based on the ADOP can be given as

$$P_{s,\text{ILS}} \leq P\left(\chi^2(n, 0) \leq \frac{c_n}{\text{ADOP}^2}\right) \tag{4.37}$$

with $c_n = \frac{\left(\frac{n}{2}\Gamma\left(\frac{n}{2}\right)\right)^{2/n}}{\pi}$. This bound was introduced in [17], while the proof was given in [18].

As previously demonstrated, utilizing the integer solution $\breve{a}$ is meaningful only if the ambiguity success rate $P_a$ is sufficiently high. Otherwise, there would be an unacceptable risk of large errors in the fixed solution $\breve{b}$. This consideration leads to the following decision rule for determining the outcome of the ambiguity resolution process,

$$\text{outcome} = \begin{cases} \breve{a} \in \mathbb{Z}^n & \text{if } P(\hat{a} \in \mathcal{P}_a) \geq P_0 \\ \hat{a} \in \mathbb{R}^n & \text{otherwise} \end{cases} \tag{4.38}$$

According to this rule, the integer solution $\breve{a}$ is accepted only if the success rate is beyond a user-defined threshold $P_0$; otherwise, the float solution $\hat{a}$ is retained. This is a model-driven rule, meaning the decision depends solely on the strength of the underlying model rather than the actual float solution $\hat{a}$ itself. Instead, the PDF of $\hat{a}$, through the probability $P(\hat{a} \in \mathcal{P}_a)$, influences the decision. Alternatively, a data-driven decision rule can be adopted, where the decision is based directly on the observed float solution rather than relying purely on the model's statistical properties. Such rules are of the form

$$\text{outcome} = \begin{cases} \breve{a} \in \mathbb{Z}^n & \text{if } \mathcal{T}(\hat{a}) \geq \tau_0 \\ \hat{a} \in \mathbb{R}^n & \text{otherwise} \end{cases} \tag{4.39}$$

with testing function $\mathcal{T} : \mathbb{R}^n \mapsto \mathbb{R}^n$ and user-defined threshold value $\tau_0$. The integer solution $\breve{a}$ is accepted if the value of $\mathcal{T}(\hat{a})$ is sufficiently large; otherwise, it is rejected in favor of the float solution $\hat{a}$. This is a data-driven rule, as the decision is based on the actual value of the float solution, which is evaluated using $\mathcal{T}(\hat{a})$.

In practice, data-driven rules are often preferred in integer validation, as they offer greater flexibility in assessing the correctness of the integer solution. Various testing functions, such as the ratio test, difference test, and projector test, can be applied in this context. Each of these tests belongs to the class of integer aperture estimators, as introduced in the literature [19, 20]. A review and evaluation of these tests can be found in [21, 22].

The ratio test is a very popular validation test in practice. The ratio test is used here, i.e., accept $\breve{a}$ if:

$$\frac{\|\hat{a} - a_2\|_{Q_{\hat{a}\hat{a}}}^2}{\|\hat{a} - a\|_{Q_{\hat{a}\hat{a}}}^2} \geq \tau_0 \tag{4.40}$$

where $a_2$ is the second-best integer candidate. In many software packages a fixed value for the ratio is used, e.g., $\tau_0 = 3$. The difference test leads to acceptance of $\breve{a}$ if:

$$\|\hat{a} - a_2\|^2_{Q_{\hat{a}\hat{a}}} - \|\hat{a} - a\|^2_{Q_{\hat{a}\hat{a}}} \geq 1/\tau_0 \tag{4.41}$$

The main advantage of data-driven rules over model-driven rules lies in their increased flexibility, particularly in controlling the fail rate. With data-driven rules, users can have full control over the fail rate, independent of the strength of the underlying GNSS model. This level of control is not possible with model-driven rules.

## 4.3  Partial Ambiguity Resolution

In theory, only the ambiguities that have enough correlation with the baselines are necessary to be correctly fixed to improve the baseline precision. For the ambiguities that have marginal correlation with baseline, we do not need to strive for their fixing, particularly when they are difficult to be fixed. In the section, we will demonstrate that in an ambiguity vector, not all ambiguities have the comparable contribution to the baseline. In other words, we will show that the full ambiguity resolution (FAR) is not always necessary.

Firstly, availability of partial ambiguity resolution (PAR) is higher than FAR. For some cases, before the FAR, the PAR results can be applied, to a great extent, to achieve the satisfied baseline/ionosphere solutions that can directly provide service to users. Secondly, the PAR is more reliable than FAR. In other words, the baseline/ionosphere solution of PAR is safer than that of FAR, especially when the success-rate of FAR is not high enough. Figure 4.1 illustrates the variation of ionosphere precision over time for PAR and FAR from a simulation experiment. Obviously, the PAR is better than FAR in the ionosphere solution. Thirdly, With the accumulation of the observations, PAR turns to the FAR as long as FAR is possible. Therefore, PAR is more flexible than FAR.

Let ambiguity vector $\hat{a} = \begin{bmatrix} \hat{a}_1^{\mathrm{T}} & \hat{a}_2^{\mathrm{T}} \end{bmatrix}^{\mathrm{T}}$ and $\hat{a}_1$ be fixed to $\check{a}_1$, then the baseline solution with PAR is

$$\check{b}_{\mathrm{PAR}} = \hat{b} - Q_{\hat{b}\hat{a}_1} Q_{\hat{a}_1\hat{a}_1}^{-1} (\hat{a}_1 - \check{a}_1) \tag{4.42}$$

and its precision is

$$Q_{\check{b}_{\mathrm{PAR}}\check{b}_{\mathrm{PAR}}} = Q_{\hat{b}\hat{b}} - Q_{\hat{b}\hat{a}_1} Q_{\hat{a}_1\hat{a}_1}^{-1} Q_{\hat{a}_1\hat{b}} \tag{4.43}$$

For the remaining ambiguity $\hat{a}_2$ if it can also be fixed to $\check{a}_2$, then precision gain of baseline is

$$Q_{\hat{a}_2,\mathrm{gain}} = Q_{\check{b}_{\mathrm{PAR}}} - Q_{\check{b}_{\mathrm{FAR}}} = Q_{\hat{b}\hat{a}_{2|1}} Q_{\hat{a}_{2|1}}^{-1} Q_{\hat{a}_{2|1}\hat{b}} \tag{4.44}$$

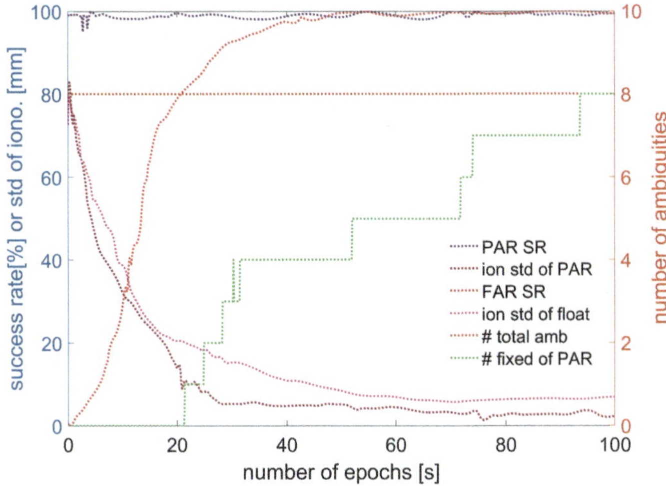


**Fig. 4.1** Variation of baseline precision over time for PAR and FAR

This quantity $Q_{\hat{a}_2,\text{gain}}$ can be used as an indicator to measure the contribution of the remaining ambiguity $\hat{a}_2$ to the baseline solution. If it is small enough, we can say the fixing of $\hat{a}_2$ has no effect on the baseline precision improvement; whereas if it is still significant, we need to strive for its fixing solution to improve the baseline precision.

In practical data processing, several PAR strategies are available for users to choose.

Success Rate-Based Selection: This strategy involves calculating the success rate for each ambiguity component, fixing only those that exceed a predefined threshold. Ambiguities with a lower success rate are kept as float solutions, while those with higher success rates are fixed. This approach ensures that only the most reliable ambiguities contribute to the final solution.

Contribution-Based Selection: Instead of relying solely on the success rate, this method focuses on the contribution each ambiguity makes to the baseline solution. Ambiguities that have a minimal impact on the baseline are left unresolved as float solutions. This helps ensure that the fixed ambiguities significantly influence the accuracy of the solution. A common approach is to analyze the contribution through the variance-covariance matrix.

Stepwise or Sequential Fixing: In this method, ambiguities are resolved one by one, starting with those that are most likely to be fixed correctly. This stepwise process reduces the chance of error propagation, as each ambiguity is carefully assessed before the next one is resolved.

Quality-Control-Based PAR: This approach involves applying statistical quality control tests, such as ratio tests and difference tests, to evaluate the reliability of each ambiguity solution. Only those ambiguities that pass these quality checks are fixed, ensuring the robustness of the final solution. Methods like R-ratio test, W-ratio test, difference test and project test are commonly used.

Subset Optimization-Based Selection: This strategy selects the optimal subset of ambiguities to resolve from a given set, with the aim of maximizing the precision and reliability of the solution. By carefully selecting the most impactful ambiguities, this method improves the overall performance of the ambiguity resolution.

These strategies provide users with various ways to balance precision, reliability, and availability when resolving ambiguities, offering flexibility in addressing different observational conditions.

## 4.4  Ambiguity Resolution with Multiple Frequencies

With the development trend of multi-frequency of GNSS, ambiguity resolution has been extended from three-frequency carrier ambiguity resolution (TCAR) to quad-frequency carrier ambiguity resolution (QCAR), penta-frequency carrier ambiguity resolution (PCAR), and even multi-frequency carrier ambiguity resolution (MCAR). In this section, MCAR is first briefly introduced, and the basic theory of MCAR is expounded, and finally the advantages and challenges of MCAR are discussed.

To generalize the scenarios, we derive the formulae based on the arbitrary number of frequencies. The derivations are analogous to our previous study for triple-frequency case [23]. The single-epoch double-differenced (DD) geometry-based (GB) model with ionospheric constraints is presented at first, from which the various models are then reduced. These reduced models can be used to simplify the ambiguity resolution and positioning under different specific situations.

Considering the residual ionospheric effects, the single-epoch DD observation equations of code and phase on $f$ frequencies read

$$\mathrm{E}\left(\begin{bmatrix} p \\ \phi \end{bmatrix}\right) = \begin{bmatrix} e_f \otimes A & \mu \otimes I_s & 0 \\ e_f \otimes A & -\mu \otimes I_s & \Lambda \otimes I_s \end{bmatrix} \begin{bmatrix} x \\ \iota \\ a \end{bmatrix} \tag{4.45}$$

where $p = [p_1^{\mathrm{T}}, \ldots, p_f^{\mathrm{T}}]^{\mathrm{T}}$ is the $f$-frequency code observations with $p_j$ the observations of frequency $f_j$. $\phi$ is the $f$-frequency phase observations with the same structure as $p$. $A$ is the design matrix to baseline parameter $x$. $\mu = [\mu_1, \ldots, \mu_f]^T$ with $\mu_j = f_1^2/f_j^2$ is the scalar vector to DD ionosphere parameters $\iota$. $\Lambda = \mathrm{diag}([\lambda_1, \ldots, \lambda_f])$ is diagonal matrix of wavelengths to DD ambiguities $a = [a_1^{\mathrm{T}}, \ldots, a_f^{\mathrm{T}}]^{\mathrm{T}}$. The subscript $s$ denotes the number of DD satellite pairs. Here we neglect the residual tropospheric effects in the DD observations since the tropospheric delays have been corrected by at least 90% with a standard tropospheric model, and the residuals have limited effects on the ambiguity resolution. Moreover, if a zenith tropospheric delay (ZTD) parameter is set up to further absorb such residuals will extend the convergence due to the strong correlation between ZTD and height component. The stochastic model of (4.45) is formulated as

$$\mathrm{D}\left(\begin{bmatrix} p \\ \phi \end{bmatrix}\right) = \mathrm{diag}([\sigma_p^2, \sigma_\phi^2]) \otimes I_f \otimes Q \tag{4.46}$$

where $\sigma_p^2$ and $\sigma_\phi^2$ are the frequency-independent variance scalars of *undifferenced* code and phase. $Q$ is an $(s \times s)$ cofactor matrix of DD observations with elevation-dependent weighting.

To make the model more general, we introduce the DD ionospheric constraints as pseudo-observations to (4.45)

$$\mathrm{E}(\iota_0) = \iota, \mathrm{D}(\iota) = \sigma_\iota^2 Q \tag{4.47}$$

where the variance $\sigma_\iota^2$ is used to model the spatial uncertainty of baseline-dependent ionosphere. Incorporating the ionospheric constraints into (4.45) and further equivalently reducing the ionospheric parameters yields the GB ionosphere-weighted (IW) model as

$$\mathrm{E}\left(\begin{bmatrix} \bar{p} \\ \bar{\phi} \end{bmatrix}\right) = \begin{bmatrix} e_f \otimes A & 0 \\ e_f \otimes A & \Lambda \otimes I_s \end{bmatrix} \begin{bmatrix} x \\ a \end{bmatrix} \tag{4.48}$$

where $\bar{p} = p - \mu \otimes \iota_0$ and $\bar{\phi} = \phi + \mu \otimes \iota_0$. Accordingly, the stochastic model is

$$\mathrm{D}\left(\begin{bmatrix} \bar{p} \\ \bar{\phi} \end{bmatrix}\right) = \begin{bmatrix} \sigma_p^2 I_f + \sigma_\iota^2 \mu\mu^{\mathrm{T}} & -\sigma_\iota^2 \mu\mu^{\mathrm{T}} \\ -\sigma_\iota^2 \mu\mu^{\mathrm{T}} & \sigma_\phi^2 I_f + \sigma_\iota^2 \mu\mu^{\mathrm{T}} \end{bmatrix} \otimes Q \tag{4.49}$$

As a special case of the GB model, the geometry-free (GF) model is formulated with $A = I_s$. In other words, the satellite-to-receiver distance is directly applied as unknowns instead of its expansion with baseline unknowns [24]. Although the GB model is the most common operation mode in most surveying applications, the GF model has its own appeal, stemming mainly from its simplicity and the exemption of complicated tropospheric variations [25]. The GF IW model follows then

$$\mathrm{E}\left(\begin{bmatrix} \bar{p} \\ \bar{\phi} \end{bmatrix}\right) = \begin{bmatrix} e_f \otimes I_s & 0 \\ e_f \otimes I_s & \Lambda \otimes I_s \end{bmatrix} \begin{bmatrix} \varrho \\ a \end{bmatrix} \tag{4.50}$$

Two extreme cases, i.e., the ionosphere-float (IFlt) model with $\sigma_\iota^2 = \infty$ and the ionosphere-fixed (IFix) model with $\sigma_\iota^2 = 0$, can be further reduced from the IW model for specific situations. Similar to [23], we can also derive the canonical formulae of the covariance matrix of the float ambiguities.

In fact, the most benefit of multi-frequency signals is to provide the more possibility to form the combinations with longer wavelength and then easier or instantaneous ambiguity resolution [26]. In this study, we select the extra-wide-lane/wide-lane (EWL/WL) combinations by transforming $f$-frequency ambiguities with a preset between-frequency transformation matrix in terms of their maximized success rates. With the covariance matrices of float ambiguities in Eq. (4.50), we solve for the optimal integer transformation matrices to minimize the variance of the transformed EWL ambiguities based on the ILS criterion. Since the GF model can already obtain the high success rate of EWL/WL ambiguity resolution (based on our extensive computations), the GF model is used to search the optimal combinations. Considering the IW model, we apply the between-frequency transformation matrix $\left(z_{\mathrm{E}}^{\mathrm{T}} \otimes I_s\right)$ with the integer vector $z_{\mathrm{E}}^{\mathrm{T}} = [z_1, \ldots, z_f]$ to transform the covariance matrix. It follows

$$\left(z_{\mathrm{E}}^{\mathrm{T}} \otimes I_s\right) Q_{\hat{a}\hat{a}}^{(\mathrm{GF,IW})} (z_{\mathrm{E}} \otimes I_s) = \sigma_{\hat{z}}^{2(\mathrm{GF,IW})} Q \tag{4.51}$$

with

$$\sigma_{\hat{z}}^{2(\mathrm{GF,IW})} = z_{\mathrm{E}}^{\mathrm{T}} \left[ \Lambda^{-1} \left( \sigma_\phi^2 I_f + \frac{\sigma_p^2 \mu \mu^{\mathrm{T}}}{\sigma_p^2 / \sigma_\iota^2 + \mu^{\mathrm{T}} \mu} \right) \Lambda^{-1} + \Theta \right] z_{\mathrm{E}} \tag{4.52}$$

Obviously, maximizing the success rate to obtain the optimal transformation vector $z_{\mathrm{E}}^{\mathrm{T}}$ corresponds minimizing the scalar $\sigma_{\hat{z}}^{2(\mathrm{GF,IW})}$, i.e., $\sigma_{\hat{z}}^{2(\mathrm{GF,IW})} = \min$. The LAMBDA method [2] can be applied to solve this minimization problem where the zero-vector plays the role of float solution while $\left[ \Lambda^{-1} \left( \sigma_\phi^2 I_f + \frac{\sigma_p^2 \mu \mu^{\mathrm{T}}}{\sigma_p^2 / \sigma_\iota^2 + \mu^{\mathrm{T}} \mu} \right) \Lambda^{-1} + \Theta \right]^{-1}$ the role of its corresponding covariance matrix. In the original version of LAMBDA software, only two optimal integer vectors are provided. Here the new version of LAMBDA software [27] was employed to provide a number of integer vectors in ascending order of their corresponding variance scalars.

The minimization problem of (4.52) is governed by the uncertainty $\sigma_\iota$ of ionospheric constraint. For the IFlt model, the variance scalar (4.52) becomes

$$\sigma_{\hat{z}}^{2(\mathrm{GF,IFlt})} = z_{\mathrm{E}}^{\mathrm{T}} [\Lambda^{-1} \left( \sigma_\phi^2 I_f + \sigma_p^2 \frac{\mu \mu^{\mathrm{T}}}{\mu^{\mathrm{T}} \mu} \right) \Lambda^{-1} + \Theta_\infty] z_{\mathrm{E}} \tag{4.53}$$

while for IFix model, the variance scalar is

$$\sigma_{\hat{z}}^{2(\mathrm{GF,IFix})} = z_{\mathrm{E}}^{\mathrm{T}} \left[ \sigma_\phi^2 \Lambda^{-2} + \sigma_p^2 \frac{\Lambda^{-1} e_f e_f^{\mathrm{T}} \Lambda^{-1}}{f} \right] z_{\mathrm{E}} \tag{4.54}$$

and its bias is accordingly transformed in a scalar case as

$$b_{\hat{z}}^{(GF,\ IFix)} = z_E^T b_{\hat{a}}^{(GF,IFix)} = -\frac{1}{f} z_E^T \Lambda^{-1} \left( f I_f + e_f e_f^T \right) \mu \iota_b = -\frac{1}{f} z_E^T \chi \iota_b \tag{4.55}$$

with $\chi = f \Lambda^{-1} \mu + \Lambda^{-1} e_f e_f^T \mu$. In the IFix model, one should consider both variance and bias of ambiguity to obtain the best transformation matrix $z_E^T$. The mean square error (MSE) is applied as a measure by capturing both the variance and bias,

$$4\sigma_{\hat{z}}^{2(GF,IFix)} + b_{\hat{z}}^{(GF,IFix)} b_{\hat{z}}^{(GF,IFix)}$$

$$= z_E^T \left[ 4\sigma_\phi^2 \Lambda^{-2} + \frac{4\sigma_p^2}{f} \Lambda^{-1} e_f e_f^T \Lambda^{-1} + \frac{\iota_b^2}{f^2} \chi \chi^T \right] z_E = \min \tag{4.56}$$

Here the factor of 4 is used to transform the ambiguity variance in DD mode and thus match the DD ionospheric bias $\iota_b$. The LAMBDA method is again applied to solve this minimization problem.

By solving the minimization problems of the IW model (4.52) with the ionospheric constraints $\sigma_\iota = 0.1, 0.2$ or $0.3$ m, the IFlt model (4.53) with $\sigma_\iota = \infty$ and the IFix model (4.54) with $\sigma_\iota = 0$ and $\iota_b = 0.1, 0.2$ or $0.3$ m, respectively, the several optimal EWL combinations for triple-, quad- and penta-frequency BeiDou Global Navigation Satellite System-3 (BDS-3) signals are displayed in Tables 4.1, 4.2 and 4.3. In the computations, we take the precisions of undifferenced code and phase as $\sigma_p = 0.2$ m and $\sigma_\phi = 3$mm. In Tables, the standard deviations (STD) are $\sigma_{\hat{z}}^{(GF,IW)}$ and $\sigma_{\hat{z}}^{(GF,IFlt)}$, and the root mean square error (RMSE) is $\sqrt{4\sigma_{\hat{z}}^{2(GF,IFix)} + b_{\hat{z}}^{(GF,IFix)} b_{\hat{z}}^{(GF,IFix)}}$. The combined EWL observation and its ionospheric coefficient, wavelength, frequency, ambiguity and variance are defined as

$$\begin{aligned}
\phi_E &= \frac{z_E^T \Lambda^{-1}}{z_E^T \Lambda^{-1} e_f} \phi \\
\mu_E &= \frac{z_E^T \Lambda^{-1} \mu}{z_E^T \Lambda^{-1} e_f} \\
\lambda_E &= \frac{1}{z_E^T \Lambda^{-1} e_f} \\
f_E &= c z_E^T \Lambda^{-1} e_f \\
a_E &= z_E^T a \\
\sigma_{\phi_E} &= \frac{\sqrt{z_E^T \Lambda^{-1} \Lambda^{-1} z_E}}{z_E^T \Lambda^{-1} e_f} \sigma_\phi = \alpha_E \sigma_\phi
\end{aligned} \tag{4.57}$$

where $c$ is the velocity of light and $\phi = [\phi_1, \ldots, \phi_f]^T$. Besides, the ionospheric constraint $\sigma_\iota$ corresponds to undifferenced observations and the ionospheric bias $\iota_b$ corresponds to DD observations.

For a given ambiguity resolution model with a specific ionospheric constraint or bias, only $(f - 1)$ EWL combinations are selected because only $(f - 1)$ EWL/WL combinations are linearly independent for an $f$-frequency system and any other EWL/

**Table 4.1**  Optimal combinations for triple-frequencies, $f_2, f_3$ and $f_4$

| Model | $z_{\mathrm{E}}^{\mathrm{T}}$ | $\mu_{\mathrm{E}}$ | $\lambda_{\mathrm{E}}$ [m] | STD/RMSE [cycle] with $\sigma_\iota$ or $\iota_b$ [m] | | |
|---|---|---|---|---|---|---|
| | | | | 0.1 | 0.2 | 0.3 |
| IW | $[0, 0, 1, -1, 0]$ | $-1.5915$ | 4.8842 | **0.0297** | **0.0303** | **0.0310** |
| | $[0, 1, -4, 3, 0]$ | $-0.6179$ | 2.7646 | **0.0816** | 0.0922 | 0.1028 |
| | $[0, 1, -3, 2, 0]$ | $-0.9698$ | 1.7654 | 0.0844 | **0.0921** | **0.1000** |
| IFix | $[0, 0, 1, -1, 0]$ | $-1.5915$ | 4.8842 | **0.0590** | **0.0594** | **0.0601** |
| | $[0, 1, -4, 3, 0]$ | $-0.6179$ | 2.7646 | **0.1560** | **0.1637** | **0.1758** |
| IFlt | $[0, 0, 1, -1, 0]$ | $-1.5915$ | 4.8842 | **0.0336** | | |
| | $[0, 1, -2, 1, 0]$ | $-1.1558$ | 1.2967 | **0.1245** | | |

**Table 4.2**  Optimal combinations for quad-frequencies, $f_1, f_2, f_3$ and $f_5$

| Model | $z_{\mathrm{E}}^{\mathrm{T}}$ | $\mu_{\mathrm{E}}$ | $\lambda_{\mathrm{E}}$ [m] | STD/RMSE [cycle] with $\sigma_\iota$ or $\iota_b$ [m] | | |
|---|---|---|---|---|---|---|
| | | | | 0.1 | 0.2 | 0.3 |
| IW | $[1, -1, 0, 0, 0]$ | $-1.009$ | 20.9323 | **0.0228** | **0.0229** | **0.0229** |
| | $[0, 0, 1, 0, -1]$ | $-1.6631$ | 3.2561 | **0.0365** | **0.0389** | **0.0410** |
| | $[0, 1, -3, 0, 2]$ | $-0.5575$ | 2.7646 | **0.0654** | **0.0757** | 0.0841 |
| | $[0, 1, -2, 0, 1]$ | $-1.0652$ | 1.4952 | 0.0762 | 0.0801 | **0.0835** |
| IFix | $[1, -1, 0, 0, 0]$ | $-1.009$ | 20.9323 | **0.0454** | **0.0455** | **0.0456** |
| | $[0, 0, 1, 0, -1]$ | $-1.6631$ | 3.2561 | **0.0712** | **0.0733** | **0.0766** |
| | $[0, 1, -3, 0, 2]$ | $-0.5575$ | 2.7646 | **0.1226** | **0.1320** | **0.1463** |
| IFlt | $[1, -1, 0, 0, 0]$ | $-1.009$ | 20.9323 | **0.0232** | | |
| | $[0, 0, 1, 0, -1]$ | $-1.6631$ | 3.2561 | **0.0458** | | |
| | $[-1, 2, -2, 0, 1]$ | $-1.0695$ | 1.6102 | **0.0909** | | |

WL combinations can be linearly recovered by these $(f - 1)$ EWL/WL combinations [23, 25]. The STD or RMSE of the selected EWL combination are marked in bold.

The results show that no matter what models used with varying ionospheric constraints or biases, all EWL combinations are nearly immune to the varying ionospheric constraints and all can obtain very small STD/RMSE, thus allowing instantaneous ambiguity resolution. Although the optimal EWL/WL combinations of GB model are not presented here, they are generally the same as those obtained with GF model.

**Table 4.3** Optimal combinations for penta-frequencies

| Model | $z_E^T$ | $\mu_E$ | $\lambda_E[m]$ | STD/RMSE [cycle] with $\sigma_\iota$ or $\iota_b$ [m] | | |
|---|---|---|---|---|---|---|
| | | | | 0.1 | 0.2 | 0.3 |
| IW | $[0, 0, 0, 1, -1]$ | $-1.7477$ | 9.7684 | **0.0195** | **0.0200** | **0.0204** |
| | $[1, -1, 0, 0, 0]$ | $-1.0090$ | 20.9323 | **0.0227** | **0.0228** | **0.0229** |
| | $[0, 0, 1, -1, 0]$ | $-1.6208$ | 4.8842 | **0.0257** | **0.0263** | **0.0267** |
| | $[0, 1, -3, 0, 2]$ | $-0.5575$ | 2.7646 | **0.0643** | 0.0757 | 0.0840 |
| | $[0, 1, -3, 1, 1]$ | $-0.8200$ | 2.1548 | 0.0660 | **0.0750** | **0.0817** |
| IFix | $[0, 0, 0, 1, -1]$ | $-1.7477$ | 9.7684 | **0.0385** | **0.0390** | **0.0397** |
| | $[1, -1, 0, 0, 0]$ | $-1.009$ | 20.9323 | **0.0452** | **0.0454** | **0.0457** |
| | $[0, 0, 1, -1, 0]$ | $-1.6208$ | 4.8842 | **0.0509** | **0.0514** | **0.0523** |
| | $[0, 1, -3, 0, 2]$ | $-0.5575$ | 2.7646 | **0.1189** | **0.1304** | 0.1475 |
| | $[0, 1, -3, 1, 1]$ | $-0.8200$ | 2.1548 | 0.1247 | 0.1335 | **0.1470** |
| IFlt | $[0, 0, 0, 1, -1]$ | $-1.7477$ | 9.7684 | **0.0212** | | |
| | $[1, -1, 0, 0, 0]$ | $-1.009$ | 20.9323 | **0.0232** | | |
| | $[0, 0, 1, -1, 0]$ | $-1.6208$ | 4.8842 | **0.0278** | | |
| | $[-1, 2, -2, 0, 1]$ | $-1.0695$ | 1.6102 | **0.0908** | | |

One can solve the EWL/WL ambiguities based on either the GB model or the GF model. For GB model, it is formulated by multiplying the phase observations of (4.50) with a vector $(\lambda_E z_E^T \Lambda^{-1} \otimes I_s)$ as

$$\mathrm{E}\left(\begin{bmatrix} p - \mu \otimes \iota_0 \\ \phi_E + \mu_E \iota_0 \end{bmatrix}\right) = \begin{bmatrix} e_f \otimes A & 0 \\ A & \lambda_E I_s \end{bmatrix} \begin{bmatrix} x \\ a_E \end{bmatrix} \tag{4.58}$$

where the stochastic model reads

$$\mathrm{D}\left(\begin{bmatrix} p - \mu \otimes \iota_0 \\ \phi_E + \mu_E \iota_0 \end{bmatrix}\right) = \begin{bmatrix} \sigma_p^2 I_f + \sigma_\iota^2 \mu \mu^T & -\sigma_\iota^2 \mu_E \mu \\ -\sigma_\iota^2 \mu_E \mu^T & \sigma_{\phi_E}^2 + \sigma_\iota^2 \mu_E^2 \end{bmatrix} \otimes Q \tag{4.59}$$

Once $(f - 1)$ EWL/WL ambiguities are fixed, one will then solve the narrow-lane (NL) ambiguities. Here we choose the ambiguities at the first frequency as the NL ambiguities because the ionospheric parameters are unknown for each DD observations such that the efficient wavelength is much shorter than the original one. The NL ambiguity is solved by using the IW model as

$$\begin{bmatrix} \check{p}_k - \mu \otimes \iota_0 \\ \check{\phi}_{E,k} + \mu_E \otimes \iota_0 \\ \phi_{1,k} + \mu_1 \otimes \iota_0 \end{bmatrix} = \begin{bmatrix} e_f \otimes A_k & 0 \\ e_{f-1} \otimes A_k & 0 \\ A_k & \lambda_1 I_s \end{bmatrix} \begin{bmatrix} x_k \\ a \end{bmatrix}, Q_{yy} \otimes Q_k \tag{4.60}$$

$$Q_{yy} = \sigma_\iota^2 \begin{bmatrix} \mu\mu^{\mathrm{T}} & \mu\mu_{\mathrm{E}}^{\mathrm{T}} & \mu_1\mu \\ \mu_{\mathrm{E}}\mu^{\mathrm{T}} & \mu_{\mathrm{E}}\mu_{\mathrm{E}}^{\mathrm{T}} & \mu_1\mu_{\mathrm{E}} \\ \mu_1\mu^{\mathrm{T}} & \mu_1\mu_{\mathrm{E}}^{\mathrm{T}} & \mu_1^2 \end{bmatrix} + \begin{bmatrix} \sigma_p^2 I_f & 0 & 0 \\ 0 & \sigma_\phi^2 Q_{\mathrm{E}} & \sigma_\phi^2 Q_{\mathrm{E1}} \\ 0 & \sigma_\phi^2 Q_{\mathrm{1E}} & \sigma_\phi^2 \end{bmatrix} \tag{4.61}$$

where the subscript $k$ denotes the $k$th epoch. $\check{\phi}_{\mathrm{E},k} = [\check{\phi}_{\mathrm{E}_1,k}^{\mathrm{T}}, \ldots, \check{\phi}_{\mathrm{E}_{f-1},k}^{\mathrm{T}}]^{\mathrm{T}}$ is the vector of ambiguity-corrected EWL/WL observations. $\mu_{\mathrm{E}} = [\mu_{\mathrm{E}_1}, \ldots, \mu_{\mathrm{E}_{f-1}}]^{\mathrm{T}}$, $Q_{\mathrm{E}} = \Lambda_{\mathrm{E}} Z_{\mathrm{E}}^{\mathrm{T}} \Lambda^{-2} Z_{\mathrm{E}} \Lambda_{\mathrm{E}}^{\mathrm{T}}$ and $Q_{\mathrm{E1}} = \Lambda_{\mathrm{E}} Z_{\mathrm{E}}^{\mathrm{T}} \Lambda^{-1} c_1$ with $Z_{\mathrm{E}} = [z_{\mathrm{E}_1}, \ldots, z_{\mathrm{E}_{f-1}}]$, $\Lambda_{\mathrm{E}} = \mathrm{diag}([\lambda_{\mathrm{E}_1}, \ldots, \lambda_{\mathrm{E}_{f-1}}])$ and $c_1$ is an $f$-dimensional column vector with first element of 1 and the others of 0's. The normal equations of $x_k$ and $a$ read

$$\begin{bmatrix} p Q_{\hat{x}_k}^{-1} & t A_k^{\mathrm{T}} Q_k^{-1} \\ t Q_k^{-1} A_k & \vartheta Q_k^{-1} \end{bmatrix} \begin{bmatrix} \hat{x}_k \\ \hat{a} \end{bmatrix} = \begin{bmatrix} u_{\hat{x},k+1} \\ u_{\hat{a},k} \end{bmatrix} \tag{4.62}$$

$$\begin{bmatrix} u_{\hat{x},k+1} \\ u_{\hat{a},k} \end{bmatrix} = \begin{bmatrix} e_{2f}^{\mathrm{T}} Q_{yy}^{-1} \otimes Q_k^{-1} \\ d_{2f}^{\mathrm{T}} Q_{yy}^{-1} \otimes Q_k^{-1} \end{bmatrix} \begin{bmatrix} p_k - \mu \otimes \iota_0 \\ \check{\phi}_{\mathrm{E},k} + \mu_{\mathrm{E}} \otimes \iota_0 \\ \phi_{1,k} + \mu_1 \otimes \iota_0 \end{bmatrix} \tag{4.63}$$

where $p = e_{2f}^{\mathrm{T}} Q_{yy}^{-1} e_{2f}$, $t = \lambda_1 e_{2f}^{\mathrm{T}} Q_{yy}^{-1} c_{2f}$, $\vartheta = \lambda_1^2 c_{2f}^{\mathrm{T}} Q_{yy}^{-1} c_{2f}$ and $Q_{\hat{x}_k}^{-1} = A_k^{\mathrm{T}} Q_k^{-1} A_k$. Here $c_{2f}$ and $d_{2f}$ are the $2f$-dimensional column vectors with last element of 1 and $\lambda_1$, and the others of 0's. By reducing the parameter $x_k$, one obtains the reduced normal equations of $\hat{a}$ over total of K epochs as

$$N_{\hat{a},K} \hat{a} = u_{\hat{a},K} \tag{4.64}$$

with

$$N_{\hat{a},K} = \vartheta \sum_{k=1}^{K} Q_k^{-1} - \frac{t^2}{p} \left( \sum_{k=1}^{K} Q_k^{-1} A_k Q_{\hat{x}_k} A_k^{\mathrm{T}} Q_k^{-1} \right) \tag{4.65}$$

$$u_{\hat{a},K} = \sum_{k=1}^{K} d_{2f}^{\mathrm{T}} Q_{yy}^{-1} \otimes Q_k^{-1} \begin{bmatrix} p_k - \mu \otimes \iota_0 \\ \check{\phi}_{\mathrm{E},k} + \mu_{\mathrm{E}} \otimes \iota_0 \\ \phi_{1,k} + \mu_1 \otimes \iota_0 \end{bmatrix} \tag{4.66}$$

For a short period, for instance the initialization, the satellite elevations have generally rather small variations such that they can be deemed as constants. As a consequence, we can adequately take $Q_k = Q$ for $k = 1, \ldots, K$. The LS estimate of float solution and its covariance matrix read

$$\hat{a} = Q_{\hat{a},K} d_{2f}^{\mathrm{T}} Q_{yy}^{-1} \otimes Q^{-1} \sum_{k=1}^{K} \begin{bmatrix} p_k - \mu \otimes \iota_0 \\ \check{\phi}_{\mathrm{E},k} + \mu_{\mathrm{E}} \otimes \iota_0 \\ \phi_{1,k} + \mu_1 \otimes \iota_0 \end{bmatrix} \tag{4.67}$$

$$Q_{\hat{a},K} = \frac{1}{K\vartheta}Q + \frac{1}{K\vartheta}\frac{t^2}{p}Q\left(\left(\frac{1}{K\vartheta}\sum_{k=1}^{K}A_k Q_{\hat{x}_k}A_k^{\mathrm{T}}\right)^{-1}Q - \frac{t^2}{p}I_s\right)^{-1} \tag{4.68}$$

Since all $(f-1)$ independent EWL/WL combinations are applied in (4.60), the selection of EWL/WL combinations does not affect the NL ambiguity resolution. To intuitively show how float NL ambiguities are improved by the increasing epochs, we take a special case of GF model, i.e., $A_k = I_s$, for $k = 1, \ldots, K$. As a result, the covariance matrix of (4.68) reduces to

$$Q_{\hat{a},K}^{\mathrm{GF}} = \frac{1}{K}\frac{p}{\vartheta p - t^2}Q \tag{4.69}$$

The covariance matrix of float NL ambiguity estimate is inversely proportional to the number of epochs, which means that the precision of float estimate is improved and become easier to be fixed by adding more epochs of data. Giving $\sigma_p = 0.2$ m, $\sigma_\phi = 3$ mm and $\sigma_t = 0.1$ m, the scalar $\frac{p}{\vartheta p - t^2}$ can be computed for the triple-, quad- and penta-frequency cases, where the employed frequencies are the same as in Tables 4.1, 4.2 and 4.3. The results reveal a numerical understanding how the additional frequency bands improve the precision of float NL ambiguities. In addition, although the more frequencies involved can improve the NL ambiguity resolution, it is not necessarily to fix all ambiguities due to the between-ambiguity quality diversity. In real applications, one often prefers to fix partial ambiguities in reliability.

As a reduced alternative, one can also resolve the NL ambiguities in a both geometry-free and ionosphere-free (GIF) model for simplification, which was proposed in [28] for triple-frequency NL ambiguity resolution. We analogously form the GIF model for $f$-frequency NL ambiguity resolution of one DD satellite pair as

$$\hat{a} = \frac{\check{\phi}_1 - b^{\mathrm{T}}\check{\phi}_{\mathrm{E}}}{\lambda_1}, \quad \sigma_{\hat{a}}^2 = 4\frac{\gamma^2\sigma_\phi^2}{\lambda_1^2} \tag{4.70}$$

where $\gamma^2 = b^{\mathrm{T}}Q_{\mathrm{E}}b - Q_{1\mathrm{E}}b - b^{\mathrm{T}}Q_{\mathrm{E}1} + 1$ and $\check{\phi}_{\mathrm{E}} = [\check{\phi}_{\mathrm{E}_1}, \ldots, \check{\phi}_{\mathrm{E}_{f-1}}]^{\mathrm{T}}$ is the $(f-1)$ ambiguity-fixed EWL/WL observations of one DD satellite pair. The coefficient vector, $b = [b_1, \ldots, b_{f-1}]^{\mathrm{T}}$, is obtained by solving the constrained minimization problem as

$$\gamma^2 = \min, \text{ s.t. } e_f^{\mathrm{T}}b = 1, b^{\mathrm{T}}\mu_{\mathrm{E}} = \mu_1 \tag{4.71}$$

By using the EWL/WL observations in Tables 4.1, 4.2 and 4.3 for triple-, quad- and penta-frequency signals, the coefficients are computed in Table 4.4. Obviously, although the more frequency signals can significantly enhance the GIF ambiguity resolution, multiple epochs are still needed to average down the ambiguity STD and then improve the success of ambiguity resolution. Once NL ambiguities are fixed to

                                                    

**Table 4.4**  Coefficients of GIF NL ambiguity resolution and the corresponding ambiguity precisions with $\sigma_\phi = 2$ mm

|         | $b_1$    | $b_2$    | $b_3$    | $b_4$   | $\sigma_{\hat{a}}$ [cycle] |
|---------|----------|----------|----------|---------|---------------------------|
| $f = 3$ | $-4.7650$ | $5.6750$ |          |         | 4.239                     |
| $f = 4$ | $0.5396$  | $-3.4313$ | $3.8917$ |         | 2.611                     |
| $f = 5$ | $-1.1947$ | $0.5381$  | $-2.2250$ | $3.8816$ | 2.609                     |

their integers, the ambiguities of all raw frequencies can be recovered from the fixed EWL/WL and NL ambiguities.

# References


1. Teunissen PJG (1993) Least squares estimation of the integer GPS ambiguities. Proceedings of the IAG general meeting, pp. 1-16
2. Teunissen PJG (1995) The least-squares ambiguity decorrelation adjustment: a method for fast GPS integer ambiguity estimation. J Geod 70:65–82
3. Teunissen PJG (1999) An optimality property of the integer least-squares estimator. J Geod 73:587 593
4. Euler HJ, Schaffrin B (1991) On a measure for the discernibility between different ambiguity solutions in the static-kinematic GPS-mode. Kinematic Syst Geodesy Surv Remote Sens 107:285–295
5. Wang J, Stewart M, Tsakiri M (1998) A discrimination test procedure for ambiguity resolution on-the-fly. J Geod 72:644–653
6. Tiberius CCJM, De Jonge PJ (1995) Fast positioning using the LAMBDA method, Proceedings of DSNS-95
7. Han S (1997) Quality control issues relating to instantaneous ambiguity resolution for real-time GPS kinematic positioning. J Geod 71:351–361
8. Verhagen S, Li B, Teunissen PJG (2013) Ps-LAMBDA: ambiguity success rate evaluation software for interferometric applications. Comput Geosci 54:361–376
9. Frei E, Beutler G (1990) Rapid static positioning based on the fast ambiguity resolution approach. Manuscr Geodaet 15:344–352
10. Mader GL (1992) Rapid static and kinematic Global Positioning System solutions using the ambiguity function technique. J Geophys Res 97:2371–2383
11. Lachapelle G, Canon ME, Erickson C, Falkenberg W (1992) High-precision C/A code technology for rapid static DGPS surveys. Proceedings of the 6th international geodetic symposium on satellite positioning, pp 1–6
12. Hatch D (1990) Instantaneous ambiguity resolution. Proceedings of kinematic systems in geodesy, surveying, and remote sensing, pp. 299–308
13. De Jonge PJ, Tiberius C (1996) The LAMBDA method for integer ambiguity estimation: implementation aspects, Delft University of Technology
14. Chang X, Yang X, Zhou T (2005) MLAMBDA: a modified LAMBDA method for integer least-squares estimation. J Geod 79:552–565
15. Teunissen PJG (1998) Success probability of integer GPS ambiguity rounding and bootstrapping. J Geod 72:606–612

16. Teunissen PJG, Kleusberg A (1998) GPS for geodesy. Springer, Berlin
17. Hassibi A, Boyd S (1998) Integer parameter estimation in linear models with applications to GPS. IEEE Trans Signal Process 46:2938–2952
18. Teunissen PJG (2000) ADOP based upper-bounds for the bootstrapped and the least-squares ambiguity success rates. Artif Satellites 35:171–179
19. Teunissen PJG (2003) Integer aperture GNSS ambiguity resolution. Artif Satellites 38:79–88
20. Teunissen PJG (2004) Penalized GNSS ambiguity resolution. J Geod 78:235–244
21. Verhagen S (2004) Integer ambiguity validation: an open problem? GPS Solut 8:36–43
22. Verhagen S, Teunissen PJG (2013) The ratio test for future GNSS ambiguity resolution. GPS Solut 17:535–548
23. Li B, Li Z, Zhang Z, Tan Y (2017) ERTK: extra-wide-lane RTK of triple-frequency GNSS signals. J Geod 91:1031–1047
24. Li B (2016) Stochastic modeling of triple-frequency BeiDou signals: estimation, assessment and impact analysis. J Geod 90:593–610
25. Li B, Feng Y, Gao W, Li Z (2015) Real-time kinematic positioning over long baselines using triple-frequency BeiDou signals. IEEE Trans Aerosp Electron Syst 51:1–16
26. Li B (2018) Review of triple-frequency GNSS: ambiguity resolution, benefits and challenges. J Glob Position Syst 16:10
27. Verhagen S, Li B (2012) LAMBDA software package: Matlab implementation, Version 3.0, Delft University of Technology and Curtin University
28. Li B, Feng Y, Shen Y (2010) Three carrier ambiguity resolution: distance-independent performance demonstrated using semi-generated triple frequency GPS signals. GPS Solut 14:177–184

# Chapter 5
# Cycle Slip Detection and Repair

## 5.1 Introduction

Carrier phase observations are fundamental to high-precision Global Navigation Satellite System (GNSS) positioning, such as real-time kinematic (RTK) and precise point positioning (PPP), because they offer millimeter-level accuracy once integer ambiguities are resolved. However, these observations are susceptible to cycle slips, which are abrupt discontinuities in the integer ambiguity caused by factors such as multipath effects, ionospheric disturbances, signal obstructions, and electromagnetic interference. In challenging environments with prevalent signal blockage, cycle slips of varying magnitudes frequently occur and may even cause simultaneous data gaps across all satellites. Undetected cycle slips can introduce uncontroled errors into the positioning solution, while even detected slips pose significant challenges for accurate correction. In RTK, an unrepaired cycle slip necessitates re-fixing the corresponding ambiguity, and in PPP it increases the number of unknowns, both of which deteriorate positioning results. Moreover, if the cycle slips are not correctly repaired, additional time is required for the ambiguities to reconverge, potentially taking from several seconds to minutes in RTK or even tens of minutes in PPP, thus severely impairing the system availability and continuity. Therefore, effective processing that reliably detects and accurately repairs cycle slips is indispensable for maintaining continuous high-precision GNSS positioning without incurring time-consuming reinitializations.

For the problem of multi-frequency cycle slip processing, in the past few decades, the extensive investigations have been done for effective cycle slip detection and repair. The single-differenced (SD) and double-differenced (DD) cycle slips are much easier to be detected since many systematic errors are greatly reduced or eliminated, for instance, the clock errors of satellites or/and receiver, the satellite orbital errors and the atmospheric delays [1]. However, the cycle slip detection methods designed for undifferenced observations, which maintain more useful information, still have their advantages for those applications with standalone GNSS receiver such



© The Author(s) 2025
B. Li et al., *GNSS Real-Time Kinematic Positioning*, Navigation: Science and Technology 17, https://doi.org/10.1007/978-981-96-9116-6_5

as PPP and PPP-RTK. Most of existing studies are based on dual-frequency signals. The TurboEdit algorithm developed by Blewitt is one of the most popular methods dedicated for dual-frequency undifferenced cycle slip processing [2]. It employs the Hatch-Melbourne-Wübbena combination [3–5] together with the geometry-free (GF) combination. The TurboEdit algorithm has been implemented in many famous softwares such as PANDA, GIPSY-OASIS II and Bernese. However it could be inefficient in case of active ionospheric condition with large biases and quick variations. The improvements to TurboEdit have been made in many literatures. Liu employed the ionospheric total electron content rate (TECR) instead of the GF combination to implement dual-frequency cycle slip detection [6]. However, one requirement is that the GNSS data are recorded at 1 Hz or even higher rate in order to detect small cycle slips as stated by author. A forward and backward moving window averaging algorithm integrated with a second-order time-differenced phase ionospheric residual algorithm were presented based on TurboEdit [7]. It is unfortunately not suitable for real-time applications. Motivated by the extensive 1-cycle slips found in low-elevation BeiDou Geostationary Earth Orbit (GEO) satellites, Ju et al. jointly used a polynomial fit algorithm and a generalized autoregressive conditional heteroscedastic (GARCH) model to provide an adaptive threshold for the GF combination in TurboEdit [8]. In addition, de Lacy et al. applied the Bayesian theory to detect the cycle slips as outliers, while its efficiency is constrained by high sampling rate of GNSS data [9]. A similar rationale can be found in [10].

Along with the gradual construction of BeiDou Navigation Satellite System (BDS) and modernization of the Global Positioning System (GPS), the triple-frequency signals become available, triggering a new upsurge for triple-frequency cycle slip estimation. The triple-frequency signals in theory can improve the cycle slip estimation [11]. One benefit of triple-frequency signals lies in the formation of various useful linear combinations, for instance, the extra-wide-lane (EWL) and the wide-lane (WL) combinations, which retain integer nature of cycle slips but with small errors in cycle thanks to their long wavelengths [12–14]. de Lacy et al. employed five GF linear combinations to detect cycle slips in three cascading steps, but the performance has been tested only with 1 Hz triple-frequency GPS data from a moderate multipath environment [15]. Dai et al. applied two GF phase combinations in cycle slip detection and their integer candidates are searched by least-squares ambiguity decorrelation adjustment (LAMBDA) method. All these methods mentioned above have a common premise that the between-epoch ionosphere variation is so small that its effect in GF combination can be ignored. This premise is not necessarily true in case of the active ionospheric condition and the large sampling interval or even data gap. Without properly compensating these misspecifications, one will obtain the wrong cycle slip estimation. Zhao et al. and Li et al. took into account the influence of between-epoch ionospheric biases in narrow-lane (NL) cycle slip estimation of triple-frequency signals [14, 16].

It is noted that all GF methods are implemented on a satellite-by-satellite basis where the coordinate parameters are eliminated. In this case, the correlations between satellites are completely neglected [6, 15, 16]. As a result, the satellites whose cycle slips have been correctly fixed cannot contribute to the cycle slip estimation of remained satellites. Especially, when multiple GNSS systems are applied, the satellites from a system without cycle slip can never help the cycle slip estimation of other systems. In other words, the key shortcoming of GF cycle slip estimation is to ignore the link between satellites embodied by coordinate parameters. However, the fact is that multi-frequency and multi-GNSS application is booming for high-precision applications as all BDS and Galileo satellites and part of GPS satellites are transmitting triple- or even more frequency signals (penta-frequency signals served by Galileo). To make best use of correlations between satellites, between frequencies and between systems, a geometry-based (GB) model is thus in demand, which involves the receiver coordinate parameters and possesses of stronger model strength compared to the GF model. Banville and Langley proposed a GB model based cycle slip correction procedure, and discussed the model tolerance to the ionospheric biases [17]. It is advised that the ionospheric biases are limited to a few centimeters with rigorously processing other non-dispersive systematic components.

In this contribution, we proposed a new geometry-based ionosphere-weighted (GBIW) method dedicating to efficiently estimating the cycle slips in scenarios of active ionospheric condition and connecting the phase data with a certain data gap. In GBIW method, the phase and code observations of all satellites are processed simultaneously in an integrated adjustment to achieve mathematically stronger model strength and then better solutions than the satellite-wise basis mode in GF model. To compensate the increased between-epoch ionospheric variations with the prolonged sampling intervals or data gaps, we predict them by using consecutive historical data of a sliding window with an adaptive polynomial order and sliding window length. Once the float solution of cycle slips is computed, its optimal integer solution is searched via LAMBDA method. In addition, because it is impossible to always fix all cycle slips correctly due to different quality from individual cycle slips, we further propose a partial cycle slip resolution (PCSR) strategy to successively fix the cycle slip based on a so-called bootstrapping procedure [18, 19]. In fact, once the sufficient number of cycle slips is correctly fixed, one can already maintain the continuous precise positioning and then improve the availability of precision solutions. In summary, the proposed method is universal and applicable to various scenarios specified by the arbitrary number of frequencies and systems, the static and kinematic modes, the quiet and active ionosphere conditions. Moreover, the proposed method can be easily reduced to its two special models, i.e., geometry-fixed (GFI) and GF models that are widely used so far.

The previous discussion primarily focused on the multi-frequency scenario. Now, we turn our attention to the single-frequency case for further analysis. Lots of algorithms dedicated to cycle slip detection and correction have been proposed in the past a few years. Some of them are universal for all situations with single- or multi-frequency signals in both static and kinematic modes, while some are limited to only specific situations. The statistical testing based methods are universal, where the cycle slips are treated as outliers and detected via testing the relevant statistics often constructed with observation residuals [20]. However, few redundancy always brings in the strong correlation amongst testing statistics and then the error transfer amongst observation residuals, which inevitably causes wrong decisions [21, 22].

The other two universal methods are the between-epoch high-order phase difference and the polynomial fitting of phase observation series [9, 23]. They are both widely known and easy-implemented. But the former one amplifies the random noise so much that the small cycle slips cannot be identified. The latter one is hindered by the need of normal historical observations within a fitting window, and it is troublesome to determine the window length and polynomial order. The reliability of cycle slip correction using measurement polynomial fitting is improved, if the kinematics of the satellite position and satellite clock are subtracted from the measurements before the polynomial fitting. In addition, some studies employed the Doppler measurements and/or inertial navigation system (INS) to improve the cycle slip detection and correction. But the accuracy of Doppler integration is limited, and meanwhile adding an INS system to GNSS increases the cost and complexity.

Most of the cycle slip detection and correction methods focus on the multi-frequency signals and are generally based on the combination of multi-frequency signals, for instance, the WL and GF combinations [2, 6, 14, 15]. Although some of them are incapable of special cycle slips and vulnerable to code noise and iono-spheric variations, they are proved to be generally efficient in cycle slip detection and have been widely implemented in many software. Moreover, with the emerging of triple-frequency signals, the different schemes of combinations among three frequencies are developed which makes it easier to effectively detect and repair the cycle slips [14, 16, 24]. Obviously, the specific methods with multi-frequency signals such as the cycle slip sensitive combinations are not available in single-frequency case. Extensive researches have been done in single-frequency cycle slip detection as well. Carcanague [25] presented a GB single-frequency cycle slip resolution, where the cycle slips are solved by using Doppler and phase observations and fixed via the LAMBDA method [26]. Unfortunately, the Doppler is not necessarily available for different types of receivers. All these characteristics multiply the challenges of single-frequency cycle slip estimation. In summary, the cycle slip estimation is still a challenging and open problem in single-frequency GNSS data processing.

In traditional measurement-based polynomial fitting (MPF), the cycle slip detection is more likely affected by the systematic errors of measurements, e.g. multipath and atmospheric delay, and also by some abnormality, e.g., ionospheric scintillations. Besides, the fitting accuracy degrades if the satellite signals are discontinued due to the loss of lock and data interruption. Therefore, compared with this satellite-by-satellite MPF cycle slip detection, detecting the cycle slips of all satellites simultaneously via an integrated adjustment with the GB model can provide mathematically better results [27]. However, in the GB method, both of baseline and cycle slips are solved as unknowns, which results in a relatively weak model and hinder the success rate of cycle slip estimation. Nevertheless, if there is accurate prior baseline information available, the efficiency of cycle slip estimation will be definitely improved. Motivated by the fact that the more accurate prior baseline leads to the more efficient cycle slip estimation, we propose a position polynomial fitting (PPF) method to construct the accurate prior baseline information with a series of positions, which is referred to as position-domain constraint. As well-known in GNSS navigation, an object motion is reasonably well subject to a low-order polynomial over a short period

[28, 29]. Thus, the positions of multiple historical epochs in a given time-window are employed to establish a low-order polynomial to characterize the object motion behavior and then predict the position of the current epoch. Afterwards, this kinematic constraint is incorporated into the GB cycle slip estimation model to improve the model strength and then the efficiency of cycle slip estimation.

## 5.2 Multi-frequency Cycle Slip Processing

This section explores the cycle slip detection and repair method for multi-frequency GNSS data. The method integrates SD ionospheric biases to strengthen the model, with extensive experiments confirming its high success rates across different sampling intervals and system combinations. The method ensures continuous positioning even with data gaps, particularly benefiting BDS applications.

The general GB mathematical model of cycle slip estimation is firstly presented. Based on which, the ionosphere constraint is introduced to enhance the model strength. We start with the single-epoch observation equations of undifferenced (UD) phase and code

$$\mathrm{E}\big(\boldsymbol{\Phi}_j\big) = \boldsymbol{Gx} + \boldsymbol{e}_n\delta t_j - \boldsymbol{\delta t}_{s,j} + \boldsymbol{\tau} - \mu_j\boldsymbol{\iota} + \lambda_j\boldsymbol{a}_j \tag{5.1}$$

$$\mathrm{E}\big(\boldsymbol{P}_j\big) = \boldsymbol{Gx} + \boldsymbol{e}_n dt_j - \boldsymbol{dt}_{s,j} + \boldsymbol{\tau} + \mu_j\boldsymbol{\iota} \tag{5.2}$$

where the subscript $j$ denotes the frequency $f_j$, corresponding to the wavelength $\lambda_j$, which is used to emphasize the frequency-specific terms. Assuming that $n$ satellites are tracked simultaneously, the vectors, $\boldsymbol{\Phi}_j = \left[\Phi_j^1, \ldots, \Phi_j^n\right]^{\mathrm{T}}$ and $\boldsymbol{P}_j = \left[P_j^1, \ldots, P_j^n\right]^{\mathrm{T}}$, denote the phase and code observations in meters. $\Phi_j^i$ and $P_j^i$ are the phase and code observations of the $i$th satellite on frequency $f_j$. $\boldsymbol{G}$ is an $(n \times 3)$ design matrix pertaining to the receiver coordinate parameters $\boldsymbol{x}$. $\delta t_j$ and $dt_j$ are the receiver clock errors for phase and code, while $\boldsymbol{\delta t}_{s,j} = \left[\delta t_{s,j}^1, \ldots, \delta t_{s,j}^n\right]^{\mathrm{T}}$ and $\boldsymbol{dt}_{s,j} = \left[dt_{s,j}^1, \ldots, dt_{s,j}^n\right]^{\mathrm{T}}$ are the satellite clock errors of $n$ satellites for phase and code, respectively. $\boldsymbol{\tau} = \left[\tau^1, \ldots, \tau^n\right]^{\mathrm{T}}$ is the slant tropospheric delay vector. $\boldsymbol{\iota} = \left[\iota^1, \ldots, \iota^n\right]^{\mathrm{T}}$ is the ionospheric delay vector on frequency $f_1$ with $\mu_j = f_1^2/f_j^2$. $\boldsymbol{u}_j - \left[u_j^1, \ldots, u_j^n\right]^{\mathrm{T}}$ is the ambiguity vector with the $k_{\mathrm{th}}$ element $a_j^k = \varphi_j(t_0) - \varphi_{s,j}^k(t_0) + z_j^k$, where $z_j^k$ and $\varphi_{s,j}^k(t_0)$ are the integer ambiguity and initial satellite phase bias for satellite $k$, and $\varphi_j(t_0)$ is the initial receiver phase bias.

Different from the outlier, the GNSS cycle slip has two properties, i.e., integer and continuity. The integer is an inherent property of cycle slip. The continuity property means that the same integer jump is introduced afterwards from the epoch where the cycle slip happens. Therefore, the between-epoch observations are often applied to isolate the cycle slips. The between-epoch SD model reads

$$E\left(\Delta\boldsymbol{\Phi}_j\right) = \Delta(\boldsymbol{Gx}) + e_n\Delta\delta t_j - \Delta\boldsymbol{\delta t}_{s,j} + \Delta\boldsymbol{\tau} - \mu_j\Delta\iota + \lambda_j\Delta z_j \qquad (5.3)$$

$$E\left(\Delta\boldsymbol{P}_j\right) = \Delta(\boldsymbol{Gx}) + e_n\Delta dt_j - \Delta\boldsymbol{dt}_{s,j} + \Delta\boldsymbol{\tau} + \mu_j\Delta\iota \qquad (5.4)$$

where the difference operator $\Delta(*) = (*)_{k+1} - (*)_k$, and the subscripts $k$ and $k + 1$ refer to two adjacent epochs. In the SD model, the initial phase biases, $\varphi_j(t_0)$ and $\varphi_{s,j}^k(t_0)$, of both receiver and satellite are completely removed thanks to their stability over a certain time. Then the integer difference, $\Delta z_j = \Delta a_j$, between two epochs is defined as the cycle slip.

Let us analyze the terms in Eqs. (5.3) and (5.4). With the receiver clock errors as an example, it follows $\Delta\delta t_j = \delta t_j(k + 1) - \delta t_j(k)$ and $\Delta dt_j = dt_j(k + 1) - dt_j(k)$ for phase and code, respectively. The inter-observation-type bias between phase and code is rather stable over time [30], which results in

$$\Delta\delta t_j = \Delta dt_j \qquad (5.5a)$$

The receiver clock errors rigorously differ from frequencies due to the inter-frequency-bias (IFB). The clock errors on frequency $j$ can be expressed as $\delta t_j = \delta t_1 + \text{IFB}_{1,j}$, where $\text{IFB}_{1,j}$ is the IFB between frequencies $f_1$ and $f_j$. Since the IFB is also very stable for a while, it follows

$$\Delta\delta t_1 = \Delta\delta t_j \qquad (5.5b)$$

with (5.5a) and (5.5b), we have

$$\Delta\delta t_j = \Delta dt_j \overset{\Delta}{=} \Delta\delta t \qquad (5.5c)$$

for all $f$ frequencies. $\Delta\delta t$ is the between-epoch SD receiver clock error for phase on $f_1$. It is exactly the same case for satellite clock errors, namely, $\Delta\boldsymbol{\delta t}_{s,j} = \Delta\boldsymbol{dt}_{s,j} \overset{\Delta}{=} \Delta\boldsymbol{\delta t}_s$.

The between-epoch SD slant tropospheric delay $\Delta\boldsymbol{\tau}$ is typically very small owing to very similar propagation paths over a short time. However, it cannot be ignored in case of lower elevation angles and is thus corrected with a tropospheric model, e.g., the New Brunswick 3 (i.e., UNB3) model. For the between-epoch SD geometric term, it follows

$$\Delta(\boldsymbol{Gx}) = \boldsymbol{G}_{k+1}\boldsymbol{x}_{k+1} - \boldsymbol{G}_k\boldsymbol{x}_k = \frac{\boldsymbol{G}_{k+1} + \boldsymbol{G}_k}{2}\boldsymbol{b} + \frac{\boldsymbol{G}_{k+1} - \boldsymbol{G}_k}{2}\boldsymbol{b} \approx \overline{\boldsymbol{G}}\boldsymbol{b} \qquad (5.6)$$

where $\overline{\boldsymbol{G}} = (\boldsymbol{G}_{k+1} + \boldsymbol{G}_k)/2$ and $\boldsymbol{b} = \boldsymbol{x}_{k+1} - \boldsymbol{x}_k$ is defined as the between-epoch baseline parameter. The reason of ignoring the term $(\boldsymbol{G}_{k+1} - \boldsymbol{G}_k)\boldsymbol{b}/2$ is explained as follows. Due to the high altitude of GNSS satellite orbits, the design matrix changes very slowly over time for normal kinematic applications. To give an insight, we simply compute the values of $(G_{k+1} - G_k)$ for all BDS and GPS satellites in a static observation mode with sampling interval of 60 s. Related study reveal that the maximum amplitudes of $(G_{k+1} - G_k)$ are in order of $10^{-3}$ for Medium Earth Orbit (MEO) satellites of BDS and GPS. These values can be smaller for smaller sampling interval. It is therefore adequate to ignore the term $(G_{k+1} - G_k)b/2$. Besides, the satellite clock errors are available from the ephemeris data and thus treated as known.

For sake of simplicity of expressions, we omit the epoch-difference operator $\Delta$ without any confusion. Then the between-epoch SD observation is finally symbolized as

$$\mathrm{E}\left(\boldsymbol{\Phi}_j + \delta t_s\right) = \overline{\boldsymbol{G}}\boldsymbol{b} + e_n \delta t - \mu_j \iota + \lambda_j z_j \tag{5.7}$$

$$\mathrm{E}\left(\boldsymbol{P}_j + \delta t_s\right) = \overline{\boldsymbol{G}}\boldsymbol{b} + e_n \delta t + \mu_j \iota \tag{5.8}$$

Collecting all observations of $f$ frequencies yields

$$\mathrm{E}(\boldsymbol{\Phi}) = \left(e_f \otimes \overline{\boldsymbol{G}}\right)\boldsymbol{b} + e_{fn}\delta t - \boldsymbol{\mu} \otimes \iota + (\boldsymbol{\Lambda} \otimes \boldsymbol{I}_n)z \tag{5.9}$$

$$\mathrm{E}(\boldsymbol{P}) = \left(e_f \otimes \overline{\boldsymbol{G}}\right)\boldsymbol{b} + e_{fn}\delta t + \boldsymbol{\mu} \otimes \iota \tag{5.10}$$

where $\boldsymbol{\Phi} = \left[\boldsymbol{\Phi}_1^{\mathrm{T}}, \ldots, \boldsymbol{\Phi}_f^{\mathrm{T}}\right]^{\mathrm{T}}$ is the observed-minus-computed vector corrected with satellite clock errors. $\boldsymbol{P}$ has the same structure as $\boldsymbol{\Phi}$. $\boldsymbol{\mu} = \left[\mu_1, \ldots, \mu_f\right]^{\mathrm{T}}$, $z = \left[z_1^{\mathrm{T}}, \ldots, z_f^{\mathrm{T}}\right]^{\mathrm{T}}$ and $\boldsymbol{\Lambda} = \mathrm{diag}\left(\lambda_1, \ldots, \lambda_f\right)$. The compact form is

$$\mathrm{E}(\boldsymbol{y}) = \left(e_{2f} \otimes \boldsymbol{H}\right)\boldsymbol{b} + (v \otimes \boldsymbol{I}_n)\iota + (\boldsymbol{\Gamma} \otimes \boldsymbol{I}_n)z \tag{5.11}$$

where $\boldsymbol{y} = \left[\boldsymbol{\Phi}^{\mathrm{T}}, \boldsymbol{P}^{\mathrm{T}}\right]^{\mathrm{T}}$, $v = \left[-\boldsymbol{\mu}^{\mathrm{T}}, \boldsymbol{\mu}^{\mathrm{T}}\right]^{\mathrm{T}}$, $\boldsymbol{\Gamma} = [\boldsymbol{\Lambda}, 0]^{T}$. $\boldsymbol{H} = [\overline{\boldsymbol{G}}, e_n]$ is augmented matrix incorporating the clock error and correspondingly the baseline parameter vector is extended to, i.e., $\boldsymbol{b} \triangleq \left[\boldsymbol{b}^{T}, \delta t\right]^{T}$. The variance matrix of single-epoch UD observations is generalized

$$\boldsymbol{Q}_E = \boldsymbol{Q}_f \otimes \boldsymbol{Q}_e \tag{5.12}$$

where $\boldsymbol{Q}_e$ is the elevation-dependent cofactor matrix of UD observations. $\boldsymbol{Q}_f = \mathrm{blkdiag}\left(\boldsymbol{Q}_\Phi, \boldsymbol{Q}_P\right)$ captures the frequency and observation-type specific precisions. If we assume that the observations of all $f$ frequencies are of the unique precision for each observation type, it follows $\boldsymbol{Q}_\Phi = \boldsymbol{I}_f \sigma_\Phi^2$ and $\boldsymbol{Q}_P = \boldsymbol{I}_f \sigma_P^2$. Here $\sigma_\Phi^2$ and $\sigma_P^2$ are the variance scalars of UD phase and code, respectively. Without time correlations, the variance matrix of SD observations is ultimately expressed as

$$Q_y = 2Q_E \tag{5.13}$$

In real data processing, in case of small sampling interval/data gap, the SD ionospheric variations can be so small to be ignored and accordingly the ionospheric parameters in (5.11) can be omitted. Reducing the ionospheric parameters in the adjustment system can improve the model strength. However, when the SD ionospheric variations are significant for instance in case of large sampling interval/ data gap, ignoring the ionospheric parameters in (5.11) will definitely lead to biased parameter estimates. In the event, one should conservatively apply model (5.11) although the model strength will be very weak.

To enhance the model strength of (5.11) and meantime without introducing biases in estimates, the pseudo-observations of prior SD ionospheric variations $\iota^0$ are often applied as constraints [31]

$$E(\iota^0) = \iota, \quad D(\iota^0) = \sigma_\iota^2 \otimes Q_e \tag{5.14}$$

where the variance $\sigma_\iota^2$ is used to model the uncertainty of prior SD ionospheric constraints. Integrating the observation model with ionospheric constraints yields the GBIW method, which is equivalent to

$$E(\bar{y}) = (e_{2f} \otimes H)b + (\Gamma \otimes I_n)z \tag{5.15a}$$

$$Q_{\bar{y}} = [2Q_f + \sigma_\iota^2 v v^{\mathrm{T}}] \otimes Q_e = Q \otimes Q_e \tag{5.15b}$$

where $\bar{y} = y - (v \otimes I_n)\iota^0$ is the SD observation vector corrected with prior ionospheric constraints. The smaller variance $\sigma_\iota^2$ generates the stronger model, which needs more precise prior ionospheric constraints. How to obtain the precise prior SD ionospheric constraints is therefore an important issue. In this chapter, we will predict the SD ionospheric biases of current epoch by fitting the polynomial of foregoing data within a time window, which will be detailed in the methodology section.

We simply discuss two reduced models, i.e., GFI and GF models, from the GB model, which corresponding to extreme situations. For the GFI model, the between-epoch baseline parameter is completely known, which is the case for static applications of which continuously operating reference stations (CORS) application is more special. The GFI model is reduced from (5.15a) by setting $H = e_n$, where only the cycle slips and receiver clock error need to be solved. However, the GF model directly takes the between-epoch satellite-to-receiver ranges $\varrho$ as unknowns instead of between-epoch baseline parameters by setting $H = [I_n, e_n]$. Obviously, the GF model is rank-deficient since the receiver clock error $\delta t$ and satellite-to-receiver ranges $\varrho$ are completely dependent. One of full rank GF models can be obtained by setting $H = I_n$ in (5.15a). It is apparent that the model strength is ordered as GFI > GB > GF, which will be numerically demonstrated in the results section. It is worth pointing out that most of existing cycle slip estimation methods are based on GF model with satellite-by-satellite processing.

As aforementioned, for a sufficient small sampling interval/data gap or quiet ionosphere condition, the SD ionospheric biases will be so small that they can be ignored, i.e., $\iota^0 = \mathbf{0}$ in (5.15a) and $\sigma_\iota^2 = 0$ in (5.15b), or their possible discrepancies are conservatively captured with a small $\sigma_\iota^2$ to avoid the potential biases in float cycle slip solution; while it is not the case for large sampling interval/data gap or active ionosphere, where the SD ionospheric biases can reach several decimeters. It is noticed that some of BDS satellites have very small SD ionospheric biases. Those satellites are all GEO and Inclined Geosynchronous Satellite Orbit (IGSO) satellites because their smaller location variations than MEO satellites with respect to the ground stations.

If phase observations at two adjacent epochs are free of cycle slips or the cycle slips have been correctly fixed, the between-epoch SD ionospheric bias $\iota$ can be derived by using GF combination of two frequency phase observations as

$$\iota = \frac{\Phi_i - \Phi_j}{\mu_j - \mu_i} \tag{5.16}$$

where subscripts $i$ and $j$ denote two frequencies. Based on the strong temporal correlation characteristic of ionospheric biases between adjacent epochs [32, 33], we apply a simple polynomial fitting to model the SD ionospheric biases of forgoing epochs within a time window and then predict those of the current epoch. Two important parameters are involved in this processing, namely, the window length and the polynomial order. Let the window length and polynomial order be $n$ and $m$, respectively, the polynomial model of SD ionospheric biases reads

$$\begin{aligned}
\iota_{(t-n)} &= \sum_{r=0}^{m} a_r (t-n)^r \\
\iota_{(t-n+1)} &= \sum_{r=0}^{m} a_r (t-n+1)^r \\
&\vdots \\
\iota_{(t-1)} &= \sum_{r=0}^{m} a_r (t-1)^r
\end{aligned} \tag{5.17}$$

where $t$ indicates the current epoch number. $\iota_{(t-k)}$ $(k = 1, \ldots, n)$ is the SD ionospheric biases computed with (5.16). $a_r (r = 0, \ldots, m)$ is the polynomial coefficient to be estimated via the least squares adjustment. Once the coefficients are determined, the SD ionospheric bias of current epoch can be predicted

$$\hat{\iota}_t = \sum_{r=0}^{m} \hat{a}_r t^r \tag{5.18}$$

where $\hat{a}_r$ denotes the estimate of $a_r$. This predicted SD ionospheric bias serves as a prior constraint in GBIW method.

For a better contribution to enhancing the GBIW model strength, one always expects more accurate prior SD ionospheric biases, i.e., the discrepancies between the real SD ionospheric biases and their predictions are as small as possible. The polynomial model is specified by both the window length $n$ and the polynomial order $m$. They would differ from the sampling intervals/data gaps or ionosphere activity conditions. Hence, the optimal parameters $n$ and $m$ of the polynomial model should be determined for each dataset to get a prediction effect as accurate as possible. The term "optimal" usually refers to the balance between the window length $n$ along with the polynomial order $m$, namely, the number of parameters need to be estimated and the goodness of fitting. Akaike introduced the Akaike's Information Criterion (AIC) to measure the goodness of an estimated statistical model and select an optimal model from a set of candidate models [34]. For polynomial model (5.17), AIC is defined as [34]

$$\text{AIC}(M_q) = -2\ln(f(\iota|M_q)) + 2(m_q + 2) \tag{5.19}$$

where "ln" indicates the natural logarithm. $M_q$ is the $q_{\text{th}}$ candidate model specified by the model parameter $\alpha_q = \{n_q, m_q\}$ with $n_q$ and $m_q$ the window length and polynomial order. $f(\iota|M_q)$ is the probability density function of SD ionospheric biases $\iota$ conditioned on the $q_{\text{th}}$ model. It is assumed of normal distribution as

$$f(\iota|M_q) = \left(\frac{1}{\sqrt{2\pi\sigma_{\iota|M_q}^2}}\right)^{n_q} \exp\left(-\frac{1}{2\sigma_{\iota|M_q}^2} \sum_{k=1}^{n_q}\left(\sum_{r=0}^{m_q} a_r(t-k)^r - \iota_{(t-k)}\right)^2\right) \tag{5.20}$$

with the error variance conditioned on the $q_{\text{th}}$ model $M_q$ as

$$\sigma_{\iota|M_q}^2 = \frac{1}{n_q}\sum_{k=1}^{n_q}\left(\sum_{l=0}^{m_q} a_r(t-k)^r - \iota_{(t-k)}\right)^2 \tag{5.21}$$

Inserting (5.20) into (5.19) yields

$$\text{AIC}(M_q) = n_q(\ln 2\pi + 1) + n_q\ln\sigma_{\iota|M_q}^2 + 2(m_q + 2) \tag{5.22}$$

Several candidate models are ranked in terms of their associated AIC values where the variance $\sigma_{\iota|M_q}^2$ is replaced by its posterior estimate computed by (5.21) with least squares estimates of polynomial coefficients and in consideration of the degree of freedom. The model with the smallest AIC is selected as optimal model.

The integer cycle slip estimation of model (5.15a, 5.15b) consists of four steps. In the first step, the so-called float solution $\hat{z}$ of cycle slips and its variance-covariance (VC) matrix $Q_{\hat{z}\hat{z}}$ are computed via the least squares adjustment. Then the optimal integer cycle slip solution $\check{z}$ is efficiently obtained by employing the LAMBDA method in the second step. Once the integer solution is obtained, a validation is executed in the third step to confirm whether it is reliable enough to accept $\check{z}$ or not, since the acceptance of a wrong solution, without notice, will result in severe positioning error. In the present contribution, the model-driven integer bootstrapping (IB) success rate [27] and the data-driven ratio test [35, 36] will be adopted to validate the reliability of the integer cycle slip solution. The IB success rate is defined as [27]

$$P_B = \prod_{i=1}^{m} \left( 2\Phi\left(\frac{1}{2\sigma_{\hat{z}_{i|I}}}\right) - 1 \right) \tag{5.23}$$

with $\Phi(x) = \int_{-\infty}^{x} \frac{1}{\sqrt{2\pi}} \exp\{-\frac{1}{2}t^2\}dt$ and $\sigma_{\hat{z}_{i|I}}$ is the standard deviation of the $i$th decorrelated float solution, conditioned on the cycle slips $I = \{i+1, \ldots, n\}$. The IB success rate is a sharp lower bound of the integer least squares (real) success rate [37]. Obviously, the IB success rate considers only the precisions and partial correlations of float solution that is model-driven. In this study, the user-defined success rate is set to $P_0 = 99.5\%$. To further control the reliability of integer solution, the data-driven ratio test is applied. The key is to determine a proper threshold $R_0$. Commonly the fixed threshold of for instance, 2 or 3, is applied. Verhagen and Teunissen proposed to adapt this threshold in terms of quality of float solution itself and the fixed failure-rate ratio-test (FF-RT) was advised [40]. However, as noticed by Li et al. [31], once float solution with higher success rate than $P_0$, its associated FF-RT threshold will be always almost 1. It means that the integer solution does not need further validation. Hence, based on the fact that the FF-RT threshold becomes smaller for larger success rate and/or higher dimension, we used a set of dimension-dependent thresholds advised by Li et al. [31]. They are 2, 1.5, 1.3 and 1.2 for integer dimensions of 2, 3–4, 5–7 and larger than 7, respectively.

In real data analysis, one cannot sometimes fix all cycle slips simultaneously. In another words, the vector of integer cycle slip estimates cannot pass the validation of IB success rate or/and ratio test. Close analysis reveals that the failure of fixing all cycle slips is mainly due to the poor quality of only some cycle slips. Hence, we turn to the fixing of partial cycle slips that can be reliably fixed. Here, fixing all cycle slips refers to as the full cycle slip resolution (FCSR) while fixing part of them refers to as PCSR. Completely analogous to partial ambiguity resolution provided in Wang et al. [38], we select the subset of cycle slips based on the successively increased elevations considering the fact that lower elevation corresponds to the poor quality of float cycle slip estimate.

Let the following partitions of float cycle slip solution $\hat{z}$ and its VC matrix $Q_{\hat{z}\hat{z}}$ as

$$\hat{z} = \begin{bmatrix} \hat{z}_1 \\ \hat{z}_2 \end{bmatrix}, \quad Q_{\hat{z}\hat{z}} = \begin{bmatrix} Q_{\hat{z}_1\hat{z}_1} & Q_{\hat{z}_1\hat{z}_2} \\ Q_{\hat{z}_2\hat{z}_1} & Q_{\hat{z}_2\hat{z}_2} \end{bmatrix} \tag{5.24}$$

where the cycle slip vector $\hat{z}_1$ is the subset assumed to be reliably fixed to its integer counterpart $\check{z}_1$. Stemmed from the bootstrapped ambiguity resolution, the remaining cycle slip subset $\hat{z}_2$ is then updated depending on its correlation with $\hat{z}_1$ by

$$\check{z}_2 = \hat{z}_2 - Q_{\hat{z}_2\hat{z}_1} Q_{\hat{z}_1\hat{z}_1}^{-1} (\hat{z}_1 - \check{z}_1) \tag{5.25a}$$

$$Q_{\check{z}_2\check{z}_2} = Q_{\hat{z}_2\hat{z}_2} - Q_{\hat{z}_2\hat{z}_1} Q_{\hat{z}_1\hat{z}_1}^{-1} Q_{\hat{z}_1\hat{z}_2} \tag{5.25b}$$

As the low-elevation cycle slips are more prone to be affected by the observation anomaly, the subset $\hat{z}_1$ is selected according to the successively increased elevations. If the FCSR fails, namely the user-defined thresholds of ratio test or/and IB success rate are not satisfied, the PCSR is therewith invoked. Let the elevations of cycle slips in an ascending order as $\theta_1 < \theta_2 < \cdots < \theta_n$, we start the PCSR by removing the cycle slips of the lowest elevation. The remained cycle slips with elevations higher than $\theta_1$ form the selected subset, denoted by $\hat{z}_1(\theta_2, \ldots, \theta_n)$. Then we check the fixing of $\hat{z}_1(\theta_2, \ldots, \theta_n)$. If yes, we fix them to $\check{z}_1$ and update the remaining subset $\hat{z}_2$ by (5.25a, 5.25b). Otherwise, we further remove the cycle slips of the second lowest elevation. This process is repeated until the selected cycle slips can be successfully fixed or the vector $\hat{z}_1$ is null. Once the remaining subset $\hat{z}_2$ is updated, we try to fix them with the same PCSR procedure.


## 5.3   Single-Frequency Cycle Slip Processing

This section comprehensively addresses cycle slip estimation for single-frequency RTK applications using the PPF method. The method leverages positional polynomial constraints to enhance cycle slip detection and correction. The PCSR strategy further improves both the success and fix rates, demonstrating significant advancements over traditional methods in various scenarios.

First, the observation models for cycle slip estimation are discussed. The observation equations of single-frequency code and phase read

$$P_r^s = \varrho_r^s + c\left(dt_r - dt^s\right) + \tau + \iota + \varepsilon_p \tag{5.26}$$

$$\Phi_r^s = \varrho_r^s + c\left(\delta t_r - \delta t^s\right) + \lambda a_r^s + \tau - \iota + \varepsilon_\phi \tag{5.27}$$

where the superscript $s$ and subscript $r$ denote the satellite and receiver, respectively. $P$ and $\Phi$ are the phase and code observations, and $\varepsilon_P$ and $\varepsilon_\phi$ are their corresponding random noises. $\varrho$, $\tau$ and $\iota$ are the satellite-to-receiver geometric range, the tropospheric delay and the ionospheric delay. $dt_r$ and $\delta t_r$ are the clock errors of code and phase receiver, while $dt^s$ and $\delta t^s$ are for satellite. $a_r^s = z_r^s + \varphi_r(t_0) - \varphi^s(t_0)$ is the ambiguity with wavelength $\lambda$, where $z_r^s$ and $\varphi^s(t_0)$ are the integer ambiguity and initial satellite phase bias, and $\varphi_r(t_0)$ is the initial receiver phase bias.

Besides the integer property, the cycle slip has an important property of continuity. It is that the same integer is introduced for all epochs afterward once a cycle slip occurs. To isolate the cycle slip only at the epoch it occurs, the between-epoch difference model is always formed

$$\Delta P_r^s = \Delta \varrho_r^s + c\left(\Delta dt_r - \Delta dt^s\right) + \Delta \tau + \Delta \iota + \Delta \varepsilon_P \tag{5.28}$$

$$\Delta \Phi_r^s = \Delta \varrho_r^s + c\left(\Delta \delta t_r - \Delta \delta t^s\right) + \lambda \Delta z_r^s + \Delta \tau - \Delta \iota + \Delta \varepsilon_\phi \tag{5.29}$$

where the symbol $\Delta$ denotes the between-epoch difference operator. $\Delta z_r^s = \Delta a_r^s$ denotes the integer cycle slip and it is zero if no cycle slips occur since both $\varphi_r(t_0)$ and $\varphi^s(t_0)$ are constant and eliminated. The atmospheric errors vary so slowly that their between-epoch variations can be basically neglected, especially for small sampling interval [39]. Thus, a cycle slip will lead to a sudden jump in $\Delta \Phi_r^s$ if the other errors can be properly modelled.

In RTK processing, the between-satellite and -receiver DD model is widely used to eliminate the errors of receiver and satellite dependence. In case of short baselines, the residual DD tropospheric delays after corrections with standard tropospheric model and the ionospheric delays can be already ignored. If the between-epoch difference is further applied, i.e., triple-differenced (TD), all systematic errors are eliminated, and the associated equations follows

$$\Delta P_{br}^{ik} = \Delta \varrho_{br}^{ik} + \Delta \varepsilon_{br}^{ik} \tag{5.30}$$

$$\Delta \Phi_{br}^{ik} = \Delta \varrho_{br}^{ik} + \lambda \Delta z_{br}^{ik} + \Delta \varepsilon_{br}^{ik} \tag{5.31}$$

where the DD operator $(*)_{br}^{ik} = \left( *_r^k - *_r^i \right) - ( *_b^k - *_b^i )$. In fact, in RTK applications, one does not need to estimate the cycle slip of undifference observation, instead the DD cycle slip $\Delta z_{br}^{ik}$. Here, we give a brief comment on the multipath that is a troublesome error factor and include both deterministic (systematic) and random components. The multipath could be still significant on the DD observations especially for the DD pseudorange observations. However, in the TD observations, its deterministic component can be adequately eliminated after between-epoch time difference. Regarding the random component, it is commonly captured with the stochastic modelling by giving a relatively enlarged variance.

In the following review of single-frequency detection methods, three typical methods of single-frequency cycle slip detection are discussed. i.e., the between-epoch high-order difference, the MPF [23, 40] and the TD residual-based snooping (TRS) [20]. Since the high-order difference scheme amplifies the random noise, it is not able to detect the small cycle slips and thus almost incapable in practice. Thus, in this section, we will discuss latter two methods. Note the following discussions focus on the real-time applications.

The principle of MPF method is to fit a polynomial with the historical phase observations of a few epochs and then to extrapolate the observation of the current detection epoch. By comparing with the real phase observation, the decision of whether the cycle slip exists is made [23]. For a given satellite $s$, we express the $k_{\text{th}}$ epoch observation with a polynomial function as

$$\mathrm{E}(\Phi_k^s) = a_0 + a_1 k + a_2 k^2 + \cdots + a_m k^m \tag{5.32}$$

where E denotes the expectation operator. Collecting the observations of $n$ epochs together yields

$$E(\boldsymbol{\Phi}^s) = A\boldsymbol{a} \tag{5.33}$$

where $\boldsymbol{\Phi}^s = [\Phi_1^s, \Phi_2^s, \ldots, \Phi_n^s]^T$, $A = \begin{bmatrix} 1 & 1 & 1 & \cdots & 1 \\ 1 & 2 & 2^2 & \cdots & 2^m \\ \vdots & \vdots & \vdots & \vdots & \vdots \\ 1 & n & n^2 & \cdots & n^m \end{bmatrix}$ and $\boldsymbol{a} = [a_0, a_1, \ldots, a_m]^T$. In case of $n > (m + 1)$, taking the unit weight matrix, the least-squares solution of polynomial coefficients reads

$$\hat{\boldsymbol{a}} = (A^T A)^{-1} A^T \boldsymbol{\Phi}^s \tag{5.34}$$

The formal standard deviation (STD) of unit weight is computed as

$$\hat{\sigma}_0 = \sqrt{\frac{\boldsymbol{v}^T \boldsymbol{v}}{n - m - 1}} \tag{5.35}$$

with $\boldsymbol{v} = \boldsymbol{\Phi}^s - A\hat{\boldsymbol{a}}$. Then the cycle slip is detected by individually testing the residual

$$|v_i| > \kappa \hat{\sigma}_0 \tag{5.36}$$

where $\kappa$ is a positive scalar that can be determined in terms of a certain significance.

In general, a low-order polynomial function is sufficient to fit phase series. As an alternative to (5.33), some specific polynomials, such as, for instance, Lagrange interpolation polynomial and Chebyshev polynomial, can also be applied. The MPF method is simple and easy-to-implement. However, some drawbacks still exist. It is observed from (5.36) that the cycle slip detection is relevant to both $\kappa$ and $\hat{\sigma}_0$. Firstly, the estimation of $\hat{\sigma}_0$ is associated with the number of redundancies, namely, the length $n$ of fitting window and fitting order $m$. Too long window with a given polynomial order or too high-order polynomial with a given window cannot obtain the stable estimate of $\hat{\sigma}_0$. Hence, it is critical to determine the reasonable window length and the polynomial order. In addition, the threshold $\kappa$ is empirically given. If $\kappa$ is too large, the cycle slips will be missed; while if $\kappa$ is too small, the normal observation will be wrongly detected as cycle slip. Besides, the MPF method is applied for individual satellite, the correlated information amongst different satellites is fully neglected. It means that observations from satellites without cycle slips cannot be used to assist cycle slip detection of the other satellites.

Following TD observation Eqs. (5.30) and (5.31) with assumption that no cycle slip exists, the linearized single-epoch TD observation equations read

$$\boldsymbol{l} = G\boldsymbol{x} + \boldsymbol{\epsilon} \tag{5.37}$$

where $\boldsymbol{l} = [\boldsymbol{P}^T, \boldsymbol{\Phi}^T]^T$ with $\boldsymbol{P}$ and $\boldsymbol{\Phi}$ denoting the single-epoch TD code and phase observation vector. $G$ is the design matrix to TD baseline $\boldsymbol{x}$. Let the cofactor matrix of DD observations be denoted by $Q$ and the variances of DD code and phase be denoted by $\sigma_P^2$ and $\sigma_\Phi^2$, the covariance matrix of single-epoch DD observations is $Q_{DD} = \mathrm{diag}\big(\big[\sigma_P^2, \sigma_\Phi^2\big]\big) \otimes Q$. It follows then that the covariance matrix of single-epoch TD observations is $Q_{ll} = 2\mathrm{diag}\big(\big[\sigma_P^2, \sigma_\Phi^2\big]\big) \otimes Q$ with assumption that the between-epoch observations are independent. The least-squares TD baseline solution follows

$$\hat{x} = (G^\mathrm{T} Q_{ll}^{-1} G)^{-1} G^\mathrm{T} Q_{ll}^{-1} l \tag{5.38}$$

The cycle slips are iteratively detected based on the TD residuals following the outlier detection theory. In this case, the underlying assumption is that the outliers are solely driven by cycle slips. In general, one first checks the compatibility of model (5.33) by using the overall test statistic as [41, 42]

$$T_q = \frac{\hat{\epsilon}^\mathrm{T} Q_{ll}^{-1} \hat{\epsilon}}{q} \tag{5.39}$$

where $\hat{\epsilon} = l - G\hat{x}$ is the TD residual vector and $q = 2s - 3$ is the degree of freedom with $2s$ the number of TD code and phase observations and $s$ for number of TD observations of either code or phase. Hence, if the misspecification is detected with $T_q < F_{1-\alpha}(q, \infty)$ by giving a significance level $\alpha$, one needs then to identify the occurrence of cycle slip. Here $F$ denotes the Fisher distribution. Usually, one starts with testing the cycle slips for individual observations by using $w$-test. The $w$-test statistic of the $i_{th}$ observation reads [42, 43]

$$w_i = \frac{c_i^\mathrm{T} Q_{ll}^{-1} \hat{\epsilon}}{\sqrt{c_i^\mathrm{T} Q_{ll}^{-1} Q_{\hat{\epsilon}\hat{\epsilon}} Q_{ll}^{-1} c_i}} \tag{5.40}$$

where $c_i$ is an $2s$-column vector with all elements of 0 except the $i_{th}$ element of $\lambda$. Here $i$ is taken from the set $\{s + 1, \ldots, 2s\}$ considering the order of code and phase observations in $l$. $Q_{\hat{\epsilon}\hat{\epsilon}} = Q_{ll} - G(G^\mathrm{T} Q_{ll}^{-1} G)^{-1} G^\mathrm{T}$ is the covariance matrix of residuals $\hat{\epsilon}$. With a significance level $\alpha$, the $i_{th}$ observation is decided as an outlier if its $w$-statistic has the largest $|w_i|$ of all $s$ phase alternatives and meantime $|w_i| > N_{1-\alpha/2}$. Then float solution of cycle slip is estimated as

$$\hat{z}_i = \frac{c_i^\mathrm{T} Q_{ll}^{-1} \hat{\epsilon}}{c_i^\mathrm{T} Q_{ll}^{-1} Q_{\hat{\epsilon}\hat{\epsilon}} Q_{ll}^{-1} c_i}, \quad \sigma_{\hat{z}_i}^2 = \big(c_i^\mathrm{T} Q_{ll}^{-1} Q_{\hat{\epsilon}\hat{\epsilon}} Q_{ll}^{-1} c_i\big)^{-1} \tag{5.41}$$

Then one can fix the integer value of cycle slip completely as the integer ambiguity resolution does. For a scalar case, the rounding method is used with a given criterion. One may refer to Li et al. [44] for the failure rate controllable rounding scheme.

In any statistical hypothesis test, one has to encounter the type I error of false alarm and the type II error of wrong detection, namely, the error of rejecting a correct hypothesis and the error of accepting a wrong hypothesis. In such case, the corresponding detection power, $\gamma = 1 - \beta$ with $\beta$ the probability of the type II error, can be computed under $H_a$. Thus in w-test, the drawbacks mainly lie in three terms. (1) It is observed from (5.40) that the w-statistics would be correlated due to the correlation of $\hat{\epsilon}$, which will bring in error transfer amongst observation residuals and lead to the type III errors besides the type I and type II errors [21]. (2) In fact, the TRS method often fails in case of simultaneous multiple cycle slips. To avoid the error transfer, the cycle slip is often detected satellite by satellite until there is no cycle slip. But it cannot always work well in practice. (3) Even though the cycle slip can be correctly detected, it is still difficult to exactly estimate the integer values of cycle slip by using its corresponding adjustment residual since the correlation amongst cycle slips are totally ignored.

To fully consider the influence of potential cycle slips with each other, one does not need to fix the cycle slips individually. Instead, once all cycle slips were detected for specific observations, we introduce the cycle slips to those observations as unknowns in TD model

$$l = Gx + Bz + \epsilon \tag{5.42}$$

where $z$ is the unknown cycle slip vector with $t$ dimension ($t \leq s$). $B$ is a ($2s \times t$) design matrix to $z$. Then the float cycle slip vector is derived with least squares criterion as

$$\hat{z} = Q_{\hat{z}\hat{z}}^{-1} B^{\mathrm{T}} Q_{ll}^{-1} \hat{\epsilon}, \, Q_{\hat{z}\hat{z}} = \left( B^{\mathrm{T}} Q_{ll}^{-1} Q_{\hat{\epsilon}\hat{\epsilon}} Q_{ll}^{-1} B \right)^{-1} \tag{5.43}$$

where $B = \begin{bmatrix} \mathbf{0}_{s \times t} \\ C_{s \times t} \end{bmatrix}$. All elements of $C$ matrix are zeros except for some elements of $\lambda$ at the positions associated to cycle slip parameters. Then with float solution $\hat{z}$ and its covariance matrix $Q_{\hat{z}\hat{z}}$, the integer cycle slips can be obtained by using integer estimation method. In this case, the correlations amongst cycle slips are fully taken into account. The model (5.42) is the basis for processing the multiple cycle slips, based on which, the additional constraints can be integrated. In this model, the normal phase measurements over several consecutive epochs are not required, which enables the real-time applications. Moreover, different from the satellite-by-satellite detection methods, the model (5.42) has mathematically stronger model strength since the all measurements are used in an integrated adjustment.

Method with position polynomial constraint, we elaborate our new single-frequency cycle slip estimation method, which is realized by imposing a position polynomial constraint on the model (5.42). Starting with the mathematic formation of the new method, we derive out its two solutions.

A time-window that contains the current epoch and ($n - 1$) consecutive historical epochs are selected. For estimating the cycle slips of satellites at current epoch, the observations of those satellites should be free of cycle slips over all previous ($n - 1$) epochs. Assigning the subscript to denote the epoch number, the mathematic model for cycle slip estimation with a window observation reads

$$
\begin{bmatrix} l_1 \\ \vdots \\ l_n \end{bmatrix} = \begin{bmatrix} G_1 & \cdots & 0 & 0 \\ \vdots & \ddots & \vdots & \vdots \\ 0 & \cdots & G_n & B \end{bmatrix} \begin{bmatrix} x_1 \\ \vdots \\ x_n \\ z \end{bmatrix} + \begin{bmatrix} \epsilon_1 \\ \vdots \\ \epsilon_n \end{bmatrix} \tag{5.44}
$$

where all symbols have the exactly same meanings as those in (5.37) and (5.42) except that an epoch number is assigned. The compact form of (5.44) follows

$$
y = \begin{bmatrix} G & F \end{bmatrix} \begin{bmatrix} x \\ z \end{bmatrix} + \epsilon \tag{5.45}
$$

where $y = [l_1^T, \ldots, l_n^T]^T$, $G = \mathrm{blkdiag}(G_1, \ldots, G_n)$, $F = c_n \otimes B$ with $c_n$ the $n$-column zero vector except for the $n_{\mathrm{th}}$ element of 1. The baseline vector $x = [x_1^T, \ldots, x_n^T]^T$ collects the baseline vectors of all $n$ epochs. The vector $\epsilon = [\epsilon_1^T, \ldots, \epsilon_n^T]^T$ is the random noise for observations of all $n$ epochs. Here we would like to give a comment on the mathematical model of cycle slip estimation. First of all, we emphasize that model (5.44) is for the cycle slip estimation but not for ambiguity resolution and positioning. For estimating the cycle slips of the satellites at current epoch, the observations of those satellites should be without cycle slips over all previous $(n - 1)$ epochs. It implies that not all observed satellites at current epoch are included in (5.44) for cycle slip estimation, for instance, for the newly tracked satellites at current epoch and the satellites whose cycle slips are not successfully repaired at all previous $(n - 1)$ epochs. For those satellites, the corresponding new ambiguities will be set up and their observations will not include in (5.44) for cycle slip estimation. After cycle slip processed, the ambiguities for those satellites will be estimated together with position parameters. Regarding the problem how to confirm that the cycle slips at the previous epochs have been successfully repaired, one can completely apply the validation method used for validating integer ambiguity resolution, like success rate, ratio test etc. In the step of cycle slip estimation, it is advised to use relatively tight validation thresholds since even if the cycle slip estimation fails, the corresponding new ambiguities can be set up for further processing.

Regarding the stochastic model, for a short time span, it is adequate to assume that the covariance matrices of DD observations are the same for all epochs, i.e., $Q_{DD}$. Then the covariance matrix of TD observations of consecutive $n$ epochs is derived via the error propagation law:

$$
Q_{yy} = \begin{bmatrix} 2 & -1 & 0 & \cdots & & 0 \\ -1 & 2 & -1 & \cdots & & 0 \\ \vdots & \ddots & \ddots & & \ddots & \vdots \\ 0 & \cdots & -1 & 2 & & -1 \\ 0 & \cdots & 0 & & -1 & 2 \end{bmatrix} \otimes Q_{DD}
$$

Based on the general observation model, we introduce the kinematic constraint. Assume that the position of the rover station is subjected to a polynomial

$$
r_i = a_0 + a_1 i + a_2 i^2 + \cdots + a_m i^m \tag{5.46}
$$

where $i$ denotes the epoch number; $r_i$ stands for the three-dimensional coordinates of the rover station. $a_k\,(k = 0, \ldots, m)$ is the polynomial coefficients for three coordinate components. For the sake of convenience, the epoch time $t_i$ is replaced by the epoch number within the time window. Accordingly, the TD baseline between two adjacent epochs can be expressed as:

$$
x_i = r_i - r_{i-1} = a_1 + (2i - 1)a_2 + \cdots + (i^m - (i - 1)^m)a_m \tag{5.47}
$$

Taking into account all TD baselines of $n$ epochs within the time-window, Eq. (5.44) can be written in a compact matrix format as:

$$
x = Sa \tag{5.48}
$$

where $S = \begin{bmatrix} 1 & 1 & \cdots & & 1 \\ \vdots & \vdots & \ddots & & \vdots \\ 1 & 2n - 1 & \cdots & n^m - (n - 1)^m \end{bmatrix} \otimes I_3$ and $a = [a_1^{\mathrm{T}}, \ldots, a_m^{\mathrm{T}}]^{\mathrm{T}}$. Inte-

grating the position polynomial constraint (5.48) with the observation model (5.45) yields the cycle slip estimation model

$$
\begin{cases} y = [G\ F]\begin{bmatrix} x \\ z \end{bmatrix} + \epsilon, Q_{yy} \\ x = Sa \end{cases} \tag{5.49}
$$

This model is referred to as the PPF model. In this model, the solutions of TD baselines $x$ and cycle slips $z$ relay on the observations of not only current epoch but also $(n - 1)$ historical epochs. Moreover, introducing the constraint (5.48) can significantly improve the accuracy and reliability for cycle slip estimation.

The first solution of (5.49) is derived by substituting the second equation of (5.49) into the first equation, namely, using $a$ instead of $x$. The least squares normal equations follows

$$\begin{bmatrix} S^T G^T Q_{yy}^{-1} GS & S^T G^T Q_{yy}^{-1} F \\ F^T Q_{yy}^{-1} GS & F^T Q_{yy}^{-1} F \end{bmatrix} \begin{bmatrix} \hat{a} \\ \hat{z} \end{bmatrix} = \begin{bmatrix} S^T G^T Q_{yy}^{-1} y \\ F^T Q_{yy}^{-1} y \end{bmatrix} \tag{5.50}$$

Then the float cycle slip vector $\hat{z}$ can be solved via reducing the normal equation as

$$\hat{z} = Q_{\hat{z}\hat{z}} u \tag{5.51}$$

$$Q_{\hat{z}\hat{z}}^{-1} = F^T Q_{yy}^{-1} F - F^T Q_{yy}^{-1} GS \left( S^T G^T Q_{yy}^{-1} GS \right)^{-1} S^T G^T Q_{yy}^{-1} F \tag{5.52}$$

$$u = F^T Q_{yy}^{-1} y - F^T Q_{yy}^{-1} GS \left( S^T G^T Q_{yy}^{-1} GS \right)^{-1} S^T G^T Q_{yy}^{-1} y \tag{5.53}$$

With float solutions, one can then apply the LAMBDA method to search for the most likely candidate of integer cycle slips. In this study, a new version of the LAMBDA software (version 3.0) is applied with a more efficient search strategy [45]. One may consult Chang et al. [46] for the modified LAMBDA, and also other researchers for more efficient ambiguity resolution method.

However, the direct solution needs to the observations of all historical epochs within the time window to form normal equations, which increases calculation burden. As an alternative but efficient implementation, an indirect solution is presented where the predicted position of a current epoch is fused with the observations of the current epoch. Firstly, the positions of the forgoing epochs are adopted to fit the polynomial, with which the position of a current epoch is predicted. Taking the predicted position as virtual observations, the model (5.49) is modified as

$$\begin{cases} \hat{\xi} = S_1 a + \epsilon_{\hat{\xi}}, Q_{\hat{\xi}} \\ l_n = G_n x_n + Bz + \epsilon_n, Q_{l_n} \\ x_n = S_2 a \end{cases} \tag{5.54}$$

where $\hat{\xi} = \left[ \hat{x}_1^T, \ldots, \hat{x}_{n-1}^T \right]^T$ is the vector of all estimated baselines of $(n-1)$ epochs and $Q_{\hat{\xi}} = \text{blkdiag} \left( Q_{\hat{x}_1}, \ldots, Q_{\hat{x}_{n-1}} \right)$ is its covariance matrix. The matrix $S$ is partitioned to two parts as

$$S_1 = \begin{bmatrix} 1 & 1 & \cdots & 1 \\ \vdots & \vdots & \ddots & \vdots \\ 1 & 2n-3 & \cdots & (n-1)^m - (n-2)^m \end{bmatrix} \otimes I_3$$

and

$$S_2 = \begin{bmatrix} 1 & 2n-1 & \cdots & n^m - (n-1)^m \end{bmatrix} \otimes I_3$$

We first use the first equation of (5.54) to solve the polynomial coefficients $\boldsymbol{a}$, with which as virtual observations, the equivalence of Eq. (5.54) follows

$$\begin{cases} \boldsymbol{l}_n = \boldsymbol{G}_n\boldsymbol{S}_2\boldsymbol{a} + \boldsymbol{B}\boldsymbol{z} + \boldsymbol{\epsilon}_n, \boldsymbol{Q}_{l_n} \\ \tilde{\boldsymbol{a}} = \boldsymbol{a}, \boldsymbol{Q}_{\tilde{a}\tilde{a}} \end{cases} \tag{5.55}$$

where $\tilde{\boldsymbol{a}} = \boldsymbol{Q}_{\tilde{a}\tilde{a}}\boldsymbol{S}_1^{\mathrm{T}}\boldsymbol{Q}_{\hat{\xi}}^{-1}\hat{\boldsymbol{\xi}}$ and $\boldsymbol{Q}_{\tilde{a}\tilde{a}} = (\boldsymbol{S}_1^{\mathrm{T}}\boldsymbol{Q}_{\hat{\xi}}^{-1}\boldsymbol{S}_1)^{-1}$ are the polynomial coefficients and its covariance matrix solved by $(n-1)$ historical epochs. For this equation system, the least squares normal equations are derived as

$$\begin{bmatrix} \hat{\boldsymbol{a}} \\ \hat{\boldsymbol{z}} \end{bmatrix} = \begin{bmatrix} \boldsymbol{Q}_{\hat{a}\hat{a}} & \boldsymbol{Q}_{\hat{a}\hat{z}} \\ \boldsymbol{Q}_{\hat{z}\hat{a}} & \boldsymbol{Q}_{\hat{z}\hat{z}} \end{bmatrix} \begin{bmatrix} \boldsymbol{S}_2^{\mathrm{T}}\boldsymbol{G}_n^{\mathrm{T}}\boldsymbol{Q}_{l_n}^{-1}\boldsymbol{l}_n + \boldsymbol{S}_1^{\mathrm{T}}\boldsymbol{Q}_{\hat{\xi}}^{-1}\hat{\boldsymbol{\xi}} \\ \boldsymbol{B}^{\mathrm{T}}\boldsymbol{Q}_{l_n}^{-1}\boldsymbol{l}_n \end{bmatrix} \tag{5.56a}$$

$$\begin{bmatrix} \boldsymbol{Q}_{\hat{a}\hat{a}} & \boldsymbol{Q}_{\hat{a}\hat{z}} \\ \boldsymbol{Q}_{\hat{z}\hat{a}} & \boldsymbol{Q}_{\hat{z}\hat{z}} \end{bmatrix} = \begin{bmatrix} \boldsymbol{S}_2^{\mathrm{T}}\boldsymbol{G}_n^{\mathrm{T}}\boldsymbol{Q}_{l_n}^{-1}\boldsymbol{G}_n\boldsymbol{S}_2 + \boldsymbol{Q}_{\tilde{a}\tilde{a}}^{-1} & \boldsymbol{S}_2^{\mathrm{T}}\boldsymbol{G}_n^{\mathrm{T}}\boldsymbol{Q}_{l_n}^{-1}\boldsymbol{B} \\ \boldsymbol{B}^{\mathrm{T}}\boldsymbol{Q}_{l_n}^{-1}\boldsymbol{G}_n\boldsymbol{S}_2 & \boldsymbol{B}^{\mathrm{T}}\boldsymbol{Q}_{l_n}^{-1}\boldsymbol{B} \end{bmatrix}^{-1} \tag{5.56b}$$

Finally, the float solution of cycle slips reads

$$\hat{\boldsymbol{z}} = \boldsymbol{Q}_{\hat{z}\hat{z}}\boldsymbol{B}^{\mathrm{T}}\boldsymbol{Q}_{l_n}^{-1}\boldsymbol{l}_n + \boldsymbol{Q}_{\hat{z}\hat{a}}(\boldsymbol{S}_2^{\mathrm{T}}\boldsymbol{G}_n^{\mathrm{T}}\boldsymbol{Q}_{l_n}^{-1}\boldsymbol{l}_n + \boldsymbol{S}_1^{\mathrm{T}}\boldsymbol{Q}_{\hat{\xi}}^{-1}\hat{\boldsymbol{\xi}}) \tag{5.57}$$

Again with float solutions, one can employ the LAMBDA software to fix the integers of cycle slips.

In PPF method, it does not require that for individual satellite the observations within time window are all normal without discontinuity. Instead, it requires only that the positions within time window are all available. This is very promising in real applications with low-cost single-frequency receivers where the discontinuity would frequently happen for some satellites. Besides, the position polynomial fitting in PPF method is more reliable than the measurement polynomial fitting in MPF method since processing all observations in an integrated adjustment in PPF provides a mathematically stronger solution than satellite-by-satellite testing. Moreover, the PPF is immune to the number of observation redundancies because the strong PPF constraint is applied. The PPF method is capable of processing the multiple cycle slips, which enables a more reliable and stable performance in GNSS-adverse environments.

## 5.4 Results and Discussion

To demonstrate the effectiveness of the proposed related methods, three different experiments under complex observation conditions were carried out. Specifically, they are the harsh environment, single-frequency low-cost receiver, and real-time kinematic situation.

### 5.4.1  Analysis in the Harsh Environment

This section adopts single- and dual-frequency mixed baseline dataset No. 1, of which the baseline length is 22.65 km. The 24-h GPS and BDS observations were collected with a sampling interval of 30 s. Hence, the ionospheric delays at this time cannot be totally eliminated by using the between-epoch differencing. The reference and rover stations are both located in Hong Kong, China. Besides, the data used here is from the day during the summer vacation. The sum of Kp indices of that day is 18, thus indicating the ionosphere is relatively active. For a comprehensive analysis, RTK and static PPP modes are both carried out. When dealing with the cycle slips, the broadcast ephemeris and precise ephemeris are used in RTK and PPP, respectively.

To verify the performance of the cycle slip processing in terms of positioning, Figs. 5.1 and 5.2 show the positioning results of the TurboEdit and proposed method based on the state-domain-aided GFI model in RTK and PPP, respectively. Three directions, including east (E), north (N), and up (U) directions, are all presented. The orange and purple points denote the float and fixed solutions in Fig. 5.1. The corresponding statistics including, STD and bias of positioning results are also computed. In this study, since the observation condition is not very good, the ratio when fixing the ambiguities is set to 1.5 in RTK mode. The LAMBDA method is adopted if the integer ambiguities need to be fixed. The success rates of the TurboEdit and proposed methods are 66.0%, and 89.4%, respectively. It indicates that 23.4% improvements of the proposed method can be obtained. In Fig. 5.2, due to the impact of cycle slips, the solution of the TurboEdit method is frequently initialized and cannot be converged. In contrast, the solution of the proposed method converges to centimeter-level after about 140 min. Taking a closer look at Figs. 5.1 and 5.2, and combining the estimated

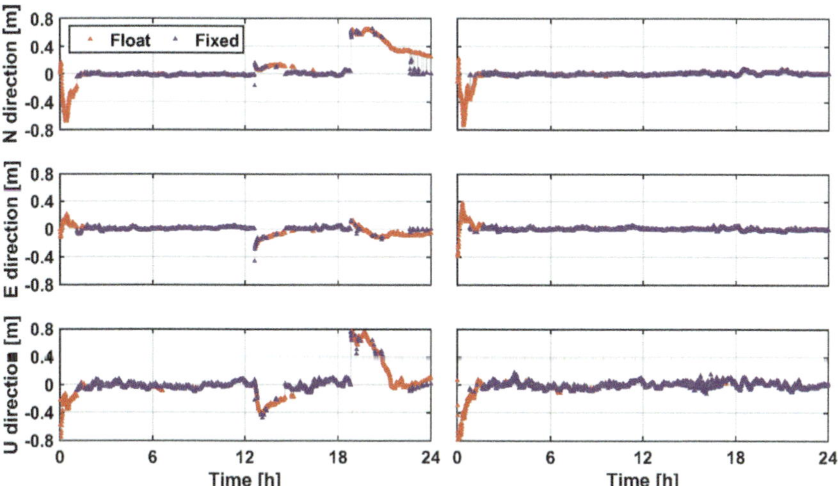


**Fig. 5.1** Positioning errors of three directions by using the TurboEdit method (left) and the GFI method (right) in RTK. The orange and purple points denote the float and fixed solutions, respectively

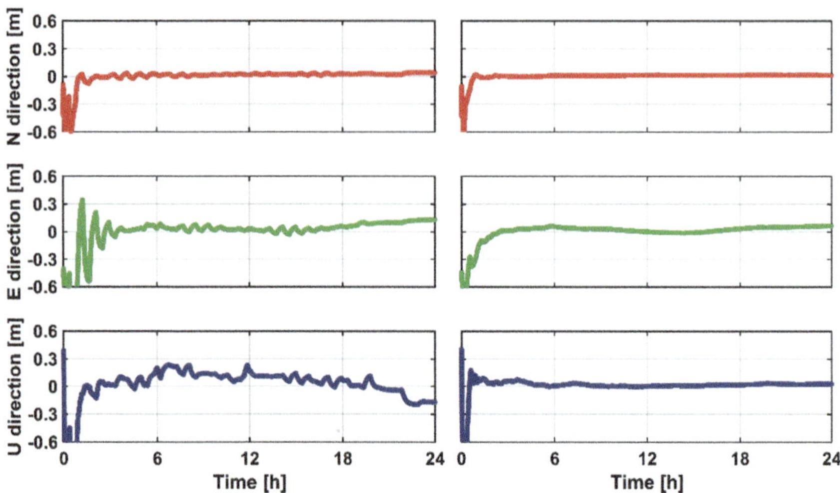

**Fig. 5.2** Positioning errors of three directions by using the TurboEdit method (left) and the GFI method (right) in PPP

STD and bias, it can be found that the proposed method is more precise. Specifically, for the STD, approximately 78.4, 48.8, and 56.4% improvements can be obtained in E, N, and U directions in RTK mode, and 55.9, 42.1 and 86.0% in PPP mode. Similarly, for the bias, 57.1, 26.3, and 30.0% improvements can be obtained by using the proposed method in RTK mode, and 54.1, 39.6 and 86.1% in PPP mode. Therefore, the cycle slips must be handled carefully in harsh environments, and the proposed method is practical and trustworthy to a great extent.

## 5.4.2  Analysis in the Single-Frequency Low-Cost Receiver

This section collects 24-h dataset No. 2 from the single-frequency low-cost receiver. The model of the low-cost receiver only costs a few hundred USD. For miniaturization purposes, a built-in GNSS full-band antenna and low-cost board are integrated into the receiver. Here 5-s single-frequency GPS/BDS observations were adopted in a 54.50-m baseline. A short baseline is used because the observation condition is not good enough where the unmodeled errors inevitably exist. Hence, the atmospheric delays need to be eliminated better. Data analysis indicates that the signal reception is disturbed, which is most likely due to the internal low-cost receiver and external challenging environment. Therefore, the quality of the observations of dataset No. 2 is not very good, which is sufficiently representative. As aforementioned, the proposed method based on the GB model is used in this section.

To certify the effectiveness of the coordinate-domain-aided approach, three-dimensional time-differenced coordinate components based on the GB model are illustrated in Fig. 5.3. We can easily find that the time-differenced coordinate components in three directions are highly concentrated around 0 m. It is reasonable that the rover and reference stations are both relatively static during this period, hence the time-differenced coordinate components are nearly 0 m. It proves that the proposed coordinate-domain-aided GB method is highly reliable. Moreover, the threshold of time-differenced coordinate components can be set (e.g., 0.1 m in this study). The coordinate-domain aided approach will be not be used if the resolved time-differenced coordinate components are larger than the above threshold. The accuracy and reliability of the coordinate-domain aided GB method can be further improved.

Figure 5.4 shows the float solutions of the cycle slip parameters resolved in the proposed method of dataset No. 2. It can be clearly seen that most float cycle slip parameters are nearly 0 cycles, thus indicating that there are no cycle slips at this time. Also, most of the other cycle slip parameters are usually close to an integer. Hence, the GB model is effective to a great extent. Taking a closer look at Fig. 5.4, the absolute values of the float cycle slip parameters are all smaller than 10 cycles since the potential large cycle slips (i.e., larger than 10 cycles) have been processed by the polynomial fitting before. In addition, there are small cycle slip parameters, especially those with an absolute value equal to 1 cycle, thus proving that the proposed method can handle small cycle slips.

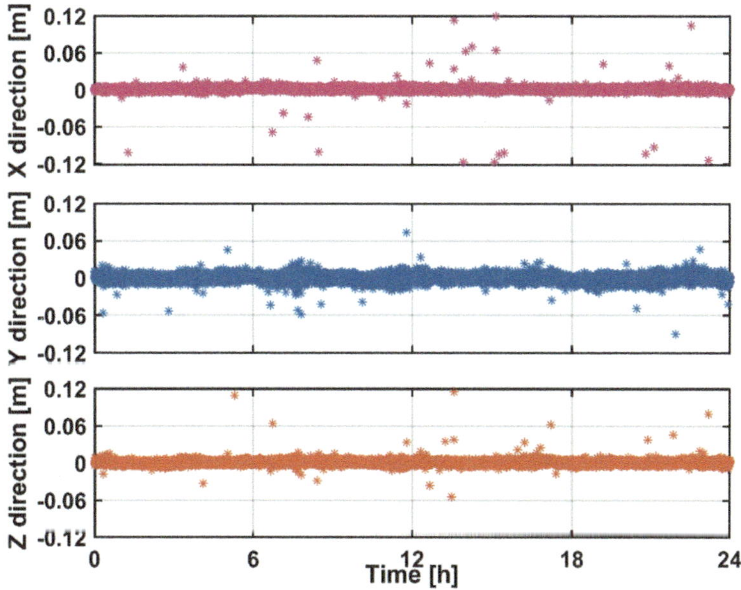

**Fig. 5.3** Three-dimensional time-differenced coordinate components resolved in the GB method of dataset No. 2

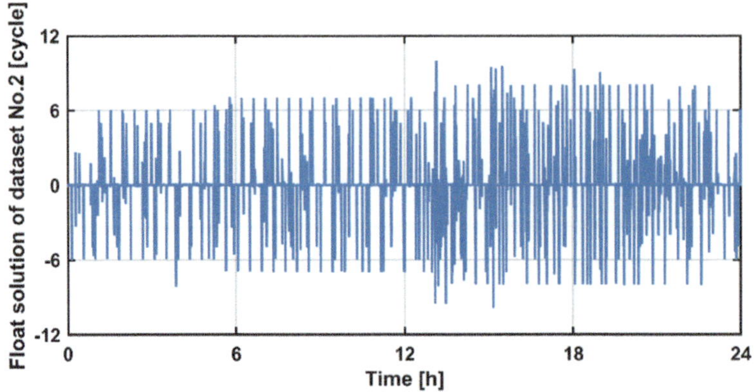

**Fig. 5.4** Float solutions of the cycle slip parameters resolved in the GB method of dataset No. 2

### *5.4.3   Analysis in the Real-Time Kinematic Situation*

The real-time kinematic data is adopted in this section named dataset No. 3. The 1-s dual-frequency vehicle data was collected from a high-end receiver. The experiment site was located in the urban area, where the signals may be obstructed, reflected, etc. The experiment lasted 1 h and 21 min. Although real cycle slips inevitably exist in a such kinematic situation, several simulated cycle slips, including small, multiple, insensitive, and even large cycle slips, are added. Hence the performance of the proposed method can be validated better. Specifically, small cycle slips from 2 to 3 cycles are added on G04 at L1, G29 at L2, C04 at B2, and C05 at B1 every 20 s from 02:17:09; multiple cycles from 4 to 5 cycles on more than 60% of satellites except reference satellite every 200 s from 02:17:49; insensitive cycle slips of GF plus ionospheric-biased and GF plus ionospheric-free combinations are added on G03 at L1 (1 cycle) and L2 (1 cycle), G22 at L1 (77 cycles) and L2 (60 cycles) every 20 s from 02:17:14; large cycle slips from 100 to 225 cycles are added on G31 at L2, G32 at L1, C01 at B1, and C02 at B2 every 20 s from 02:17:19. The proposed method based on GF and GB models is applied here.

Figures 5.5 and 5.6 show the float solutions and corresponding fractional parts of cycle slips based on the GB method alone and the proposed GB combined GF method, respectively. Here the fractional part means the difference between the float solution of the cycle slip and its nearest integer. Based on the relatively concentrated fractional parts in Fig. 5.5, the GB method alone is effective to some extent. However, the GB method can only work some of the time. The reason may be that there are not enough redundant observations. Hence, the GB method alone cannot work well in case of multiple cycle slips. It can be confirmed in Fig. 5.6 that the fractional parts are more concentrated since some significant cycle slips are processed in advance by the proposed method. Therefore, it proves the effectiveness of the proposed method.

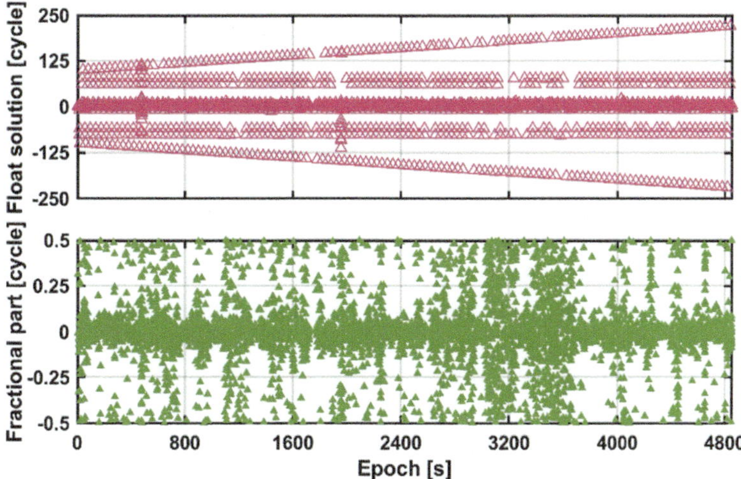

**Fig. 5.5**   Float solutions and corresponding fractional parts of cycle slips based on the GB method alone

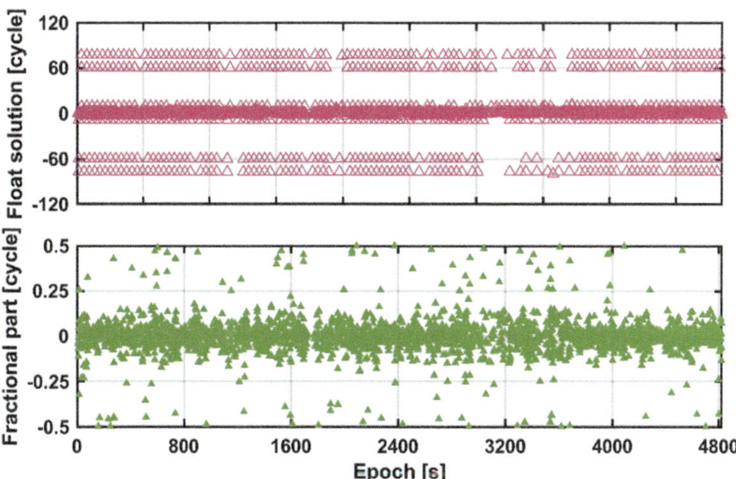

**Fig. 5.6**   Float solutions and corresponding fractional parts of cycle slips based on the GB combined GF method

## 5.5   Conclusion

The cycle slip estimation is a conventional but rather important hotspot in precision GNSS applications with phase observations. The efficient cycle slip processing can avoid re-initialization and maintain precise solutions continuously. In high-precision GNSS applications, the frequent occurrences of cycle slips and data gaps contaminate carrier phase observations, triggering the time-consuming re-initialization, which otherwise will result in positioning errors and degradation of positioning continuity. In this monograph, a GBIW method was proposed for integer cycle slip and data gap repair, which can be easily reduced to the GF and GFI models. In this method, cycle slips are processed simultaneously in an integrated adjustment, and a key portion is to provide the SD ionospheric biases as prior constraints to enhance the model strength in case of large sampling intervals or data gaps. To further improve the reliability of solutions, a PCSR strategy is employed. It should be clear that the proposed method should perform better in real applications since the simulation for cycle slips and data gaps was run at every epoch, which is a scenario unlikely to occur in practical situations. Although this monograph mainly focuses on the universal GB model, the results should, in theory, be better for the GFI model in static mode, where the station coordinates are already precisely known.

Essentially, cycle slips can be considered as fixed integer ambiguities, but their form is expressed as differential ambiguities. Compared to traditional integer ambiguity resolution, cycle slip resolution has its own characteristics. By leveraging this property, cycle slip correction can be conducted more efficiently, facilitating continuous and reliable high-precision positioning. In addition, it is a great challenge to efficiently process cycle slips in single-frequency RTK applications. The impact of cycle slip handling in such scenarios is particularly significant, as it directly affects the positioning accuracy and reliability. This is even more crucial in low-cost receivers and challenging environments, where frequent signal blockages and multipath effects exacerbate the difficulty of maintaining continuous and precise positioning solutions. To address this issue, we have proposed a new method with PPF aimed at single-frequency cycle slip estimation. The key aspect of the PPF method is the imposition of a positional polynomial constraint. Its essence lies in accurately predicting the position of the current epoch using the positions of several historical epochs. By leveraging the useful information gained from the PPF constraint, the accuracy of cycle slip estimation is significantly improved. Moreover, with the presented PCSR strategy, both the success rate and fix rate can be further enhanced, ensuring more reliable and robust positioning performance. Finally, we clarify that the PPF method, although originally motivated by single-frequency RTK applications, can be directly extended to multi-frequency and multi-GNSS scenarios. In principle, even better results should be achieved in such cases, given the additional redundancy and stronger observational geometry provided by multiple frequencies and constellations.

# References


1. Chen D, Ye S, Zhou W, Liu Y, Jiang P, Tang W, Yuan B, Zhao L (2016) A double-differenced cycle slip detection and repair method for GNSS CORS network. GPS Solut 20:439–450
2. Blewitt G (1990) An automatic editing algorithm for GPS data. Geophys Res Lett 17:199–202
3. Wübbena G (1985) Software developments for geodetic positioning with GPS using TI-4100 code and carrier measurements. Proceedings of the first international symposium on precise positioning with the global positioning system, pp 403–412
4. Melbourne WG (1985) The case for ranging in GPS-based geodetic systems. Proceedings of the first international symposium on precise positioning with the Global Positioning System, pp 373–386
5. Hatch R (1983) The synergism of GPS code and carrier measurements. Proceedings of the International geodetic symposium on satellite Doppler positioning, pp 1213–1231
6. Liu Z (2011) A new automated cycle slip detection and repair method for a single dual-frequency GPS receiver. J Geod 85:171–183
7. Cai C, Liu Z, Xia P, Dai W (2013) Cycle slip detection and repair for undifferenced GPS observations under high ionospheric activity. GPS Solut 17:247–260
8. Ju B, Gu D, Chang X, Herring TA, Duan X, Wang Z (2017) Enhanced cycle slip detection method for dual-frequency BeiDou GEO carrier phase observations. GPS Solut 21:1227–1238
9. de Lacy MC, Reguzzoni M, Sansò F, Venuti G (2008) The Bayesian detection of discontinuities in a polynomial regression and its application to the cycle-slip problem. J Geod 82:527–542
10. Zhang Q, Gui Q (2013) Bayesian methods for outliers detection in GNSS time series. J Geod 87:609–627
11. Zhang X, Li P (2016) Benefits of the third frequency signal on cycle slip correction. GPS Solut 20:451–460
12. Feng Y (2008) GNSS three carrier ambiguity resolution using ionosphere-reduced virtual signals. J Geod 82:847–862
13. Feng Y, Li B (2008) A benefit of multiple carrier GNSS signals: regional scale network-based RTK with doubled inter-station distances. J Spatial Sci 53:135–147
14. Li B, Qin Y, Li Z, Lou L (2016) Undifferenced cycle slip estimation of triple-frequency BeiDou signals with ionosphere prediction. Mar Geodesy 39:348–365
15. de Lacy MC, Reguzzoni M, Sansò F (2012) Real-time cycle slip detection in triple-frequency GNSS. GPS Solut 16:353–362
16. Zhao Q, Sun B, Dai Z, Hu Z, Shi C, Liu J (2015) Real-time detection and repair of cycle slips in triple-frequency GNSS measurements. GPS Solut 19:381–391
17. Banville S, Langley RB (2013) Mitigating the impact of ionospheric cycle slips in GNSS observations. J Geod 87:179–193
18. Dong D, Bock Y (1989) Global Positioning System Network analysis with phase ambiguity resolution applied to crustal deformation studies in California. J Geophys Res: Solid Earth 94:3949–3966
19. Blewitt G (1989) Carrier phase ambiguity resolution for the Global Positioning System applied to geodetic baselines up to 2000 km. J Geophys Res: Solid Earth 94:10187–10203
20. Teunissen PJG (1998) Minimal detectable biases of GPS data. J Geod 72:236–244
21. Yang L, Wang J, Knight NL, Shen Y (2013) Outlier separability analysis with a multiple alternative hypotheses test. J Geod 87:591–604
22. Li B, Zhang L, Verhagen S (2017) Impacts of BeiDou stochastic model on reliability: overall test, w-test and minimal detectable bias. GPS Solut 21:1095–1112
23. Xu G (2007) GPS: theory, algorithms and applications. Springer, Berlin
24. Li B, Qin Y, Liu T (2019) Geometry-based cycle slip and data gap repair for multi-GNSS and multi-frequency observations. J Geod 93:399–417
25. Carcanague S (2012) Real-time geometry-based cycle slip resolution technique for single-frequency PPP and RTK. Proceedings of the ION GNSS 2012, pp 1136–1148
26. Teunissen PJG (1995) The least-squares ambiguity decorrelation adjustment: a method for fast GPS integer ambiguity estimation. J Geod 70:65–82

27. Teunissen PJG (1998) Success probability of integer GPS ambiguity rounding and bootstrapping. J Geod 72:606–612
28. Zhou Z, Shen Y, Li B (2010) A windowing-recursive approach for GPS real-time kinematic positioning. GPS Solut 14:365–373
29. Zhou Z, Li B (2015) GNSS windowing navigation with adaptively constructed dynamic model. GPS Solut 19:37–48
30. Zhou Z, Li B (2017) Optimal Doppler-aided smoothing strategy for GNSS navigation. GPS Solut 21:197–210
31. Li B, Verhagen S, Teunissen PJG (2014) Robustness of GNSS integer ambiguity resolution in the presence of atmospheric biases. GPS Solut 18:283–296
32. Zhang X, Li X (2012) Instantaneous re-initialization in real-time kinematic PPP with cycle slip fixing. GPS Solut 16:315–327
33. Dai L, Wang J, Rizos C, Han S (2003) Predicting atmospheric biases for real-time ambiguity resolution in GPS/GLONASS reference station networks. J Geod 76:617–628
34. Akaike H (1974) A new look at the statistical model identification. IEEE Trans Autom Control 19:716–723
35. Wang J, Stewart MP, Tsakiri M (1998) A discrimination test procedure for ambiguity resolution on-the-fly. J Geod 72:644–653
36. Verhagen S, Teunissen PJG (2013) The ratio test for future GNSS ambiguity resolution. GPS Solut 17:535–548
37. Verhagen S, Li B, Teunissen PJG (2013) Ps-LAMBDA: ambiguity success rate evaluation software for interferometric applications. Comput Geosci 54:361–376
38. Wang S, Li B, Li X, Zang N (2018) Performance analysis of PPP ambiguity resolution with UPD products estimated from different scales of reference station networks. Adv Space Res 61:385–401
39. Freda P, Angrisano A, Gaglione S, Troisi S (2015) Time-differenced carrier phases technique for precise GNSS velocity estimation. GPS Solut 19:335–341
40. Zangeneh-Nejad F, Amiri-Simkooei AR, Sharifi MA, Asgari J (2017) Cycle slip detection and repair of undifferenced single-frequency GPS carrier phase observations. GPS Solut 21:1593–1603
41. Koch KR (1999) Parameter estimation and hypothesis testing in linear models. Springer, Berlin
42. Teunissen PJG (2006) Testing theory: an introduction, 2nd edn. Delft University Press, Delft
43. Baarda W (1968) A testing procedure for use in geodetic networks, Netherlands Geodetic Commission
44. Li B, Teunissen PJG (2014) GNSS antenna array-aided CORS ambiguity resolution. J Geod 88:363–376
45. Verhagen S, Li B (2012) LAMBDA software package: Matlab implementation, version 3.0
46. Chang X-W, Yang X, Zhou T (2005) MLAMBDA: a modified LAMBDA method for integer least-squares estimation. J Geod 79:552–565

# Chapter 6
# Stochastic Modeling

## 6.1 Introduction

In geodetic data processing, two models are essential: the functional model and stochastic model. The functional model establishes the relationship between measurements and unknown parameters, typically via linear or nonlinear equations, while the stochastic model characterizes the random errors by describing their accuracy and inter-correlations through a covariance matrix. Although any arbitrarily positive definite and symmetric matrix may serve as a weight matrix in a least squares (LS) adjustment to yield unbiased estimates, only the correct stochastic model can guarantee minimum variance estimators and reliable statistical tests. Given that the true stochastic properties of measurements are often not adequately known, methods such as variance and covariance component estimation (VCE) have been developed to better capture the actual dispersion and improve overall parameter estimation and reliability in data adjustment systems.

Significant research efforts have been devoted to VCE in geodetic data processing over the past century, resulting in many elegant formulations. In a typical adjustment, the covariance matrix of the observations is known only up to a scale factor, namely, the variance components, which must be accurately estimated to achieve optimal parameter estimation and reliable statistical testing. Starting with Helmert's quadratic estimation of unbiased variance components using LS residuals in a Gauss-Markov (GM) model (and its subsequent extension and simplification [1]), Grafarend extended this method to conditional adjustment models (the Gauss-Helmert model). Building on these foundations, Rao proposed the minimum norm quadratic unbiased estimator (MINQUE) [2], which is equivalent to Helmert's estimate under normality assumptions [3]; Sjöberg further extended MINQUE to conditional adjustment models [4] and Koch introduced the best invariant quadratic unbiased estimator (BIQUE) with a minimal variance objective [5]. In the general Gauss-Helmert framework, Yu derived a generalized BIQUE [6], showing that both Rao's MINQUE [2] and Koch's BIQUE [5] are special cases of his approach. Concurrently, maximum

© The Author(s) 2025

B. Li et al., *GNSS Real-Time Kinematic Positioning*, Navigation: Science and Technology 17, https://doi.org/10.1007/978-981-96-9116-6_6

likelihood estimation (MLE) methods have been explored extensively: Kubik derived an MLE VCE for the case of one variance per observation group [7], Koch formulated an MLE for unbiased estimates in the GM model [8], and Ou provided an MLE formula equivalent to Koch's and Helmert's methods [9]; moreover, a general MLE is derived that encompasses the methods of Kubik [7] and Koch [8]. Modern geodetic applications have since adopted popular VCE methods including MINQUE [2], BIQUE [3, 5], the restricted maximum likelihood estimator (RMLE) [7, 8, 10], and the least-squares variance component estimator (LS-VCE) [11–14]. Additional studies have focused on robust estimation [15], the use of VCE in collocation [16] and ill-posed problems [17, 18], and the development of efficient computational algorithms [19]. In addition, Xu et al. discussed the estimability of variance and covariance components, indicating that at most $r(r + 1)/2$ independent components are estimable for a redundancy of $r$ [20], while Teunissen and Amiri-Simkooei have demonstrated the benefits of LS VCE [21].

In general, the iterative procedure must be employed to gradually approximate the converged estimates of variance and covariance components, because the unknown covariance matrix is involved to compute the inputs of VCE. For instance, in Helmert estimation with the GM model, the LS residuals as the inputs for VCE are computed iteratively with the updated covariance matrix, since the covariance matrix is used to compute the LS residuals. Thus, in real application the bottleneck problem of VCE is the huge computation burden, particularly when many variance and covariance components are involved for many heterogeneous observations. Theoretically, the measurements can be used to extract at most their discrepancies with each other besides to estimate the unknowns in linear model and these discrepancies are the essential inputs for VCE. Any orthogonal complement matrix of coefficient matrix in the linear GM model can be used to compute these discrepancies. Traditionally, the orthogonal complement matrix is constructed by using both the coefficient and covariance matrices. Thus, the constructed orthogonal complement matrix and its derived discrepancies must be updated in the whole iterative VCE procedure, which is the key limitation for VCE computation efficiency.

The objective of this section is to develop a new method that allows efficient estimation of variance and covariance components to overcome the VCE bottleneck problem of a huge computation burden. The core of the new method is to construct the orthogonal complement matrix only by using the coefficient matrix exempting the involvement of covariance matrix. Therefore, the constructed matrix and its derived discrepancies are invariant for the whole VCE procedure. As a result, the computation efficiency is significantly improved.

For the fully-populated stochastic modeling, in Global Navigation Satellite Systems (GNSS) applications, an adequate stochastic model is required for reliable ambiguity resolution and for precise positioning. However, our knowledge of an adequate stochastic model is still at a rather rudimentary level in contrast to the functional model. Hence, refining the GNSS stochastic model is a worthy aspiration. Studies of GNSS stochastic models in earlier times were based on the elevation dependence of random observation errors [22], and later took into account time and cross correlations [11, 18, 23–25]. In addition, the signal-to-noise ratio based elevation-dependent models, i.e., SIGMA-ε and SIGMA-Δ, were developed by [26, 27]. This SIGMA-type model was further extended to incorporate the physical correlations [26, 27] based on turbulence theory [28]. However, in most existing GNSS software packages, only the elevation dependence of observation variances is implemented due to its simplicity and small computation burden [29].

BeiDou Navigation Satellite System (BDS) is the first GNSS that broadcasts triple-frequency signals in the full satellite constellation. Since its Interface Control Document (ICD) was released at the end of 2012, many investigators have intensified their efforts to understand BDS capabilities in ambiguity resolution and positioning [30–32]. However, all such investigations are based on a very empirical stochastic model borrowed from the knowledge of Global Positioning System (GPS) stochastic models, where homoscedastic or empirical elevation-dependent weighting is applied and often both time and cross correlations are disregarded. But we expect differences between the stochastic properties of BDS and GPS observations due to differences in their constellations and signal quality. Hence, it is necessary to study appropriate stochastic models for BDS observations.

The contribution of this study is threefold. First, we study estimation theory in the context of triple-frequency BDS stochastic models, where a very sophisticated covariance matrix is formulated to allow estimating satellite-specific variance, cross correlation between two arbitrary frequencies, and time correlation of phase and code observations per frequency. To purely reflect the level of random observation noise, the between-receiver single-difference (SD) geometry-free functional model is used to eliminate geometric errors. Second, six BDS data sets from four brands of receivers are analyzed to demonstrate BDS stochastic properties. The stochastic models between short and zero-length baselines are compared in detail. Finally, the impacts of elevation-dependent weighting, cross correlation and time correlation on integer ambiguity resolution and positioning are numerically analyzed to emphasize the importance of using a realistic stochastic model.

## 6.2  Variance and Covariance Component Estimation

This section comprehensively introduces variance and VCE including the efficient method. We introduce the methodology of the VCE from two parts: VCE based on the LS residuals and efficient estimation of variance and covariance components.

For the part of VCE based on the LS residuals. Starting from the linear (linearized) GM model

$$y = Ax + \varepsilon \tag{6.1}$$

where $y$ is an $n \times 1$ vector of measurements, $A$ is an $n \times t$ $(n > t)$ design matrix of full column rank, $x$ is a $t \times 1$ vector of unknown parameters to be estimated, and $\varepsilon$ is the $n \times 1$ vector of measurement errors. In general, $\varepsilon$ is random with mean of zero and covariance matrix $D_{\varepsilon\varepsilon} = D_{yy}$. The LS solution of Eq. (6.1) is

$$
\begin{cases}
\hat{x} = \left(A^{\mathrm{T}} D_{yy}^{-1} A\right)^{-1} A^{\mathrm{T}} D_{yy}^{-1} y \\
v = Ry = R\varepsilon \\
R = I_n - A\left(A^{\mathrm{T}} D_{yy}^{-1} A\right)^{-1} A^{\mathrm{T}} D_{yy}^{-1}
\end{cases}
\tag{6.2}
$$

where $v$ is the vector of the LS residuals; $I_n$ is the $n \times n$ identity matrix and $R$ is an idempotent matrix satisfying with

$$
RR = R, \quad RA = 0, \quad R^{\mathrm{T}} D_{yy}^{-1} = D_{yy}^{-1} R
\tag{6.3}
$$

and

$$
\mathrm{tr}(R) = \mathrm{rk}(R) = r
\tag{6.4}
$$

Here, $\mathrm{tr}(\bullet)$ and $\mathrm{rk}(\bullet)$ are the mathematical operators for computing trace and rank of a matrix, respectively. $r = n - t$ is the redundancy. The matrix equations for VCE are established by using the LS residuals as

$$
R D_{yy} R^{\mathrm{T}} - R \mathrm{E}\left(\varepsilon\varepsilon^{\mathrm{T}}\right) R^{\mathrm{T}} - \mathrm{E}\left(vv^{\mathrm{T}}\right)
\tag{6.5}
$$

where $\mathrm{E}(\bullet)$ denotes the mathematical expectation of a variable. Obviously, the iterative procedure must be employed to estimate covariance matrix $D_{yy}$ because it is involved to compute $v$ and $R$. Ignoring the expectation and giving an appropriate initial covariance matrix $D_0$, the fundamental matrix equations for iterative VCE reads

$$
R_0 D_{yy} R_0^{\mathrm{T}} = v_0 v_0^{\mathrm{T}}
\tag{6.6}
$$

where $v_0 = R_0 y$ and $R_0 = I_n - A\left(A^{\mathrm{T}} D_0^{-1} A\right)^{-1}$. Without loss of generality, we use the general structure of $D_{yy}$ as

$$
D_{yy} = U_1 \theta_1 + U_2 \theta_2 + \cdots + U_m \theta_m = \sum_{i=1}^{m} U_i \theta_i
\tag{6.7}
$$

where $\theta_i$ is the $i$th unknown variance or covariance; $U_i$ is the given (semi-)definite matrix for variance component of $\theta_i$ and the symmetrical matrix for covariance component. Since $v_0 v_0^{\mathrm{T}}$ in Eq. (6.6) is a matrix, we transform the matrix equations into vector equations as

$$
\mathrm{vec}\left(R_0 \hat{D}_{yy} R_0^{\mathrm{T}}\right) = \mathrm{vec}\left(v_0 v_0^{\mathrm{T}}\right)
\tag{6.8}
$$

where $\text{vec}(\bullet)$ denotes the vector operator that converts a matrix to a column vector by stacking one column of this matrix underneath the previous one. Substituting Eq. (6.7) into Eq. (6.8), we have

$$M_0\hat{\theta} = \text{vec}(v_0 v_0^T) \tag{6.9}$$

with $M_0 = \left[\, \text{vec}(R_0 U_1 R_0^T) \,\cdots\, \text{vec}(R_0 U_m R_0^T) \,\right]$. We use the LS criterion on Eq. (6.9) with the weight matrix $P = D_0^{-1} \otimes D_0^{-1}$, the VCE normal equations yields

$$M_0^T P M_0 \hat{\theta} = N\hat{\theta} = M_0^T P \text{vec}(v_0 v_0^T) = q \tag{6.10}$$

After the derivation, the explicit expressions are

$$n_{ij} = \text{tr}(W_0 U_i W_0 U_j), \quad q_i = v_0^T W_0 U_i W_0 v_0 \tag{6.11}$$

with $W_0 = D_0^{-1} R_0 = R_0^T D_0^{-1} = W_0^T$, which is the same to the expressions of MINQUE given by Rao [2] and the LS VCE formulae given by Teunissen and Amiri-Simkooei [21]. Moreover, it is rather easy to prove that Helmert's formulae are the same [14]. Therefore, all VCE methods mentioned above can be derived from the fundamental matrix Eq. (6.6) and the equivalent solutions are trivially achievable.

For the part of efficient estimation of variance and covariance components. It is observed that the fundamental equations are correlated due to the rank defect of $R$. In other words, there are $r(r+1)/2$ independent equations and $n^2 - r(r+1)/2$ equations do never contribute to the estimation of variances and covariances [20]. In this section, we will derive VCE from the independent discrepancies of measurements. Theoretically, any orthogonal complement matrix of the coefficient matrix $A$ can be used to compute the discrepancies. We invent an efficient VCE method by constructing an invariant orthogonal complement matrix only using the coefficient matrix itself. Additionally, equivalent solutions from the correlated LS residuals and the independent discrepancies are proven.

For the VCE based on the independent discrepancies. The essential inputs for VCE are not all $n$ LS residuals due to their correlation, but $r = n - t$ independent observation discrepancies. In other words, the most information extracted from the measurements is their discrepancies except the information used for parameter estimation. Thus, the observation vector can be transformed to the vector comprised by parameters and discrepancies

$$\begin{bmatrix} \hat{x} \\ u \end{bmatrix} = \begin{bmatrix} \left(A^T D_{yy}^{-1} A\right)^{-1} A^T D_{yy}^{-1} \\ B \end{bmatrix} y \tag{6.12}$$

where $u$ is the $r \times 1$ vector of the discrepancies; $B$ is an $r \times n$ matrix with rank number of $r$ and orthogonal with $A$, namely, $BA = 0$. In other words, $B$ is an orthogonal complement matrix of $A$ in the $n$-dimensional real-valued space $\mathrm{R}^n$. From the mathematical point of view, $y \in \mathrm{R}^n$, $\hat{x} \subset \mathrm{R}(A^T)$ and $u \subset \mathrm{R}(B)$, and $\mathrm{R}(B) =$

$N(A^T)$. Here $R(\bullet)$ denotes the space spanned by all column vectors of a matrix and $N(\bullet)$ the null space of a matrix.

Starting with the equations of discrepancies

$$u = By = B(Ax + \varepsilon) = B\varepsilon \tag{6.13}$$

we establish alternative fundament matrix equations for VCE similar to Eq. (6.5) as

$$BD_{yy}B^T = uu^T \tag{6.14}$$

Due to the symmetry of matrix $uu^T$, there are $r(r+1)/2$ equations are independent. All independent equations are extracted using "vech" product as

$$M\hat{\theta} = \text{vech}(uu^T) \tag{6.15}$$

where $M = \left[\text{vech}(BU_1B^T) \cdots \text{vech}(BU_mB^T)\right]$ The product "vech$(\bullet)$" has a similar definition to vec$(\bullet)$ except the elements on and below the diagonal of a matrix is used. It is emphasized here that the inverse of covariance matrix of vech$(uu^T)$ is used as weight matrix, otherwise the solution is not optimal. When $y$ is of normal distribution, the covariance matrix of vech$(uu^T)$ is computed as [21]

$$\sum_{vh} = D^+(D_{uu} \otimes D_{uu})D^{+T}/2 \tag{6.16}$$

where $D_{uu} = BD_{yy}B^T$, $D$ is the duplication matrix defined by the property that $D\text{vech}(S) = \text{vec}(S)$ with $S$ being a symmetrical matrix, and $D^+ = (D^TD)^{-1}D^T$ is its pseudo inverse. Given the prior covariance matrix $D_0$, namely $D_{u_0} = BD_0B^T$ and $D_{vh}^0 = D^+(D_{u_0} \otimes D_{u_0})D^{+T}/2$, the elements of LS normal equation is derived similar to Eq. (6.11)

$$n_{ij} = \text{tr}(\tilde{U}_i \Sigma_{u_0}^{-1} \tilde{U}_j \Sigma_{u_0}^{-1}), \quad q_i = u^T \Sigma_{u_0}^{-1} \tilde{U}_i \Sigma_{u_0}^{-1} u \tag{6.17}$$

with $\tilde{U}_i = BU_iB^T$.

For the Constructing the orthogonal complement matrix for invariant discrepancies. In principle, any orthogonal complement matrix of $A$ can be used as $B$ in Eq. (6.17), and the equivalent solution is achievable. But the computation efficiency is significantly various for the different choices of orthogonal complement matrices. As a geodesist, we are very familiar with that the coefficient matrix in the condition adjustment model can be directly used as matrix $B$. However, in the most of data processing problems, the GM function model is preferred due to its easier formation, particularly in the case of huge observations with many observation types. Hence, it is crucial to efficiently determine an orthogonal complement matrix based on the information from the GM model.

In this monograph, we construct the orthogonal complement matrix purely from coefficient matrix $A$ itself such that the constructed $B$ matrix is invariant and never necessary to be updated in the whole iterations. The $y$, $A$ and $\varepsilon$ in Eq. (6.1) are blocked as

$$y = \begin{bmatrix} y_1 \\ y_2 \end{bmatrix}, \quad A = \begin{bmatrix} A_1 \\ A_2 \end{bmatrix}, \quad \varepsilon = \begin{bmatrix} \varepsilon_1 \\ \varepsilon_2 \end{bmatrix} \tag{6.18}$$

where $y_1$ and $\varepsilon_1$ are of $t \times 1$ dimension and $y_2$ and $\varepsilon_2$ of $r \times 1$ dimension; $A_1$ is a $t \times t$ invertible matrix and $A_2$ a $r \times t$ matrix. Obviously, the parameter $x$ can be uniquely determined by $y_1$ without any redundancy

$$\hat{x} = A_1^{-1} y_1 \tag{6.19}$$

The remaining equations are totally redundant. If the measurements are free of error contamination, then

$$y_2 = A_2 A_1^{-1} y_1 \tag{6.20}$$

holds exactly true. In fact, all measurements are inevitably uncertain and thus their discrepancies with each other are computed as

$$u = A_2 A_1^{-1} y_1 - y_2 = \left[ A_2 A_1^{-1} \ -I_r \right] y = By \tag{6.21}$$

with [33]

$$B = \left[ A_2 A_1^{-1} \ -I_r \right] \tag{6.22}$$

Apparently, the matrix $B$ is orthogonal and complementary with $A$. Substituting it into Eq. (6.17), we obtain the formula of efficient estimation of variance and covariance components, where $\tilde{U}_i$ and $u$ are invariant and computed before the iteration procedure, and we just need to update $D_{u_0}$. However, in the traditional methods, both $R_0$ and $v_0$ have to been updated besides $D_{u_0}$.

For the Equivalent solutions from the correlated LS residuals and independent discrepancies. We have well known that the correlated equations can never contribute to the parameters and then we wonder whether the solution from independent discrepancies is equivalent to that from correlated LS residuals. The answer is positive and the proof is given as follows.

According to the equivalence theory of [34], the elimination of the partial parameters can never affect the solution of other parameters and corresponding accuracy. Therefore, the estimated variance of unit weight from the original Eq. (6.1) is equal to that from equivalent Eq. (6.21), namely, we have the following equation

$$\hat{\sigma}_0^2 = \frac{(y - A\hat{x})^{\mathrm{T}} \sum_0^{-1} (y - A\hat{x})}{n - t} = \frac{u^{\mathrm{T}} \sum_{u_0}^{-1} u}{n - t} \tag{6.23}$$

Substituting the LS solution of $\hat{x}$ and Eq. (6.21) into Eq. (6.23) yields

$$y^{\mathrm{T}}R_0^{\mathrm{T}}D_0^{-1}R_0 y = y^{\mathrm{T}}B^{\mathrm{T}}D_{u_0}^{-1}By \tag{6.24}$$

Since Eq. (6.24) holds true for arbitrary observation vector $y$, Eq. (6.24) is equivalent to

$$R_0^{\mathrm{T}}D_0^{-1}R_0 = B^{\mathrm{T}}D_{u_0}^{-1}B \tag{6.25}$$

It is rather easy to derive the same expression to Eq. (6.11) via substituting Eq. (6.25) into Eq. (6.17), namely, the equivalent solution is achievable from the efficient method.

## 6.3   Fully-Populated Stochastic Modeling

As an important part of Stochastic modeling, the fully-populated stochastic modeling is studied comprehensively in this section. Then, the estimation procedure of a fully-populated stochastic model is introduced.

First, the SD geometry-free functional model is studied. As defined by Teunissen in his canonical theory for short GPS baselines, the geometry-free model parameterizes the observation equations in terms of the receiver-satellite ranges instead of the baseline components as in a geometry-based model. The double-differenced (DD) geometry-free model has been extensively studied for DD ambiguity resolution by Delft researchers [35] due to its advantages: for instance, as mentioned in [11], the linearity of the observation model, the independence of satellite orbit information and tropospheric delay, and the capability of ambiguity resolution with only one pair of satellites.

To retrieve the stochastic properties of pure random noise of triple-frequency BDS signals, one has to eliminate the systematic errors contained in the observations. We will use the between-receiver SD, geometry-free model to remove such errors. On ultra-short (shorter than 10 m as usual) or zero-length baselines, the systematic errors can be completely eliminated such that only pure random errors remain in the resulting observations. Moreover, since no mathematical correlation is introduced in the SD model, it is more suitable for estimating the satellite-specific variances compared to the DD model, though they are equivalent for estimating the stochastic model [11].

Eliminating the systematic errors on ultra-short or zero-length baselines, the single-epoch, geometry-free, SD observation equations on frequency $j$ read

$$\mathrm{E}\big(\boldsymbol{\Phi}_j\big) = \delta\varrho + e_s\delta t_{r,j} + \lambda_j e_s \varphi_j + \lambda_j \boldsymbol{\alpha}_{r,j} \tag{6.26}$$

$$\mathrm{E}\big(\boldsymbol{P}_j\big) = \delta\varrho + e_s \mathrm{d}t_{r,j} \tag{6.27}$$

where $\boldsymbol{\Phi}_j = \left[\Phi_j^1, \ldots, \Phi_j^s\right]^{\mathrm{T}}$ is the SD observation vector of $s$ satellites for phase on frequency $j$, and $\boldsymbol{P}_j$ for code has the same structure as $\boldsymbol{\Phi}_j$. $\delta\boldsymbol{\varrho} = \left[\delta\varrho^1, \ldots, \delta\varrho^s\right]^{\mathrm{T}}$ is the vector of SD receiver-satellite ranges; $\boldsymbol{a} = \left[a^1, \ldots, a^s\right]^{\mathrm{T}}$ is the SD integer ambiguity vector with wavelength $\lambda_j$; $\varphi_j$ is the SD initial phase bias of receiver; $\delta t_{r,j}$ and $\mathrm{d}t_{r,j}$ are the SD receiver clock errors for phase and code, respectively.

Obviously, Eq. (6.26) has a rank deficiency of 1 since the coefficients of the unknown parameters $\delta t_{r,j}$, $\varphi_j$ and $\boldsymbol{\alpha}_{r,j}$ satisfy

$$\left[\boldsymbol{e}_s, \lambda_j\boldsymbol{e}_s, \lambda_j\boldsymbol{I}_s\right]\begin{bmatrix} \lambda_j & \lambda_j \\ -1 & 0 \\ \boldsymbol{0}_{s\times 1} & -\boldsymbol{e}_s \end{bmatrix} = \begin{bmatrix} \boldsymbol{0}_{s\times 1}, & \boldsymbol{0}_{s\times 1} \end{bmatrix} \tag{6.28}$$

This indicates that the SD phase clock $\delta t_{r,j}$ and SD initial bias $\varphi_j$ are linearly dependent, and also further dependent on the SD ambiguity parameters $\boldsymbol{\alpha}_{r,j}$. To eliminate this rank deficiency, we reparameterize the equations as:

$$\delta t_{r,j} =: \delta t_{r,j} + \lambda_j\varphi_j + \lambda_j a_{r,j}^1 \quad z_j = [a_{r,j}^2 - a_{r,j}^1, \ldots, a_{r,j}^s - a_{r,j}^1]^{\mathrm{T}} \tag{6.29}$$

where the same symbol $\delta t_{r,j}$ is used to denote the reparameterized variable. The vector $z_j$ contains DD ambiguities. The full-rank version of observation Eqs. (6.26) and (6.27) reads

$$\mathrm{E}\begin{pmatrix} \boldsymbol{\Phi}_j \\ \boldsymbol{P}_j \end{pmatrix} = \begin{bmatrix} \boldsymbol{I}_s & \boldsymbol{e}_s & \boldsymbol{0} & \lambda_j\boldsymbol{\Lambda} \\ \boldsymbol{I}_s & \boldsymbol{0} & \boldsymbol{e}_s & \boldsymbol{0} \end{bmatrix}\begin{bmatrix} \delta\boldsymbol{\varrho} \\ \delta t_{r,j} \\ \mathrm{d}t_{r,j} \\ z_j \end{bmatrix} \tag{6.30}$$

where $\boldsymbol{\Lambda} = \left[\boldsymbol{0}_{(s-1)1}, \boldsymbol{I}_{s-1}\right]^{\mathrm{T}}$. In general, the geometry-free model has lower success in ambiguity resolution compared to a geometry-based model. However, it still allows reliable ambiguity resolution over multiple epochs in our short baseline case. After the DD ambiguities $z_j$ are fixed, one can move them to the left side of the equations. Collecting the observations of all $f$ frequencies yields:

$$\mathrm{E}(\boldsymbol{y}) = \left[ \left(\boldsymbol{e}_{2f} \otimes \boldsymbol{I}_s\right) \left(\boldsymbol{I}_{2f} \otimes \boldsymbol{e}_s\right) \right]\begin{bmatrix} \delta\boldsymbol{\varrho} \\ \Delta\boldsymbol{t} \end{bmatrix} \tag{6.31}$$

where $\boldsymbol{y} = \left[\boldsymbol{\Phi}^{\mathrm{T}}, \boldsymbol{P}^{\mathrm{T}}\right]^{\mathrm{T}}$ with the ambiguity-corrected phase $\boldsymbol{\Phi} = \left[\Phi_1^{\mathrm{T}}, \ldots, \Phi_f^{\mathrm{T}}\right]^{\mathrm{T}}$ and the same structure for code $\boldsymbol{P}$; $\Delta\boldsymbol{t} = \left[\delta\boldsymbol{t}^{\mathrm{T}}, \ldots, \mathrm{d}\boldsymbol{t}^{\mathrm{T}}\right]^{\mathrm{T}}$ with $\delta\boldsymbol{t} = \left[\delta t_{r,1}, \ldots, \delta t_{r,f}\right]^{\mathrm{T}}$ and $\mathrm{d}\boldsymbol{t} = \left[\mathrm{d}t_{r,1}, \ldots, \mathrm{d}t_{r,f}\right]^{\mathrm{T}}$.

From the design matrix of Eq. (6.31), it is easy to see that

$$\left[\,(e_{2f}\otimes I_s)\,(C\otimes e_s)\,\right]\begin{bmatrix} e_s \\ e_{2f}\end{bmatrix}=0$$

which reveals that Eq. (6.31) is rank-deficient with a deficiency of 1. We therefore make an independent parameterization by fixing $dt_{r,1}$, i.e., $\Delta \bar{t}=\left[\delta\bar{t},\,d\bar{t}\right]$ with $\delta\bar{t}=\delta t-e_f dt_{r,1}$ and $d\bar{t}=[dt_{r,2}-dt_{r,1},\ldots,dt_{r,f}-dt_{r,1}]^{\mathrm{T}}$; $\delta\overline{\varrho}=\delta\varrho+e_s dt_{r,1}$. Finally, the single-epoch, $f$-frequency, geometry-free, SD observation equations of full rank read

$$\mathrm{E}(y)=\left[\,(e_{2f}\otimes I_s)\,(C\otimes e_s)\,\right]x \tag{6.32}$$

where $x=[\delta\overline{\varrho}^{\mathrm{T}},\,\Delta\bar{t}^{\mathrm{T}}]^{\mathrm{T}}$, $C=\mathrm{blkdiag}\big(I_f,\,[0,I_{f-1}]^{\mathrm{T}}\big)$. The observation equations of $K$ epochs are

$$\mathrm{E}(\ell)=(I_K\otimes B)\beta=A\beta \tag{6.33}$$

where $B=\left[(e_{2f}\otimes I_s),(C\otimes e_s)\right]$, $\ell=[y_1^{\mathrm{T}},\ldots,y_K^{\mathrm{T}}]^{\mathrm{T}}$ and $\beta=[x_1^{\mathrm{T}},\ldots,x_K^{\mathrm{T}}]^{\mathrm{T}}$ with the subscripts denoting the epoch number.

Second, the Formulation of the stochastic model is studied. To make the stochastic model sufficiently sophisticated, the following assumptions are made. Firstly, to address the satellite-specific variance and its elevation dependence, an unknown variance is assigned to individual satellites over a short period of $K$ epochs during which the satellite elevation is nearly invariant. Secondly, the cross correlation is assumed to be present between two different frequencies for phase and code observations, respectively, while it is assumed to be absent between phase and code of common frequency. Thirdly, for a given time lag, the time correlation coefficients are assigned respectively to phase and code per frequency. Note that the between-satellite and between-station correlations are neglected in the stochastic model.

It is emphasized that our covariance matrix is not a fully unknown covariance matrix though it is very sophisticated. That is, the number of unknown (co)variance components is much smaller than the number of elements in the full covariance matrix. Hence, our covariance matrix is mathematically (stepwisely) estimable. For more information on estimability of covariance matrices, see [20].

The single-epoch stochastic model can be derived. Based on the foregoing assumptions for our stochastic model, the observation variances are specified by the observation types, the satellites and the frequencies, while the cross correlations are associated with the between-frequency observations of phase and code, respectively. In the case of phase, the single-epoch stochastic model reads

$$Q_\phi=Q_\phi^{[\sigma]}+Q_\phi^{[c]} \tag{6.34}$$

where $Q_\phi^{[\sigma]}=\mathrm{blkdiag}\big(Q_{\phi_1}^{[\sigma]},\ldots,Q_{\phi_f}^{[\sigma]}\big)$ captures the observation variances of all $f$ frequencies, while $Q_{\phi_j}^{[\sigma]}=\mathrm{diag}\big(\big[\sigma_{\phi_{r,j}^1}^2,\ldots,\sigma_{\phi_{r,j}^s}^2\big]\big)$ the variances of all $s$ satellites on frequency $j$, with $\sigma^2_{\phi^i_{r,j}}$ being the variance of the $i$th satellite. The matrix $\boldsymbol{Q}^{[c]}_\phi = \left\{ \varrho^{[c]}_{\phi_i\phi_j} \boldsymbol{Q}^{[\sigma]}_{\phi_i\phi_j} \right\}$ captures the cross correlations with $\varrho^{[c]}_{\phi_i\phi_j}$, which is the cross correlation coefficient between frequencies $i$ and $j$, and $\boldsymbol{Q}^{[\sigma]}_{\phi_i\phi_j} = \mathrm{diag}\left( \left[ \sigma_{\phi^1_{r,i}} \sigma_{\phi^1_{r,j}}, \ldots, \sigma_{\phi^s_{r,i}} \sigma_{\phi^s_{r,j}} \right] \right)$. The superscripts '$[\sigma]$' and '$[c]$' denote the terms associated with the observation variances and cross correlations, respectively. As an example with $f = 3$, the covariance matrix (6.34) reads

$$\boldsymbol{Q}_\phi = \begin{bmatrix} \boldsymbol{Q}^{[\sigma]}_{\phi_1} & \varrho^{[c]}_{\phi_1\phi_2} \boldsymbol{Q}^{[\sigma]}_{\phi_1} & \varrho^{[c]}_{\phi_1\phi_3} \boldsymbol{Q}^{[\sigma]}_{\phi_1\phi_3} \\ & \boldsymbol{Q}^{[\sigma]}_{\phi_1\phi_2} & \varrho^{[c]}_{\phi_2\phi_3} \boldsymbol{Q}^{[c]}_{\phi_2\phi_3} \\ \mathrm{symmetry} & & \boldsymbol{Q}^{[c]}_{\phi_3} \end{bmatrix} \tag{6.35}$$

One can similarly construct the covariance matrix for code observations (denoted by $\boldsymbol{Q}^{[\sigma]}_P$ and $\boldsymbol{Q}^{[c]}_P$) and then the single-epoch covariance matrix of phase and code observations as

$$\boldsymbol{Q}_E = \mathrm{D}(\boldsymbol{y}) = \boldsymbol{Q}^{[\sigma]}_E + \boldsymbol{Q}^{[c]}_E \tag{6.36}$$

where $\boldsymbol{Q}^{[\sigma]}_E = \mathrm{blkdiag}\left( \boldsymbol{Q}^{[\sigma]}_\phi, \boldsymbol{Q}^{[\sigma]}_P \right)$ and $\boldsymbol{Q}^{[c]}_E = \mathrm{blkdiag}\left( \boldsymbol{Q}^{[c]}_\phi, \boldsymbol{Q}^{[c]}_P \right)$. The covariance matrices for code, $\boldsymbol{Q}^{[\sigma]}_P$ and $\boldsymbol{Q}^{[c]}_P$, have the same structure as those for phase. There are total $2sf + f(f - 1)$ unknown (co)variance components, of which $sf$ are for variances and $f(f - 1)$ for cross correlations.

The multiple-epoch stochastic model can be derived. In the multiple-epoch case, besides block-diagonally stacking the single-epoch covariance matrices of all epochs, the time correlations are introduced for phase and code observations per frequency for all time lags. As a result, the multiple-epoch covariance matrix is formulated as

$$\boldsymbol{Q}_\ell = \boldsymbol{I}_K \otimes \boldsymbol{Q}_E + \boldsymbol{Q}^{[t]} \circ \left( \boldsymbol{e}_K \boldsymbol{e}^{\mathrm{T}}_K \otimes \boldsymbol{Q}_E \right) \tag{6.37}$$

where the symbol "$\circ$" is the Hadamard product with $\left[ \boldsymbol{A} \circ \boldsymbol{B} \right]_{ij} = \left[ \boldsymbol{A} \right]_{ij} \left[ \boldsymbol{B} \right]_{ij}$. Matrix $\boldsymbol{Q}^{[t]}$ is symmetric, and its upper triangle part is

$$\begin{bmatrix} \boldsymbol{0} & \boldsymbol{Q}^{[\sigma]}_{,1} & \boldsymbol{Q}^{[\sigma]}_{,2} & \cdots & \boldsymbol{Q}^{[\sigma]}_{,K-1} \\ & \ddots & \ddots & \ddots & \vdots \\ & & \ddots & \ddots & \boldsymbol{Q}^{[\sigma]}_{,2} \\ & & & \ddots & \boldsymbol{Q}^{[\sigma]}_{,1} \\ & & & & \boldsymbol{0} \end{bmatrix}$$

where $\boldsymbol{Q}^{[t]}_{,\tau} = \mathrm{blkdiag}\left( \boldsymbol{Q}^{[t]}_{\phi,\tau}, \boldsymbol{Q}^{[t]}_{P,\tau} \right) \otimes \boldsymbol{I}_s$ captures the time correlations of phase and code of all $f$ frequencies for time lag of $\tau$. Matrix $\boldsymbol{Q}^{[t]}_{\phi,\tau} = \mathrm{diag}\left( \left[ \rho^{[t]}_{\phi_1,\tau}, \ldots, \rho^{[t]}_{\phi_f,\tau} \right] \right)$

corresponds to phase observations, where $\rho_{\phi_j,\tau}^{[t]}$ is the time correlation coefficient on frequency $j$ with time lag $\tau$; $\boldsymbol{Q}_{P,\tau}^{[t]}$ has a similar structure as $\boldsymbol{Q}_{\phi,\tau}^{[t]}$. To show the structure of the multiple-epoch stochastic model, an example is given for phase observations with $K = 3$ and $f = 2$ and without cross correlation:

$$
\begin{bmatrix}
\boldsymbol{Q}_{\phi_1}^{[\sigma]} & \boldsymbol{0} & \varrho_{\phi_{1,1}}^{[t]}\boldsymbol{Q}_{\phi_1}^{[\sigma]} & \boldsymbol{0} & \varrho_{\phi_{1,2}}^{[t]}\boldsymbol{Q}_{\phi_1}^{[\sigma]} & \boldsymbol{0} \\
& \boldsymbol{Q}_{\phi_2}^{[\sigma]} & \boldsymbol{0} & \varrho_{\phi_{2,1}}^{[t]}\boldsymbol{Q}_{\phi_2}^{[\sigma]} & \boldsymbol{0} & \varrho_{\phi_{2,2}}^{[t]}\boldsymbol{Q}_{\phi_2}^{[\sigma]} \\
& & \boldsymbol{Q}_{\phi_1}^{[\sigma]} & \boldsymbol{0} & \varrho_{\phi_{1,1}}^{[t]}\boldsymbol{Q}_{\phi_1}^{[\sigma]} & \boldsymbol{0} \\
& & & \boldsymbol{Q}_{\phi_2}^{[\sigma]} & \boldsymbol{0} & \varrho_{\phi_{2,1}}^{[t]}\boldsymbol{Q}_{\phi_2}^{[\sigma]} \\
& & & & \boldsymbol{Q}_{\phi_1}^{[\sigma]} & \boldsymbol{0} \\
\text{symmetry} & & & & & \boldsymbol{Q}_{\phi_2}^{[\sigma]}
\end{bmatrix}
$$

It can be further simplified, in case of single-frequency, i.e., $f = 1$, as

$$
\begin{bmatrix}
1 & \varrho_{\phi_{2,1}}^{[t]} & \varrho_{\phi_{1,2}}^{[t]} \\
\varrho_{\phi_{1,1}}^{[t]} & 1 & \varrho_{\phi_{1,1}}^{[t]} \\
\varrho_{\phi_{1,2}}^{[t]} & \varrho_{\phi_{1,1}}^{[t]} & 1
\end{bmatrix} \otimes \boldsymbol{Q}_{\phi_1}^{[\sigma]}
$$

In general, we have $2f(K-1)$ time correlation coefficients with the factor of 2 accounting for both phase and code. Such a large number of parameters precludes efficient numerical computations.

Third, the estimation of the parameters of the stochastic model is studied. In this part of the LS variance component estimation. As aforementioned, there are many VCE methods, for instance, MINQUE [2], BIQUE [5], RMLE [7, 8, 10] and LS-VCE [12, 13]. These methods differ in the estimation principles used, as well as in the distributional assumptions that they make. They might be equivalent under certain circumstances [14]. In this study, the LS-VCE is applied owing to its superiorities identified by [21]. To describe the LS-VCE method, we write the observation model Eq. (6.33) together with its stochastic model Eq. (6.37) as a general, linear GM model:

$$
\mathrm{E}(\ell) = \boldsymbol{A}\boldsymbol{\beta}, \quad \mathrm{D}(\ell) = \boldsymbol{Q}_\ell = \boldsymbol{Q}_0 + \sum_{i=1}^{p} \sigma_i \boldsymbol{U}_i \tag{6.38}
$$

where $\ell$ is an $n$-column observation vector ($n = 2fsK$), $\boldsymbol{A} = \boldsymbol{I}_K \otimes \boldsymbol{B}$ is the $n \times t$ design matrix for the $t = (2f - 1 + s)K$ parameter vector $\boldsymbol{\beta}$. The variance matrix $\boldsymbol{Q}_\ell$ consists of a known part $\boldsymbol{Q}_0$ and unknown part specified by $p$ unknown (co)variance components $\sigma_i$ and their associated known cofactor matrices $\boldsymbol{U}_i$ ($i = 1, \ldots, p$).

The normal equations of LS-VCE are [14]

$$
\boldsymbol{N}\hat{\boldsymbol{\sigma}} = \boldsymbol{\omega} \tag{6.39}
$$

where $\hat{\boldsymbol{\sigma}} = \left[\hat{\sigma}_1, \ldots, \hat{\sigma}_p\right]^{\mathrm{T}}$. The entries of normal matrix $\boldsymbol{N}$ and vector $\boldsymbol{\omega}$ are

$$n_{kl} = \mathrm{tr}\left(\boldsymbol{U}_k\boldsymbol{Q}_\ell^{-1}\boldsymbol{P}_A^\perp\boldsymbol{U}_l\boldsymbol{Q}_\ell^{-1}\boldsymbol{P}_A^\perp\right) \tag{6.40}$$

$$\omega_k = \hat{\boldsymbol{\varepsilon}}^{\mathrm{T}}\boldsymbol{Q}_\ell^{-1}\boldsymbol{U}_k\boldsymbol{Q}_\ell^{-1}\hat{\boldsymbol{\varepsilon}} - \mathrm{tr}\left(\boldsymbol{U}_k\boldsymbol{Q}_\ell^{-1}\boldsymbol{P}_A^\perp\boldsymbol{Q}_0\boldsymbol{Q}_\ell^{-1}\boldsymbol{P}_A^\perp\right) \tag{6.41}$$

where $\boldsymbol{P}_A^\perp = \boldsymbol{I} - \boldsymbol{A}(\boldsymbol{A}^{\mathrm{T}}\boldsymbol{Q}_\ell^{-1}\boldsymbol{A})^{-1}\boldsymbol{A}^{\mathrm{T}}\boldsymbol{Q}_\ell^{-1}$ is the project matrix orthogonal to $\boldsymbol{A}$, and $\hat{\boldsymbol{\varepsilon}} = \boldsymbol{P}_A^\perp\ell$ the LS residual vector. The LS-VCE needs to be solved iteratively since the unknown covariance matrix $\boldsymbol{Q}_\ell$ is contained in the normal equations. Given the initial values for (co)variance unknowns, denoted by $\sigma_i^0 (i = 1, \cdots, p)$, the iteration continues until the difference in the computed unknowns between two consecutive iterations is sufficiently small.

In this part of estimation procedure, many unknown (co)variance components are involved in the covariance matrix described above. For number of epochs $K$, number of frequencies $f$, and number of satellites $s$, there are a total of $2fs + f(f1) + 2f(K1)$ unknowns. Here, $2fs$ is for observation variances, $f(f1)$ is for cross correlations between any two frequencies, and $2f(K1)$ is for time correlations over $K$ epochs. For instance, when $K = 50$, $f = 3$ and $s = 6$, the number of unknown (co)variance components is 396. Simultaneous estimation of such a huge number of unknowns will often lead to unreliable estimates, even producing negative variances and nonsense correlation coefficients. The reason is twofold. First, the normal matrix $N$ in Eq. (6.39) associated with the geometry-free model for estimating observation variances and cross correlations is rank deficient. It means that one cannot estimate the observation variances and cross correlations simultaneously. Secondly, the fact that the precision of phase observations are much higher than that of code will cause ill-posedness, which leads to unreliable estimates. For further discussion, see [11, 25], which concludes a theoretical proof and numerical analysis. In summary, it is not feasible to simultaneously estimate all variance and covariance components. Hence, the following three-step estimation procedure is proposed: Estimate the phase and code variances per satellite of each frequency, which will be used to analyze the elevation-dependence of individual satellite precisions; Estimate the cross correlation coefficients for phase and code observations, respectively; Estimate the time correlation coefficients for phase and code of each frequency as a function of time lags.

In steps 2 and 3 of this procedure, the estimates from the previous step are held fixed. Details of each step are discussed in the following.

Step 1: The cross and time correlations are disregarded and the unknown covariance matrix is reduced to

$$\boldsymbol{Q}_\ell = \boldsymbol{I}_K \otimes \boldsymbol{Q}_E^{[\sigma]} = \boldsymbol{I}_K \otimes \sum_{i=1}^{2fs} \sigma_i\boldsymbol{U}_i \tag{6.42}$$

where $U_i$ is the zero matrix except for its $i$th diagonal entry of 1, and $\sigma_i$ is its corresponding variance component. Inserting Eq. (6.42) into Eqs. (6.40) and (6.41), it follows, from multivariate adjustment principles, that the variance estimates of $K$ epochs are equal to the averaged single-epoch estimates over $K$ epochs [11],

$$\hat{\sigma} = \sum_{k}^{K} \hat{\sigma}_{(k)}/K \qquad (6.43)$$

where $\hat{\sigma}_{(k)} = \left[\hat{\sigma}_{1,(k)}, \ldots, \hat{\sigma}_{2fs,(k)}\right]^{\mathrm{T}}$ is the vector of variance component estimates from the $k$th data epoch only. To improve the computation efficiency, one may use the following simplified formula, resulting in an almost unbiased estimation [1].

$$\hat{\sigma}_i = \frac{\sum_{k=1}^{K} \hat{\varepsilon}_{i,(k)}^{\mathrm{T}} P_i \hat{\varepsilon}_{i,(k)}}{K r_i}, \quad i = 1, \ldots, 2fs \qquad (6.44)$$

where $\hat{\varepsilon}_{i,(k)}$ is the residual of the $i$th observation at the $k$th epoch and $r_i = 1 - \mathrm{tr}\left((\boldsymbol{B}^{\mathrm{T}}\boldsymbol{PB})^{-1}\boldsymbol{B}_i^{\mathrm{T}}P_i\boldsymbol{B}_i\right)$ is its redundancy number; $\boldsymbol{B}_i$ is the $i$th row vector of matrix $\boldsymbol{B}$ in (6.38); $P_i$ is the $i$th diagonal element of $\boldsymbol{P} = (\boldsymbol{Q}^{[\sigma]})^{-1}$.

Step 2: With variance components estimated in step 1 held fixed, we estimate the cross correlations in step 2. The unknown covariance matrix is now structured as

$$\boldsymbol{Q}_t = \boldsymbol{I}_K \otimes \boldsymbol{Q}_E^{[\hat{\sigma}]} + \boldsymbol{I}_K \otimes \boldsymbol{Q}_E^{[c]} = \boldsymbol{Q}_0 + \sum_{i=1}^{f(f-1)} \sigma_i(\boldsymbol{I}_K \otimes \boldsymbol{U}_i) \qquad (6.45)$$

where $\boldsymbol{Q}_0 = \boldsymbol{I}_K \otimes \boldsymbol{Q}_E^{[\hat{\sigma}]}$ was estimated in step 1. The scalar $\sigma_i$ is the $i$th unknown cross correlation coefficient that can be collected in the vector

$$\sigma = \left[\varrho_{\phi_1\phi_2}^{[c]}, \varrho_{\phi_1\phi_3}^{[c]}, \varrho_{\phi_2\phi_3}^{[c]}, \varrho_{P_1P_2}^{[c]}, \varrho_{P_1P_3}^{[c]}, \varrho_{P_2P_3}^{[c]}\right]^{\mathrm{T}}$$

for $f = 3$. $\boldsymbol{U}_i$ is the $(2fs \times 2fs)$ cofactor matrix associated with $\sigma_i$. For example, for $\sigma_1 = \varrho_{\phi_1\phi_2}^{[c]}$ and $\sigma_4 = \varrho_{P_1P_2}^{[c]}$, the cofactor matrices are

$$\boldsymbol{U}_1 = \begin{bmatrix} 0 & \boldsymbol{Q}_{\phi_1\phi_2}^{[\hat{\sigma}]} & 0 & 0 & 0 & 0 \\ \boldsymbol{Q}_{\phi_2\phi_1}^{[\hat{\sigma}]} & 0 & 0 & 0 & 0 & 0 \\ 0 & 0 & 0 & 0 & 0 & 0 \\ 0 & 0 & 0 & 0 & 0 & 0 \\ 0 & 0 & 0 & 0 & 0 & 0 \\ 0 & 0 & 0 & 0 & 0 & 0 \end{bmatrix}, \quad \boldsymbol{U}_4 = \begin{bmatrix} 0 & 0 & 0 & 0 & 0 & 0 \\ 0 & 0 & 0 & 0 & 0 & 0 \\ 0 & 0 & 0 & 0 & 0 & 0 \\ 0 & 0 & 0 & 0 & \boldsymbol{Q}_{P_1P_2}^{[\hat{\sigma}]} & 0 \\ 0 & 0 & 0 & \boldsymbol{Q}_{P_2P_1}^{[\hat{\sigma}]} & 0 & 0 \\ 0 & 0 & 0 & 0 & 0 & 0 \end{bmatrix} \qquad (6.46)$$

Inserting Eq. (6.45) into Eqs. (6.40) and (6.41) and performing some algebraic manipulation yields

$$n_{kl} = K\mathrm{tr}\big(U_k Q_E^{-1} P_B^{\perp} U_l Q_E^{-1} P_B^{\perp}\big) \tag{6.47}$$

and

$$\omega_k = \mathrm{tr}\Big(\hat{E}^{\mathrm{T}} Q_E^{-1} U_k Q_E^{-1} \hat{E}\Big) - K\mathrm{tr}\Big(U_k Q_E^{-1} P_B^{\perp} Q_E^{[\hat{\sigma}]} Q_E^{-1} P_B^{\perp}\Big) \tag{6.48}$$

where $P_B^{\perp} = I - B(B^{\mathrm{T}} Q_E^{-1} B)^{-1} B^{\mathrm{T}} Q_E^{-1}$, $\hat{E} = P_B^{\perp} L$ with observation matrix $L = [y_1, \ldots, y_K]$. In computations, if the initial values of unknown components are taken as $\sigma = 0$, then $Q_E = Q_E^{[\hat{\sigma}]}$ and Eqs. (6.47) and (6.48) are simplified to

$$n_{kl} = K\mathrm{tr}\big(U_k P_E^{\hat{\sigma}} P_B^{\perp} U_l P_E^{\hat{\sigma}} P_B^{\perp}\big) \tag{6.49}$$

$$\omega_k = \mathrm{tr}\Big(\hat{E}^T P_E^{\hat{\sigma}} U_k P_E^{\hat{\sigma}} \hat{E}\Big) - K\mathrm{tr}\big(U_k P_E^{\hat{\sigma}} P_B^{\perp}\big) \tag{6.50}$$

where $P_E^{\hat{\sigma}} = (Q_E^{[\hat{\sigma}]})^{-1}$. Our experience is that after the first computation of (6.49) and Eq. (6.50), the estimate update from the iteration is so marginal that the iteration is practically not necessary.

Step 3: We finally estimate the time correlations for phase and code observations of each frequency. Since the products of between-frequency residuals are not used, the cross correlations have very minor effects on estimates of time correlations. Hence, in order to reduce the computation complexity, the cross correlations are disregarded and only the observation variances estimated from the step 1 are held fixed, i.e., $Q_E = Q_E^{[\sigma]}$ and $Q_E^{[c]} = 0$. The unknown stochastic model in Eq. (6.45) reads

$$Q_\ell = I_K \otimes Q_E^{[\hat{\sigma}]} + Q^{[t]} \circ \Big(e_K e_K^{\mathrm{T}} \otimes Q_E^{[\sigma]}\Big) = Q_0 + \sum_{i=1}^{2f(K-1)} \sigma_i U_i \tag{6.51}$$

where $Q_0 = I_K \otimes Q_E^{[\hat{\sigma}]}$. The scalar $\sigma_i$ is the $i$th unknown time correlation coefficient in the vector

$$\sigma = \left[\left(\varrho_{,1}^{[t]}\right)^{\mathrm{T}}, L, \left(\varrho_{,\tau}^{[t]}\right)^{\mathrm{T}}, L, \left(\varrho_{,K-1}^{[t]}\right)^{1}\right]^{\mathrm{T}}$$

with $\varrho_{,\tau}^{[t]} = [\varrho_{\phi_1,\tau}^{[t]}, \ldots, \varrho_{\phi_f,\tau}^{[t]}, \varrho_{P_1,\tau}^{[t]}, \ldots, \varrho_{P_f,\tau}^{[t]}]^{\mathrm{T}}$. The cofactor matrix is structured as $U_i = U_{\tau_i} \otimes \big(U_{\alpha_i} \otimes O_{\alpha_i}\big)$ with the subscript $i = 2f(\tau_i - 1) + \alpha_i$ and $\alpha_i = 1, \ldots, 2f$; $\tau_i = 1, \ldots, K - 1$.

Given $i$, one can uniquely determine $\alpha_i$ and $\tau_i$. As an example, when $i = 1$, it follows that $\alpha_i = 1$ and $\tau_i = 1$; while when $i = 20$, $\alpha_i = 2$ and $\tau_i = 4$ in case of $f = 3$. $Q_{\alpha_i}$ is the $\alpha_i$-th block-diagonal matrix of $Q_E^{[\hat{\sigma}]}$. For instance, $Q_{\alpha_i} = Q_{\phi_1}^{[\hat{\sigma}]}$ for $\alpha_i = 1$ and $Q_{\alpha_i} = Q_{P_1}^{[\hat{\sigma}]}$ for $\alpha_i = f + 1$. $U_{\alpha_i}$ is a $(2f \times 2f)$ zero matrix with its $\alpha_i$-th diagonal element of 1. $U_{\tau_i}$ is a $(K \times K)$ zero matrix with the elements at its $\tau_i$-th diagonal above and below main diagonal all being 1.

Substituting Eq. (6.31) into Eqs. (6.40) and (6.41), one can iteratively compute the time correlation coefficients. Note that the iteration process may be time consuming because of so many unknowns, namely $2f(K1)$. Again, our experience indicates that iteration can hardly improve the estimates from the first computation with initial values of all time correlation coefficients being 0. Therefore, one may directly obtain the estimates without iteration. In this case, $Q_\ell = Q_0 = I_K \otimes Q_E^{[\hat{\sigma}]}$ and $P_A^\perp \equiv I_K \otimes P_B^\perp$, and the normal equations of LS-VCE become

$$n_{kl} = \mathrm{tr}\left(U_{\tau_k} U_{\tau_l}\right) \mathrm{tr}\left(\overline{U}_{\alpha_k} P_B^\perp \overline{U}_{\alpha_l} P_B^\perp\right) \tag{6.52}$$

$$\omega_k = \mathrm{tr}\left(\hat{E}^{\mathrm{T}} P_E^{\hat{\sigma}} \overline{U}_{\alpha_k} \hat{E} U_{\tau_k}\right) - \mathrm{tr}\left(U_{\tau_k}\right) \mathrm{tr}\left(\overline{U}_{\alpha_k} P_B^\perp\right) \tag{6.53}$$

where $\overline{U}_{\alpha_k} = U_{\alpha_k} \otimes I_s$. Since $\mathrm{tr}\left(U_{\tau_k}\right) = 0$, $\omega_k$ is further reduced to

$$\omega_k = \mathrm{tr}\left(\hat{E}^{\mathrm{T}} P_E^{\hat{\sigma}} \left(U_{\alpha_k} \otimes I_s\right) \hat{E} U_{\tau_k}\right) \tag{6.54}$$

Some comments are given on the stepwise estimation procedure. In step 1 of estimating observation variances, both the cross and time correlations are disregarded. In such case, the variance estimates are unbiased only if the cross and time correlations are all indeed zeros; otherwise, they are biased and the biases depend on the magnitude of correlations. To understand this point, see the example with explanations given in [11]. Analogously, if the variance estimates obtained in step 1 are biased and fixed in the next steps, the estimates of cross and time correlation coefficients will be biased as well. Unfortunately, it is often the case in actual data analysis. Here, a between-step iteration strategy is advised to possibly reduce such biases of the estimates. If the estimates of time and cross correlation coefficients are not sufficiently close to 0, the iteration starts with the step 1 by fixing the cross and time correlation coefficients to their previous estimates instead of zeros. Repeat the iteration with updated cross and time correlation coefficients until their change is sufficiently small.

## 6.4  Results and Discussion

In order to have a better understanding of GNSS stochastic modeling. Firstly, the experiment setup is described. Secondly, the spatial, cross and temporal correlations are discussed, respectively.

**Table 6.1** Data description of the ten data sets

| Data set | Receiver brand | Baseline length |
| --- | --- | --- |
| 1 | Trimble NetR9 | 0.00 m |
| 2 | Trimble NetR9 | 12.49 km |
| 3 | Trimble NetR9 | 23.48 km |
| 4 | Trimble NetR9 | 31.49 km |
| 5 | Trimble NetR9 | 42.74 km |
| 6 | Leica GR25 | 4.99 m |
| 7 | Leica GR25 | 13.31 km |
| 8 | Leica GR25 | 20.93 km |
| 9 | Leica GR25 | 34.50 km |
| 10 | Leica GR25 | 49.89 km |

## *6.4.1  Experiment Description*

Ten data sets of dual-frequency BDS code and phase observations with a sampling interval of 1 s were collected using two types of receivers. The observation session was 1 h at the same time, and the baseline lengths varied from 0 to 50 km in the same area. Table 6.1 lists the receivers' specifications and the baseline lengths. It can be seen that the attributions of receiver, multipath, and atmosphere could be taken into account because of their varying baseline lengths.

All the data sets were processed with the DD version of our own research type RTK software using two types of functional models: (A) the ionosphere-fixed model; (B) the ionosphere-free (IF) model. The Hopfield model was used to correct the tropospheric effects. Since the precise coordinates of all stations are solved by using daily data with our RTK software, the DD integer ambiguities are reliably fixed in advance. Therefore, only three unknown coordinates exist in the functional models.

For the stochastic modeling, the variance elements are estimated by the elevation-dependent model. To make sure that the physical correlations to be analyzed are correct, one must obtain the reliable DD residuals. Therefore, after estimating a priori fully populated VCM, the MINQUE method is applied, as aforementioned.

## *6.4.2  Spatial Correlation*

Based on the structure of a fully populated stochastic model as aforementioned, the spatial correlations are firstly analyzed. Since a DD observation formed by four undifferenced observations, the mathematical correlation is present in the stochastic model. The mean spatial correlations of all these six signal types are shown in Fig. 6.1. It can be seen that all the mean spatial correlation coefficients are close to + 0.5, although there exist biases especially the phase observations, where the mean correlation coefficients of B1, B2 and IF phase signals are 0.391, 0.460 and 0.389 respectively. The reason is that the variance of reference satellite is usually smaller than the one of any other common satellite, hence the mean spatial correlations are usually smaller than $+ 0.5$.

In order to accurately determine the spatial correlations, the single between-receiver differenced residuals are applied since the single-differences are mathematically uncorrelated. Based on the estimated SD residuals calculated, the mean spatial correlations from the ten data sets are listed in Table 6.2. It can be found that all the absolute correlation coefficients are lower than 0.196, thus indicating that the spatial correlations in RTK are not significant. Actually, the reason is that the DD solution can eliminate the most of spatial correlated errors to a great extent, especially when the baseline length is not very long (less than 50 km as usual).

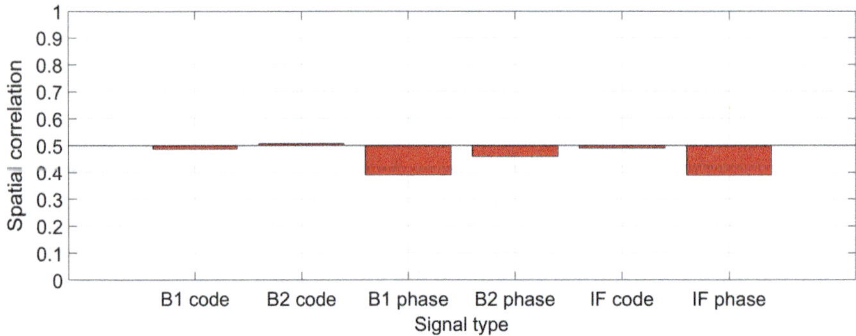

**Fig. 6.1**  Mean spatial correlations of six signal types from the ten data sets

**Table 6.2**  Mean spatial correlations of the ten data sets

| Number | Code | | | Phase | | |
|---|---|---|---|---|---|---|
| | B1 | B2 | IF | B1 | B2 | IF |
| 1 | − 0.057 | − 0.081 | − 0.065 | − 0.067 | − 0.067 | − 0.068 |
| 2 | − 0.068 | − 0.054 | − 0.068 | − 0.028 | − 0.009 | − 0.044 |
| 3 | − 0.046 | − 0.039 | − 0.049 | 0.009 | − 0.069 | − 0.036 |
| 4 | − 0.040 | − 0.047 | − 0.048 | − 0.024 | − 0.079 | − 0.017 |
| 5 | − 0.037 | − 0.037 | − 0.044 | 0.021 | 0.001 | 0.000 |
| 6 | − 0.076 | − 0.064 | − 0.075 | − 0.050 | − 0.041 | − 0.038 |
| 7 | 0.079 | 0.070 | 0.071 | − 0.079 | − 0.150 | − 0.113 |
| 8 | 0.045 | − 0.107 | 0.037 | 0.002 | − 0.112 | − 0.143 |
| 9 | 0.121 | 0.180 | 0.160 | 0.032 | − 0.028 | − 0.094 |
| 10 | − 0.068 | 0.067 | − 0.030 | 0.144 | 0.059 | − 0.062 |

### *6.4.3  Cross Correlation*

The mean cross correlations from Trimble and Leica of each data set are presented in Tables 6.3 and 6.4, respectively. It can be clearly seen that the cross correlation coefficients between B1 phase and B2 phase are significantly higher than the other types. It can be seen that these two observation types, i.e., phase and code, are generally not correlated with each other regardless of the receiver internal and external environments of receiver. However, the phase-to-phase correlations are generally correlated but not dependent on the multipath and atmospheric effects. Specifically, for the Trimble receiver, B1 phase is strongly correlated with B2 phase with the mean value of 0.859, but the correlation for the Leica receiver is weaker only with an average of 0.358. It indicates that the cross correlations can be introduced by decoding techniques employed by different receiver types. For instance, when the decoding technique is applied to obtain B2 signal that is encrypted under Anti-Spoofing, the B1 and $\Delta$B are directly obtained and $B2 = B1 + \Delta B$. As a result, it leads to a strong cross correlation between B1 and B2, like the Trimble receiver in this study.

In conclusion, this type of correlation is mainly receiver-specific and independent with the baseline length. It is worth noting that though the code-to-code cross correlations in this study are insignificant, they may be significant in some other types of receivers. In addition, the mean values can be used to substitute the specific ones since they are almost invariant.

**Table 6.3** Mean cross correlations of the five data sets from Trimble

| Number | B1 code to B2 code | B1 phase to B2 phase | B1 code to B1 phase | B1 code to B2 phase | B2 code to B1 phase | B2 code to B2 phase | IF code to IF phase |
|---|---|---|---|---|---|---|---|
| 1 | 0.129 | 0.870 | − 0.012 | − 0.018 | 0.003 | 0.008 | − 0.022 |
| 2 | 0.119 | 0.829 | − 0.001 | − 0.021 | − 0.012 | 0.022 | 0.031 |
| 3 | 0.099 | 0.823 | 0.021 | 0.009 | 0.039 | 0.025 | − 0.022 |
| 4 | 0.079 | 0.874 | − 0.009 | − 0.005 | − 0.024 | − 0.029 | − 0.001 |
| 5 | 0.069 | 0.901 | 0.000 | − 0.003 | − 0.026 | − 0.024 | − 0.010 |

**Table 6.4** Mean cross correlations of the five data sets from Leica

| Number | B1 code to B2 code | B1 phase to B2 phase | B1 code to B1 phase | B1 code to B2 phase | B2 code to B1 phase | B2 code to B2 phase | IF code to IF phase |
|---|---|---|---|---|---|---|---|
| 6 | − 0.147 | 0.291 | 0.021 | 0.014 | − 0.020 | − 0.019 | − 0.075 |
| 7 | − 0.010 | 0.262 | 0.161 | 0.040 | − 0.215 | 0.137 | 0.089 |
| 8 | 0.118 | 0.350 | 0.006 | − 0.156 | − 0.150 | 0.150 | − 0.075 |
| 9 | − 0.065 | 0.489 | − 0.030 | 0.037 | − 0.200 | − 0.135 | − 0.031 |
| 10 | − 0.130 | 0.398 | − 0.176 | 0.122 | − 0.225 | − 0.363 | − 0.080 |

### 6.4.4  *Temporal Correlation*

In this section, the temporal correlations are analyzed. Since the temporal correlation coefficients of the sampling interval will be applied into a fully populated stochastic model, the mean temporal correlations for all the observations at a lag of sampling interval (1 s) are emphasized, as shown in Tables 6.5 and 6.6. It can be seen that the correlation coefficients of code observations from Leica are larger than those from Trimble. The reason is that the preprocessed techniques (e.g., filtering and smoothing) are different for different receiver types, consequently the temporal correlations share different patterns. It proves indirectly that the temporal correlations can be influenced by the receiver. For the phase observations, the correlation coefficients become larger when the baseline length is increased. It is seen that the result of zero (No. 1) and ultra-short (No. 6) baselines seem to have little temporal correlations, particular for the zero baseline. However, the temporal correlations become more serious and can be treated significant (larger than 0.196) for longer baselines (No. 2-5 and No. 7-10). As a matter of fact, the multipath effects will exhibit in the ultra-short baseline and the atmospheric delays (including the troposphere and ionosphere) are the dominant errors in longer baselines. Therefore, the temporal correlation coefficients are both determined by the multipath and atmosphere. Moreover, the correlation coefficients of IF phase data are significantly lower than those of B1 and B2 phase data. It is reasonable that the ionospheric delays have been reduced by the IF model. Therefore, the temporal correlations are sensitive to the distance-dependent atmospheric delays.

**Table 6.5**  Mean temporal correlations with a 1-s lag of the five data sets from Trimble

| Number | Code | | | Phase | | |
|---|---|---|---|---|---|---|
| | B1 | B2 | IF | B1 | B2 | IF |
| 1 | 0.037 | 0.066 | 0.046 | 0.001 | 0.000 | 0.009 |
| 2 | 0.149 | 0.174 | 0.172 | 0.581 | 0.758 | 0.498 |
| 3 | 0.207 | 0.206 | 0.231 | 0.815 | 0.905 | 0.497 |
| 4 | 0.173 | 0.212 | 0.201 | 0.841 | 0.920 | 0.448 |
| 5 | 0.169 | 0.201 | 0.201 | 0.881 | 0.947 | 0.375 |

**Table 6.6**  Mean temporal correlations with a 1-s lag of the five data sets from Leica

| Number | Code | | | Phase | | |
|---|---|---|---|---|---|---|
| | B1 | B2 | IF | B1 | B2 | IF |
| 6 | 0.986 | 0.988 | 0.988 | 0.143 | 0.225 | 0.160 |
| 7 | 0.997 | 0.996 | 0.997 | 0.780 | 0.879 | 0.355 |
| 8 | 0.997 | 0.996 | 0.997 | 0.875 | 0.927 | 0.402 |
| 9 | 0.999 | 0.996 | 0.999 | 0.940 | 0.963 | 0.403 |
| 10 | 0.999 | 0.996 | 0.999 | 0.966 | 0.980 | 0.471 |

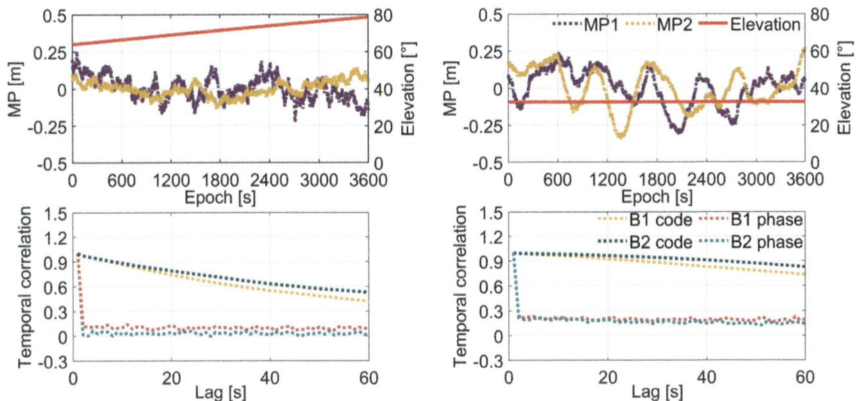


**Fig. 6.2** Code multipath effects, elevations and temporal correlations of **a** PRN 07 and **b** PRN 04 on a 5-m baseline

Another impact factor of temporal correlations, i.e., the site-specific multipath, is investigated. The data set No. 6 was applied since the multipath is the dominant error source in a 5-m baseline. To assess the strength of code multipath, the multipath combination function is applied, where the code multipath can be estimated by the peak-to-peak behaviors. In addition, the elevation-dependent code hardware variations are insignificant mainly due to the small elevation variation. Figure 6.2 illustrates the dual-frequency multipath effects, elevations, and corresponding temporal correlations of satellites PRN 07 and PRN 04, respectively. Based on the top panels, it is clear that the elevations of PRN 07 are significantly higher than PRN 04. Accordingly, the observations of PRN 04 are contaminated by multipath more seriously. Therefore, the temporal correlations of code and phase observations are higher than 0.7 and 0.2 for PRN 07, whereas for PRN 04, the corresponding ones are 0.4 and 0, respectively. It indicates that when the elevations are lower, the multipath will be contaminated more seriously and the temporal correlations will be higher.

In conclusion, the temporal correlation is an important property caused by the unmodeled errors, though it could also be influenced by the receiver. Specifically, when the multipath is the dominant error source, the temporal correlations will be positively influenced, where the elevations and satellite types are both the potential impact factors. It is noteworthy that the discrepancy of mean temporal correlation coefficients between different frequencies for the same data set is insignificant, therefore they can be treated equal.

## 6.5   Conclusion

We firstly established the fundamental matrix equations for VCE, from which the traditional VCE formula is derived. It has been theoretically proven that the essential inputs for VCE are not all correlated LS residuals but the independent discrepancies of measurements and any orthogonal complement matrix of the coefficient matrix in the GM model can be used to compute a set of independent discrepancies. We proposed to construct an orthogonal complement matrix only by using the coefficient matrix itself, such that both the constructed orthogonal complement matrix and its derived discrepancies are invariant. Consequently, they do never need to be updated in the iterative VCE procedure and the computation efficiency is significantly improved. Nowadays, many sensors are freely accessible to acquire measurements for integrated utilization. With the development of spatial geodetic technology, more and more sensors will be launched and much more plentiful measurements will be available in future. We must employ efficient estimation of variances and covariances to determine the reasonable stochastic model of the measurements from the different sensors for balancing their contributions to the fused solution, especially in the (near) real-time applications. Therefore, the new method will undoubtedly be beneficial to data fusion from multi-sensors.

In addition, we systematically studied the stochastic modeling of triple-frequency BDS observations. A very sophisticated structure of covariance matrix was designed to allow estimation of satellite-specific variances, cross correlations between two arbitrary frequencies as well as time correlations for phase and code observations per frequency. A three-step VCE procedure was presented for efficiently and stably estimating many (co)variance components. Six data sets with four brands of BDS receivers on short and zero-length baselines were used to analyze the stochastic models. Finally, with the observation variances, the cross and time-correlation coefficients estimated from real data, one can construct a stochastic model though it is still numerically challenging. The established stochastic model can largely reflect the actual observation random errors of observations. Even if it mis-specifies the stochastic model of long baselines, one can efficiently compensate for that by estimating fewer unknowns in a new unknown covariance matrix with the established stochastic model as known part, allowing for use with (near) real time applications.

## References


1. Förstner W (1979) Ein verfahren zur schätzung von varianz-und kovarianzkomponenten. Allg Vermess Nachr 86:446–453
2. Rao CR (1971) Estimation of variance and covariance components-MINQUE theory. J Multivar Anal 1:257–275
3. Crocetto N, Gatti M, Russo P (2000) Simplified formulae for the BIQUE estimation of variance components in disjunctive observation groups. J Geod 74:447–457
4. Sjoberg LE (1983) Unbiased estimation of variance-components in condition adjustment with unknowns-a MINQUE approach. ZfV 108:9

5. Koch K-R (1978) Schätzung von varianzkomponenten. Allg Vermess Nachr 85:264–269
6. Zongchou Y (1992) A generalization theory of estimation of variance-covariance components. Manuscr Geod 17:295–301
7. Kubik K (1970) The estimation of the weights of measured quantities within the method of least squares. Bull Geod 95:21–40
8. Koch KR (1986) Maximum likelihood estimate of variance components: Ideas by A.J. Pope (in memory of Allen J. Pope, 11.10.1939–29.08.1985). Bull Geod 60:329–338
9. Ou Z (1989) Estimation of variance and covariance components. Bull Geod 63:139–148
10. Yu Z (1996) A universal formula of maximum likelihood estimation of variance-covariance components. J Geod 70:233–240
11. Amiri-Simkooei AR, Teunissen PJG, Tiberius CCJM (2009) Application of least-squares variance component estimation to GPS observables. J Surv Eng 135:149–160
12. Pukelsheim F (1976) Estimating variance components in linear models. J Multivar Anal 6:626–629
13. Teunissen PJG (1988) Towards a least-squares framework for adjusting and testing of both functional and stochastic models, Geodetic Computing Centre
14. Amiri-Simkooei A (2007) Least-squares variance component estimation: theory and GPS applications. Delft University of Technology, Delft
15. Yang Y, Xu T, Song L (2005) Robust estimation of variance components with application in global positioning system network adjustment. J Surv Eng 131:107–112
16. Yang Y, Zeng A, Zhang J (2009) Adaptive collocation with application in height system transformation. J Geod 83:403–410
17. Xu P (2009) Iterative generalized cross-validation for fusing heteroscedastic data of inverse ill-posed problems. Geophys J Int 179:182–200
18. Tiberius C, Kenselaar F (2003) Variance component estimation and precise GPS positioning: case study. J Surv Eng 129:11–18
19. Li B, Shen Y, Lou L (2011) Efficient estimation of variance and covariance components: a case study for GPS stochastic model evaluation. IEEE Trans Geosci Remote Sensing 49:203–210
20. Xu P, Liu Y, Shen Y, Fukuda Y (2007) Estimability analysis of variance and covariance components. J Geod 81:593–602
21. Teunissen PJG, Amiri-Simkooei AR (2008) Least-squares variance component estimation. J Geod 82:65–82
22. Xiang Jin X, De Jong CD (1996) Relationship between satellite elevation and precision of GPS code observations. J Navig 49:253–265
23. Wang J, Satirapod C, Rizos C (2002) Stochastic assessment of GPS carrier phase measurements for precise static relative positioning. J Geod 76:95–104
24. Li B, Shen Y, Xu P (2008) Assessment of stochastic models for GPS measurements with different types of receivers. Sci Bull 53:3219–3225
25. Amiri-Simkooei AR, Tiberius CCJM (2006) Assessing receiver noise using GPS short baseline time series. GPS Solut 11:21–35
26. Schön S, Brunner FK (2008) A proposal for modelling physical correlations of GPS phase observations. J Geod 82:601–612
27. Kermarrec G, Schön S (2014) On the Mátern covariance family: a proposal for modeling temporal correlations based on turbulence theory. J Geod 88:1061–1079
28. Schön S, Brunner FK (2008) Atmospheric turbulence theory applied to GPS carrier-phase data. J Geod 82:47–57
29. Beutler G, Bock H, Dach R, Fridez P, Gäde A, Hugentobler U, Jäggi A, Meindl M, Mervart L, Prange L et al (2007) Bernese GPS software, version 5.0
30. Wang K, Rothacher M (2013) Ambiguity resolution for triple-frequency geometry-free and ionosphere-free combination tested with real data. J Geod 87:539–553
31. Odolinski R, Teunissen PJG, Odijk D (2015) Combined BDS, Galileo, QZSS and GPS single-frequency RTK. GPS Solut 19:151–163
32. Li B, Feng Y, Gao W, Li Z (2015) Real-time kinematic positioning over long baselines using triple-frequency BeiDou signals. IEEE Trans Aerosp Electron Syst 51:3254–3269

33. Mikhail E, Ackermann F (1976) Observations and least squares. Dun-Donnelley, New York
34. Koch K (1999) Parameter estimation and hypothesis testing in linear models. Springer, Berlin
35. Odijk D (2008) What does "geometry-based" and "geometry-free" mean in the context of GNSS. Inside GNSS—GNSS Solut Column 3:22–24

# Chapter 7
# Unmodeled Error Processing

## 7.1 Introduction

Global Navigation Satellite System (GNSS) has become indispensable for precise positioning across various fields such as geodesy, engineering, and artificial intelligence, with real-time kinematic positioning (RTK) playing a pivotal role. Although differencing techniques can effectively mitigate many systematic errors, such as satellite and receiver clock errors and orbital inaccuracies, residual errors inevitably persist due to the complex spatiotemporal variability of the ionosphere, troposphere, and multipath effects, as well as our limited understanding of their underlying physical mechanisms. These residual systematic errors, often termed unmodeled errors [1], cannot be entirely captured by operators of differencing and linear combination, conventional empirical models or parameterizations, thereby constraining the further improvement of positioning accuracy in real applications [2, 3]. Moreover, as the advent of multi-constellation GNSS (including Global Positioning System (GPS), BeiDou Navigation Satellite System (BDS), GLONASS, and Galileo) has enhanced signal visibility and integrity, it becomes even more critical to accurately process these errors to ensure trustworthy positioning results. Understanding the properties of unmodeled errors is therefore a prerequisite for developing effective compensation methods to enhance ambiguity resolution, data quality control, and overall positioning accuracy [4], making it essential to establish efficient procedures to test their significance and identify their components.

Since the residual systematic errors hinder the high-precision and high-reliability of GNSS positioning, many works have been carried out on how to model/reduce the systematic errors as much as possible. In recent decades, the specific residual systematic errors are focused, such as multipath, tropospheric and ionospheric effects. Firstly, for the multipath, except for choosing the favorable environment and advanced hardware, the additional data processing strategies are further applied. The most widely used approach is the sidereal filtering based on the coordinate or observation domain [5–7]. Besides, the multipath hemispherical map [8–10], signal-to-noise ratio (SNR) or carrier-to-noise power-density ratio (C/N0) [11], wavelet analysis



B. Li et al., *GNSS Real-Time Kinematic Positioning*, Navigation: Science and Technology 17, https://doi.org/10.1007/978-981-96-9116-6_7

[12], support vector regression [13] and ray-tracing approach [14] are also studied. Although the multipath can be characterized or mitigated to some extent, the strongly environment-specific multipath is still rather difficult to be totally modeled or eliminated in real applications, especially for the real-time and kinematic modes. Secondly, for the tropospheric effects, the easy-to-implement empirical troposphere correction model is often applied [15–17]. However, due to the limited accuracy of troposphere models, especially for the wet troposphere component, the residual tropospheric errors will remain in GNSS observations particularly with low elevations. To further absorb the residual tropospheric errors, a so-called zenith tropospheric delay (ZTD) parameter is introduced together with mapping function. However, it is not always effective especially in case of long-range kinematic positioning where the variation of tropospheric errors grows rapidly [18]. Thirdly, regarding the ionospheric effects, it can be compensated by parameterization or eliminated by forming ionosphere-free (IF) model combination with multiple frequency signals [19–21]. Another easier but less accurate strategy is to use the empirical model, such as Bent, IRI, Klobuchar, grid, polynomial and spherical harmonic models. Unfortunately, these methods are effective only to the first-order ionospheric errors. The second- and/or higher-order terms remain owning to their spatiotemporal complexity and predictable difficulty [3]. In conclusion, no matter how we properly model these residual systematic errors, there will inevitably leave some unmodeled errors mainly due to their complicated spatiotemporal characteristics. In theory, the unmodeled errors can be, to a great extent, compensated by introducing the additional parameters. However, introducing too many parameters would lead to an ill-posed or even inestimable. Therefore, we have to make a compromise to only mitigate the unmodeled errors that are indeed significant by introducing parameters as few as possible. It is thus urgently needed to develop a procedure to test the significance of unmodeled errors and identify their components.

So far, although there is no study directly on the unmodeled errors in GNSS community, a number of studies have indirectly suggested the existence of unmodeled errors by analyzing the statistics and stochastic characteristics of GNSS observations. For instance, many studies suggested the existence of physical correlations [1] in GNSS observations, e.g., the time correlation. The probability distribution of GNSS observations have also been extensively investigated based on the least squares (LS) derived residuals. Tiberius and Borre [22] studied the probability density of GPS observations by graphical analysis and empirical moments. They found that the normal distribution turned out to be reasonable for the zero and ultra-short (3 m) baselines, but not be suitable for the longer (approximately 13 km) baselines. Luo et al. [23] analyzed the studentized double differenced (DD) residuals of GPS phase observations mainly using four sample moments and five hypothesis tests. Then the discrepancies between the classical Gauss distribution and reality are numerically demonstrated as a function of baseline length. Finally, they attribute these discrepancies mainly to the multipath and atmospheric effects. As mentioned above, there will inevitably leave some unmodeled errors mainly due to their complicated spatiotemporal characteristics. Therefore, these findings actually imply the existence of unmodeled errors in GNSS observations.

All aforementioned studies on the unmodeled errors are all based on the LS residuals. In theory, the unmodeled errors pertain to their associated observations cannot be completely estimated from the residuals because of the their dependence. However, the residuals can, to a great extent, reflect the behaviors of unmodeled errors especially in case of the large number of redundancies with multi-frequency and multi-constellation GNSS signals. Unfortunately, little attention has realized this opportunity to study the unmodeled errors themselves. So far, very little attention has paid to the unmodeled errors themselves even in such promising area of multi-frequency and multi-constellation. Thus far, the majority of related studies mainly focus on how to capture the systematic errors that can be modeled. Only if the unmodeled errors are statistically identified with a certain significance, the further compensation methods could be applied.

In this study, a procedure will be designed for testing the significance of GNSS unmodeled errors, i.e., the Li's procedure. It is composed of the Augmented Dickey-Fuller (ADF) test, Jarque-Bera (JB) test and $t$-test. Specifically, the ADF-test is applied to test the stationarity of unmodeled errors, then the combined JB-test and $t$-test are introduced to detect the zero-mean normality of the stationary unmodeled errors. Thus three components are identified in unmodeled errors, i.e., the nonstationary signal, the stationary signal and the white noise, which can be undetstood as the deterministic signal, the clored noise and the Gaussian white noise respectively in GNSS observations. The efficiency of the testing procedure is validated by using the simulated time series and the real dual-frequency BDS observations of 10 baselines ranging between 0 and 50 km. To further verify the correctness of testing results from the proposed procedure, the Allan variance analysis and fast Fourier transform (FFT) are applied to investigate the properties of unmodeled errors from the attributions of atmosphere, multipath and receiver, respectively. It is worth mentioning that the proposed procedure allows us to test the significance of unmodeled errors individually for each satellite or satellite pair in real time. It is promising to real-time applications.

To address the challenge of unmodeled error compensation, we now explore innovative strategies aimed at mitigating their adverse effects on positioning accuracy. Thus far, many studies focus on how to model or reduce the specific unmodeled effects, such as multipath, tropospheric and ionospheric errors. First, for the multipath, one can choose the ideal environment and advanced hardware. Besides, the data processing methods are often applied, such as sidereal filtering [5–7], hemispherical map [8–10], wavelet analysis [12] and so on. Second, regarding the tropospheric delay, the tropospheric correction model including the Hopfield model [15], Saastamoinen model [16] and New Brunswick 3 (i.e., UNB3) model [17] are often applied. In addition, the ZTD parameter and mapping function are both introduced to estimate the tropospheric errors [18]. Third, for the ionospheric delay, the parameterization or empirical model can be applied. The IF combination only by using carrier phase can also be formed to mitigate these ionospheric delays if there are at least two available frequencies [21]. However, all these traditional methods cannot totally eliminate the systematic errors, and these unmodeled effects are inevitably exist in GNSS observations especially for the RTK multipath, wet tropospheric component, second- and higher-order ionospheric effects.

In essence, there are four main ideas to mitigate these residual systematic errors in GNSS applications. The first one is to make a difference or linear combination, which can mitigate or even eliminate some common systematic errors among different observations. For instance, the tropospheric and ionospheric delays can be mitigated, and the satellite and receiver clock errors can be eliminated. The second one is to apply a priori correction, empirical model or precise product [20]. This method is the most widely used in the field of atmospheric error correction. The third one is stochastic model compensation [24], where the variance and covariance elements are used to capture the residual systematic errors. The functional model compensation is the last main approach to mitigate these residual systematic errors. This method is especially suitable when these systematic errors are significant. Specifically, one can choose the adjustment model with additional systematic error parameters [25], the collocation model with additional systematic error parameters [26], or even the semiparametric estimation model [27].

However, for the unmodeled error mitigation, these traditional methods are not the best solutions. At first, theoretically, the functional model compensation is the most suitable approach among all these traditional methods since the unmodeled errors usually exhibit like a deterministic signal. Whereas, introducing too many parameters would lead to ill-conditioned or even inestimable. Therefore, we have to make a compromise to only parameterize the unmodeled errors that are indeed significant. At second, unlike the tropospheric or ionospheric errors, the unmodeled errors have their own characteristics. Specifically, the residual tropospheric errors can be parameterized a ZTD parameter together with mapping function, and the ionospheric errors can be parameterized where the parameters are related to the signal frequency and the observation type. However, the unmodeled errors are related to the elevation, azimuth, even the frequency and observation type. Therefore, the unmodeled errors are not directly estimable by the traditional parameterization. Although there exist some functional model compensation methods such as semiparametric estimation, the algorithm such as determination of smoothing parameters is a little complicated. Besides, this method cannot be applied to the real-time scenario otherwise the true coordinate component cannot be separated accurately. In conclusion, it is urgently needed to develop a functional model compensation of unmodeled effects for GNSS precise positioning, which can be conducted in real time.

Since in single-frequency multi-GNSS positioning, the first-order ionospheric delays cannot be eliminated by the IF combination in terms of two or more carrier phases with different frequencies, the problem of significant unmodeled effects that affect the positioning precision and reliability is more severe. Fortunately, there are enough redundant observations at this time, which can reveal the unmodeled errors to a great extent according to the observation residuals. In this monograph, we propose a real-time unmodeled error mitigation method in single-frequency multi-GNSS precise positioning. This method can be called as multi-epoch partial parameterization. That is, only the significant unmodeled errors are parameterized and the properties of unmodeled effects are applied to parameterize the unmodeled errors to a great extent. To evaluate the effectiveness of the proposed method, an experiment was conducted and analyzed.


## 7.2  Unmodeled Error Detection

As an important part of unmodeled error processing, the detection of unmodeled errors is studied comprehensively in this section. First, for the GNSS unmodeled error, if there are no unmodeled errors in GNSS observations, the functional and stochastic models are defined as

$$l = Ax + e \tag{7.1}$$

$$D = \sigma_0^2 Q_{ee} \tag{7.2}$$

where $l$ is an $m \times 1$ observation vector and $e$ is its corresponding noise vector with zero mean, $x$ is a $u \times 1$ parameter vector to be estimated and $A$ is its design matrix of full column rank. $D$ is the covariance matrix of the observations with $\sigma_0^2$ a variance factor and $Q_{ee}$ the cofactor matrix. It is worth noting that the Eqs. (7.1) and (7.2) can be used in the single differenced (SD) and DD observations. Typically, the noises $e$ are adequately assumed to be white noises with normal distribution (i.e., Gaussian white noises) [23, 28] for undifferenced (UD) observations. In this case, the cofactor matrix $Q_{ee}$ should be derived via error propagation law when the SD and DD observations are applied.

The LS estimator and its cofactor matrix read

$$\hat{x} = \left(A^{\mathrm{T}} Q_{ee}^{-1} A\right)^{-1} A^{\mathrm{T}} Q_{ee}^{-1} l \tag{7.3}$$

$$Q_{\hat{x}\hat{x}} = \left(A^{\mathrm{T}} Q_{ee}^{-1} A\right)^{-1} \tag{7.4}$$

The LS residual vector reads

$$\hat{v} = l - A\hat{x} = Rl = Re \tag{7.5}$$

where $R = I_m - A\left(A^{\mathrm{T}} Q_{ee}^{-1} A\right)^{-1} A^{\mathrm{T}} Q_{ee}^{-1}$ $R$ is an idempotent and rank-deficient matrix, satisfying with $RA = 0$. Because the LS solution is an unbiased estimator, i.e.,

$$E(\hat{v}) = RE(l) = RE(e) = 0 \tag{7.6}$$

where $E(\cdot)$ denotes the mathematical expectation. It means that the residuals are of zero-mean expectation if the observations are not affected by the unmodeled errors.

More specifically, the residuals are normally distributed with zero-mean if we take the common assumption for normal distribution of GNSS observation noises [23, 28].

However, if the unmodeled errors exist in observations and are ignored without notice, the LS residuals must be affected. There are three potential scenarios. Firstly, if the unmodeled errors exhibit as the deterministic signals $s$, i.e., $\mathrm{E}(l) = Ax + s$, then both LS estimator and residuals are biased. The biases are as follows

$$\Delta x_s = \mathrm{E}(\hat{x}) - x = \left(A^{\mathrm{T}}Q_{ee}^{-1}A\right)^{-1}A^{\mathrm{T}}Q_{ee}^{-1}s \tag{7.7}$$

$$\Delta v_s = R\mathrm{E}(l) = R(Ax + s) = Rs \tag{7.8}$$

Secondly, if the unmodeled errors exhibit as the colored noises $\varepsilon$, the LS solution will be not optimal anymore if the cofactor matrix $Q_{ee}$ is still applied. In this case, although the LS estimator and residuals are unbiased as well-known, their associated covariance matrices will be definitely affected. The third scenario is that the unmodeled errors include the combined deterministic signals and colored noises simultaneously. It is obvious that the LS residuals are affected by both two components of unmodeled errors. In this case, the residuals will become nonstationary. Therefore, in practice, the significance of unmodeled errors can be judged by testing the LS residuals with a given significance level.

Then, for the Li's procedure, since the unmodeled errors objectively exist as the spatiotemporal signals, they have some corresponding properties of the spatiotemporal signals, including the colored noises caused by the physical correlations [1] and the deterministic signals caused by some certain systematic errors [4]. For a typical signal, there are two types of components, i.e., the stationary term and the nonstationary term [29]. Analogously, the time series of GNSS unmodeled errors can also be subdivided into two components: (1) the stationary terms including both the white noises and colored noises. Specifically, the white noises are Gaussian white noises [23, 28], and the colored noises usually have the properties of random walk noise, flicker noise and the first order Gauss-Markov (GM) process; (2) the nonstationary terms, referred to also as the deterministic signals, including trend and periodic terms [30]. It is emphasized that in GNSS observations, the nonstationary and stationary signals can be, to a great extent, understood as the deterministic signals and colored noises. As a result, the time series of unmodeled errors $y$ can be mathematically formulated as

$$y_t = u_t + \sum_{i=1}^{k}[a_i \sin(\omega_i t) + b_i \cos(\omega_i t)] + s_t + e_t \tag{7.9}$$

where $t$ denotes the observation epoch. $u$, $s$ and $e$ denote the trend term, the colored noise and the white noise, respectively. The periodic term is expressed by the summation of $k$ harmonic functions $\sum_{i=1}^{k}[a_i \sin(\omega_i t) + b_i \cos(\omega_i t)]$ with the amplitudes $a$ and $b$, and the angular frequency $\omega$. It is noticed that separating the certain systematic effects from the signals contaminated with the colored noises is dangerous [30]. Our purpose is to test the significance of the unmodeled errors. If they are statistically found significantly, we will directly compensate their effects on the LS solutions instead of separating the systematic effects from the unmodeled errors.

In principle, if all the systematic errors are completely modeled, there remains the only white noise. If some proper hypothesis tests can be used to effectively test the significance of the colored noises and/or the deterministic signals, we can conclude the existence of unmodeled errors and consequently develop the relevant compensation methods. Therefore, we propose a procedure for testing the significance of unmodeled errors, as shown in Fig. 7.1. It is clear that the procedure starts with the time series of LS residuals and consists of two main testing steps. The first step is to test the stationarity of residual time series by applying the ADF-test, while the second step, after the time series is confirmed to be stationary, is to further test whether the colored noises are included by using the combined JB-test and $t$-test. We do not consider the situation of the combined nonstationary signals and white noise here since it is nearly inexistent based on our extensive experimental studies [1]. Once the components of unmodeled errors are identified from time series of LS residuals by hypothesis testing, the proper compensation methods should be applied if the further improvements on positioning solutions are expectable. For instance, if the stationary errors (i.e., the lumped colored noises and white noises) are justified, the stochastic model compensation should be applied to adequately assimilate the physical correlations raised by the colored noises. However, if the nonstationary signals are identified (mostly, the deterministic signals and the colored noises are simultaneously existing), one needs to assimilate the deterministic signals by proper functional model compensation (e.g., modeling or parameterization) besides the colored noises by the stochastic model compensation. Certainly, there are some other easy-to-implement compensation methods, such as deleting the observations in the functional model, down-weighting the observations in the stochastic model.

In this study, three hypothesis tests are advised, i.e., the ADF-test for identifying the deterministic signals, the combined JB-test and $t$-test for identifying the colored noises. The ADF-test is a unit root test based on the existence and uniqueness property of an autoregressive (AR) model. It is mathematically formulated as [31]

$$Y_t = \phi Y_{t-1} + \alpha_1 \Delta Y_{t-1} + \cdots + \alpha_p \Delta Y_{t-p} + E_t \tag{7.10}$$

where $Y$ is the residual time series to be tested. $\phi$ and $\alpha$ are the parameters, and $E$ is the white noise. The subscript $t$ is the observation epoch. $p$ is the number of lagged difference terms. $\Delta$ is the time differencing operator, i.e., $\Delta Y_t = Y_t - Y_{t-1}$. The parameters are estimated in terms of LS criterion. Then the test statistic of the ADF-test is constructed as

$$T_{\text{ADF}} = \frac{\hat{\phi} - 1}{\text{SE}(\hat{\phi})} \tag{7.11}$$

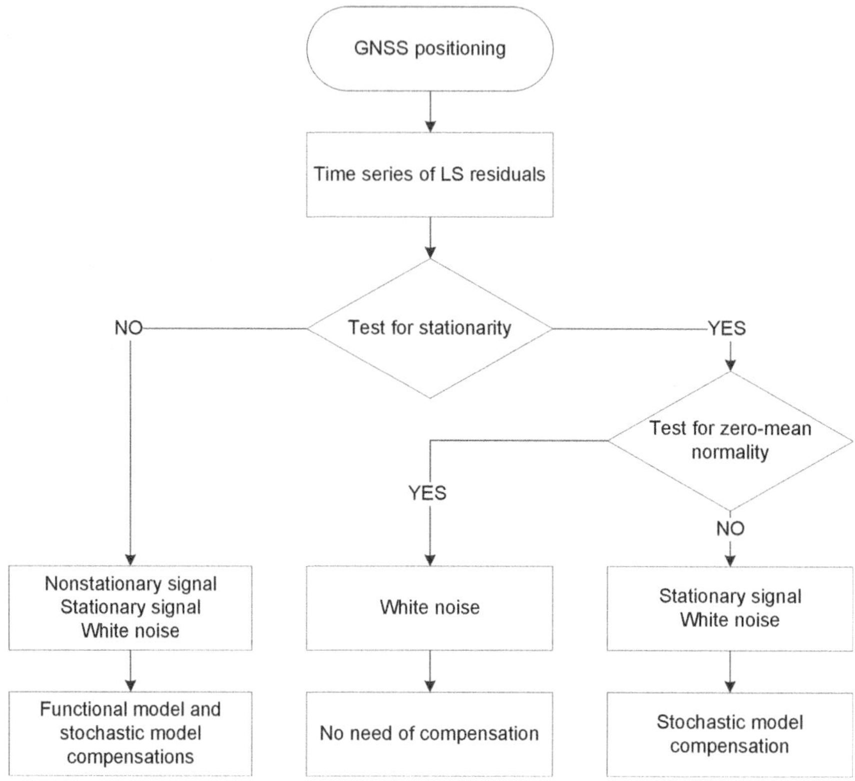


**Fig. 7.1** Flow diagram of the procedure for testing the significance of unmodeled errors

where $\text{SE}(\hat{\phi})$ denotes the standard error of LS estimator $\hat{\phi}$. Here the standard error is the standard deviation (STD) of the sampling distribution of mean. The ADF statistic follows the nonstandard distribution [32]. The critical value for $T_{\text{ADF}}$ can be generated by using the Monte-Carlo simulation. Since the ADF-test focuses on searching AR unit roots, the Kwiatkowski-Phillips-Schmidt-Shin (KPSS) test based on the moving average (MA) unit roots could also be used as an alternative choice.

If the residual time series is identified to be stationary with ADF-test, a multiple test will be further applied to test the zero-mean normality of this series. If it is negative, then one can conclude that the colored noises are contained in residuals. Specifically, the JB-test is used for testing the normality while the $t$-test is for testing the zero-mean expectation. Only when both tests pass, we can confirm that the residual time series is of zero-mean normal distribution. The JB-test is a two-sided goodness-of-fit test with statistic as [33]

$$T_{\text{JB}} = \frac{n}{6}\left[S_3^2 + \frac{(S_4 - 3)^2}{4}\right] \tag{7.12}$$

where $n$ is the sample size of residual time series to be tested. $S_3$ and $S_4$ are the third and fourth standardized central sample moments of a random distribution, i.e., the skewness and kurtosis, they are computed by

$$S_i = \frac{\sum_{j=1}^{n}\left(Y_j - \overline{Y}\right)^i / n}{\sigma^i} \tag{7.13}$$

where $\overline{Y}$ and $\sigma$ are the mean and STD of time series $Y_j$ $(j = 1, \ldots, n)$. $i$ is the order of central sample moments (i.e., 3 for skewness and 4 for kurtosis). The JB-test often uses the chi-square distribution to determine its critical value. If the sample size is less than 2000, the critical value would be better computed by Monte-Carlo simulation. Note since the JB-test is sensitive to the outliers and poorly valid for small sample sizes [23, 34], the other tests for normality like Lilliefors (LF) test [35], would be alternatively applied if the time series has a small size or the outliers probably exist. The $t$-test is applied to test the significance of zero mean of a time series. Its test statistic reads

$$T_t = \frac{\overline{Y} - u}{\sigma/\sqrt{n}} \tag{7.14}$$

where $u$ is the hypothesized population mean and $u = 0$ in our testing case. The $t$-test statistic has student's distribution with $n-1$ degrees of freedom, and its critical value can be accordingly computed.

We emphasize that the above testing procedure is not restricted to the DD observations. It can be applied to the observations in UD, SD and DD modes. Therefore, the Li's procedure can be applied for significance testing of the unmodeled errors for standalone positioning and relative positioning models, such as precise point positioning (PPP) and RTK.

## 7.3  Unmodeled Error Compensation

After introducing unmodeled error detection, this section will discuss how to compensate unmodeled errors. It is also an important part of unmodeled error processing. In theory, the unmodeled errors can be compensated by stochastic model and functional model simulatenously. Since the stochastic modeling is discussed in Chap. 6, the functional model compensation that can capture the significant unmodeled errors are discussed in this section. Specifically, the introduction, methodology, and experiments and results of unmodeled error mitigation are introduced and analyzed in turn. For the methodology of unmodeled error mitigation, we will introduce it in two parts: basic theory of unmodeled error mitigation and unmodeled error mitigation based on multi-epoch partial parameterization.

First, for the basic theory of unmodeled error mitigation, a discussion on the GNSS unmodeled error and its functional model compensation is the prerequisite of unmodeled error mitigation. Therefore, at first, the fundamental properties of the single-frequency multi-GNSS unmodeled errors are discussed. At second, the principle of functional model compensation is analyzed. In this section, the estimability of unmodeled error parameters is emphasized.

For the single-frequency multi-GNSS unmodeled error, if the GNSS mathematical model can meet the reality, the unmodeled errors do not exist. At this time, the mathematical model of single-frequency multi-GNSS observations is defined as

$$l = Ax + e \tag{7.15}$$

$$D = \sigma_0^2 Q_{ee} \tag{7.16}$$

where in functional model (7.15), $l$ and $e$ denote the observation vector and noise vector, respectively; $x$ denotes the parameters to be estimated and $A$ denotes the corresponding design matrix of full column rank. In stochastic model (7.16), $D$ denotes the variance-covariance matrix of the observation vector; $\sigma_0$ and $Q_{ee}$ (i.e., $P^{-1}$ with $P$ denoting the weighting matrix) denote the variance factor and cofactor matrix, respectively. The error equation and its adjustment criterion are as follows

$$V = Ax - l \tag{7.17}$$

$$V^{\mathrm{T}} P V = \min \tag{7.18}$$

where $V$ denotes the residual vector. As a result, the LS estimator and its cofactor matrix read

$$\hat{x} = \left(A^{\mathrm{T}} P A\right)^{-1} A^{\mathrm{T}} P l \tag{7.19}$$

$$Q_{\hat{x}\hat{x}} = \left(A^{\mathrm{T}} P A\right)^{-1} \tag{7.20}$$

However, if the GNSS mathematical model cannot meet the reality, the unmodeled errors will be existent and even significant. In case of single frequency, the ionospheric delays cannot be mitigated by using the IF combination in terms of two or more carrier phases with different frequencies. Hence, when the baseline length is long, these ionospheric delays are easily significant. Fortunately, in multi-GNSS precise positioning, the redundancies are large enough as usual. Therefore, the observation residuals can reflect the unmodeled errors including the ionospheric delays to a great extent in this situation, then several observations with significant unmodeled errors can be found out according to the behaviors of these observation residuals. Specifically, if the unmodeled errors are indeed significant for some observations, they exhibit as the deterministic signals with randomness. As usual, the priori information of the unmodeled errors can be estimated including the expectation, variance and covariance.

Since the unmodeled effects have three main types of components, i.e., the nonstationary signal, stationary signal and white noise, the GNSS unmodeled errors should be compensated by the mathematical models. It is obvious that the nonstationary signal of unmodeled errors is the main impact factor affecting the positioning precision. For the nonstationary part, i.e., the aforementioned deterministic signals, the functional model compensation is prior to be applied. Whereas, the unmodeled errors cannot be easily parameterized and estimated. The reasons are as follows. At first, the severity of unmodeled errors among different satellites is not the same due to their different elevations and azimuths. At second, unlike some specific residual systematic errors, the unmodeled effects are actually the integrated errors of observations. When the residual tropospheric delays are severe, the so-called ZTD parameter is introduced together with mapping function. That is, the tropospheric delay in any direction is considered to be related to the zenith direction, so the mapping function and the ZTD parameter can represent the residual tropospheric delay of any satellite. When the ionospheric delays are severe, the ionosphere-float model can be applied. Besides, since the ionospheric delay is related to the signal frequency and the observation type, only one ionospheric delay parameter can construct the ionospheric delay of each satellite. Therefore, unlike some specific residual systematic errors, the unmodeled errors are highly probable to be correlated with the elevation, azimuth, even the frequency and observation type. Moreover, the unmodeled errors are impacted by some site-specific errors such as multipath. In conclusion, the unmodeled errors from different sources cannot be directly estimated by one parameter.

For the classical functional model compensation, regardless of the estimability of unmodeled errors, firstly let us analyze the classical functional model compensation. The additional parameter vector $s$ is added to the (7.15), and the new functional model reads

$$l = Ax + Bs + e \tag{7.21}$$

It is worth noting that the term $s$ denotes the filtering signal, i.e., the deterministic signal with randomness. The new stochastic models read

$$\mathrm{E}(s) = \mu_s, \quad \mathrm{var}(s) = D_s \tag{7.22}$$

$$\mathrm{E}(e) = 0, \quad \mathrm{var}(e) = D_e \tag{7.23}$$

$$\mathrm{cov}(e, s) = D_{es} \tag{7.24}$$

where symbols "E($\cdot$)", "var($\cdot$)", and "cov($\cdot$)" denote the operators of prior expectation, variance and covariance, respectively. The error equations then read

$$\begin{cases} V = Ax + Bs - l \\ V_s = s - l_s \end{cases} \tag{7.25}$$

where the $V_s$ denotes the residual vectors of the filtering signal. The $l_s$ denotes the virtual observations of the filtering signal. The $l_s$ equals the prior expectation, i.e., $l_s = \mu_s$. The generalized LS criterion can be obtained

$$V^{\mathrm{T}} P V + V_s^{\mathrm{T}} P_s V_s = \min \tag{7.26}$$

Assuming the filtering signal is not correlated with the observation noise, i.e., $\mathrm{cov}(e, s) = \mathbf{0}$, the LS estimator of parameters to be estimated and the filtering signal read

$$\hat{x} = \left[ A^{\mathrm{T}} \left( B D_s B^{\mathrm{T}} + D_e \right)^{-1} A \right]^{-1} A^{\mathrm{T}} \left( B D_s B^{\mathrm{T}} + D_e \right)^{-1} (l - B l_s) \tag{7.27}$$

$$\hat{s} = l_s + D_s B^{\mathrm{T}} \left( B D_s B^{\mathrm{T}} + D_e \right)^{-1} (l - A\hat{x} - B l_s) \tag{7.28}$$

The corresponding covariance matrices read

$$D_{\hat{x}\hat{x}} = \left[ A^{\mathrm{T}} \left( B D_s B^{\mathrm{T}} + D_e \right)^{-1} A \right]^{-1} \tag{7.29}$$

$$D_{\hat{s}\hat{s}} = D_s - D_s B^{\mathrm{T}} \left( B D_s B^{\mathrm{T}} + D_e \right)^{-1} \left[ I - A D_{\hat{x}\hat{x}} A^{\mathrm{T}} \left( B D_s B^{\mathrm{T}} + D_e \right)^{-1} \right] B D_s \tag{7.30}$$

where the $I$ denotes the identity matrix.

Since the unmodeled errors can be considered to be a deterministic signal with randomness, theoretically classical functional model compensation can be used. It is worth noting that this idea is essentially one of using best linear unbiased prediction, i.e., BLUP [36]. However, according the aforementioned analysis, the above standard form of functional model compensation cannot be applied directly mainly because the unmodeled error parameters are not estimable. In order to estimate the unmodeled errors based on the functional model compensation, the properties of the unmodeled errors should be used firstly. Fortunately, the unmodeled errors have the property of temporal correlation, and the severity of unmodeled errors from different sources is not the same [1, 24]. Hence, the basic theory of functional model compensation should be based on the multi-epoch partial parameterization. Specifically, this method only parameterizes the observations with severe unmodeled errors, and uses multiple epochs to jointly estimate these unmodeled errors to a great extent, thus overcoming the problem that the unmodeled error parameters have no redundant observations.

Second, for the Unmodeled error mitigation based on multi-epoch partial parameterization, we propose a method for unmodeled error mitigation mainly based on the multi-epoch partial parameterization. At first, the methodology of multi-epoch partial parameterization is proposed. At second, a procedure of the real-time unmodeled error mitigation method is presented.

For the multi-epoch partial parameterization, take the RTK as an example to unfold the (7.21), and there are $m$ single-frequency multi-GNSS observations in the consecutive $k$ epochs, of which $n$ observations need to be parameterized ($n \leq m$). It is worth noting that since different satellite systems have their own reference satellites, there are no intersystem biases here. When the size of moving window is $k$ ($1 < k < i$), the linearized observation equations are formulated as

$$\begin{bmatrix} l_{i-k+1} \\ \vdots \\ l_i \end{bmatrix} = \begin{bmatrix} A_{i-k+1} & & \\ & \ddots & \\ & & A_i \end{bmatrix} \begin{bmatrix} x_{i-k+1} \\ \vdots \\ x_i \end{bmatrix} + \begin{bmatrix} B_{i-k+1} \\ \vdots \\ B_i \end{bmatrix} s + \begin{bmatrix} e_{i-k+1} \\ \vdots \\ e_i \end{bmatrix} \tag{7.31}$$

where $l = \left[l_{i-k+1}^{\mathrm{T}}, \ldots, l_i^{\mathrm{T}}\right]^{\mathrm{T}}$ denotes the vector of observed-minus-computed DD observations; $x = \left[x_{i-k+1}^{\mathrm{T}}, \ldots, x_i^{\mathrm{T}}\right]^{\mathrm{T}}$ denotes the vector of parameters to be estimated; $s$ denotes the vector of unmodeled errors; $A = \mathrm{blkdiag}(A_{i-k+1}, \ldots, A_i)$ denotes the design matrix to $x$; $B = \left[B_{i-k+1}^{\mathrm{T}}, \ldots, B_i^{\mathrm{T}}\right]^{\mathrm{T}}$ denotes the design matrix to $s$; $e = \left[e_{i-k+1}^{\mathrm{T}}, \ldots, e_i^{\mathrm{T}}\right]^{\mathrm{T}}$ denotes the noise vector; The symbol "blkdiag" denotes the operator of block diagonal concatenation of matrices. It can be seen that the design matrix $B$ is a rank deficient matrix since only partial observations are parameterized. Therefore, the design matrix $B$ is determined by the vector of unmodeled errors.

Since the unmodeled errors can be regarded as the deterministic signal, the priori expectation and variance should be known. Therefore, the priori information of unmodeled error parameters need to be estimated firstly when assuming the unmodeled errors are the non-random parameters. That is, the corresponding stochastic model is as follows

$$D = \sigma_0^2 Q = \sigma_0^2 P^{-1} \tag{7.32}$$

The error equation and its adjustment criterion are as follows

$$V = Ax + Bs - l \tag{7.33}$$

$$V^{\mathrm{T}} P V = \min \tag{7.34}$$

Then the normal equation reads

$$\begin{bmatrix} A^{\mathrm{T}} P A & A^{\mathrm{T}} P B \\ B^{\mathrm{T}} P A & B^{\mathrm{T}} P B \end{bmatrix} \begin{bmatrix} \hat{x} \\ \hat{s} \end{bmatrix} - \begin{bmatrix} A^{\mathrm{T}} P l \\ B^{\mathrm{T}} P l \end{bmatrix} \tag{7.35}$$

with $A^{\mathrm{T}} P A = N_{AA}$, $A^{\mathrm{T}} P B = N_{AB}$, $B^{\mathrm{T}} P A = N_{BA}$, $B^{\mathrm{T}} P B = N_{BB}$ and $M = N_{BB} - N_{BA} N_{AA}^{-1} N_{AB}$. Accordingly, the vector of unmodeled errors can be estimated as

$$\hat{s} = M^{-1} B^{\mathrm{T}} P l - M^{-1} N_{BA} N_{AA}^{-1} A^{\mathrm{T}} P l \tag{7.36}$$

The covariance matrix reads

$$Q_{\hat{s}\hat{s}} = M^{-1} \tag{7.37}$$

Finally, the priori information of the unmodeled error at the $(i + 1)$-th epoch can be obtained as follows

$$\mathrm{E}(s_{i+1}) = \hat{s} \tag{7.38}$$

$$\mathrm{var}(s_{i+1}) = M^{-1} \tag{7.39}$$

For the $(i + 1)$-th epoch, the functional model can be derived as

$$l_{i+1} = A_{i+1}x_{i+1} + B_{i+1}s_{i+1} + e_{i+1} \tag{7.40}$$

According to the generalized LS criterion, the error equations read

$$\begin{cases} V_{i+1} = A_{i+1}x_{i+1} + B_{i+1}s_{i+1} - l_{i+1} \\ V_{s_{i+1}} = s_{i+1} - l_{s_{i+1}} \end{cases} \tag{7.41}$$

where $V_{s_{i+1}}$ denotes the residual vector of the filtering signal, i.e., the unmodeled errors; $l_{s_{i+1}}$ denotes the virtual observations of the filtering signal, satisfying $l_{s_{i+1}} = \hat{s}$. The adjustment criterion then can be derived as

$$V_{i+1}^{\mathrm{T}}P_{i+1}V_{i+1} + V_{s_{i+1}}^{\mathrm{T}}P_{s_{i+1}}V_{s_{i+1}} = \min \tag{7.42}$$

where $P_{s_{i+1}} = M$ denotes the weighting matrix of the filtering signal.

Assuming there are no correlations between the filtering signal and the observation noise, i.e., $\mathrm{cov}(e, s) = 0$, the unknown parameters and unmodeled errors can be estimated as

$$\hat{x}_{i+1} = \left(A_{i+1}^{\mathrm{T}}N_{i+1}^{-1}A_{i+1}\right)^{-1}A_{i+1}^{\mathrm{T}}N_{i+1}^{-1}(l_{i+1} - B_{i+1}\hat{s}) \tag{7.43}$$

$$\hat{s}_{i+1} = \hat{s} + M^{-1}B_{i+1}^{\mathrm{T}}N^{-1}(l_{i+1} - A_{i+1}\hat{x}_{i+1} - B_{i+1}\hat{s}) \tag{7.44}$$

with $N_{i+1} = B_{i+1}M^{-1}B_{i+1}^{\mathrm{T}} + D_e$. The corresponding covariance matrices are as follows

$$D_{\hat{x}_{i+1}\hat{x}_{i+1}} = \left(A_{i+1}^{\mathrm{T}}N_{i+1}^{-1}A_{i+1}\right)^{-1} \tag{7.45}$$

$$D_{\hat{s}_{i+1}\hat{s}_{i+1}} = M^{-1} - M^{-1}B_{i+1}^{\mathrm{T}}N_{i+1}^{-1}\left(I - A_{i+1}D_{\hat{x}_{i+1}\hat{x}_{i+1}}A_{i+1}^{\mathrm{T}}N_{i+1}^{-1}\right)B_{i+1}M^{-1} \tag{7.46}$$

Similarly, the unmodeled errors of the $(i + 2)$-th epoch can be mitigated according to from (7.31) to (7.46). Apparently, the proposed method is conducted epoch-by-epoch, hence it can be applied in real-time.

For the procedure of the proposed method. Based on the preceding analysis, a procedure of real-time unmodeled error mitigation based on the multi-epoch partial parameterization is proposed. The details of the proposed method are shown in Fig. 7.2.

According to Fig. 7.2, it can be seen that the proposed method is iterative. The descriptions of these specific steps are as follows.

1. GNSS positioning.

   The positioning results and other relevant necessary data are obtained by using the conventional positioning methods, such as RTK or PPP. The code-based positioning modes such as single point positioning and code real-time differenced positioning are also suitable.

2. Observation residuals and significance testing.

   The Li's method is applied to test the residuals of GNSS observations, hence the observations with significant unmodeled errors are obtained. Specifically, a

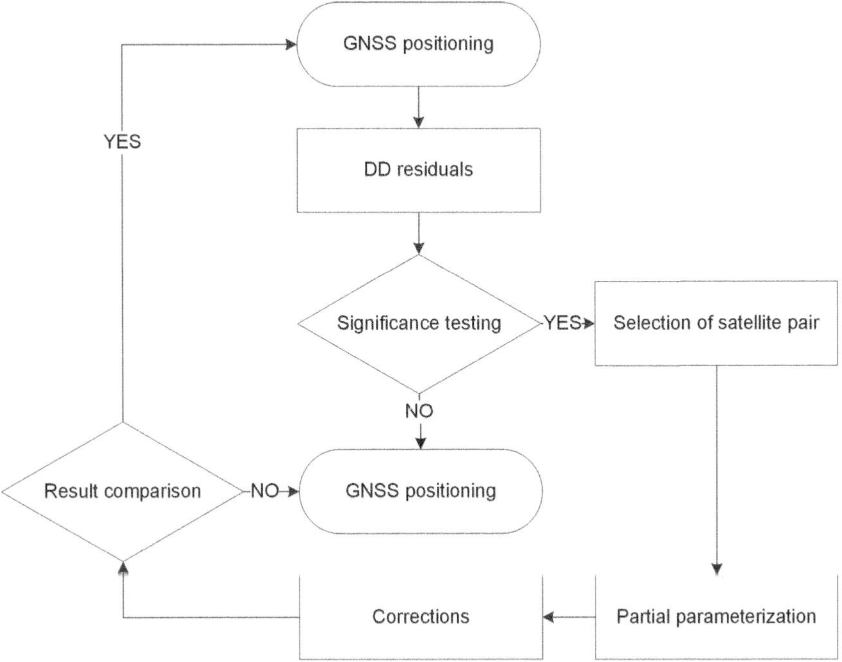


**Fig. 7.2** A procedure of unmodeled error mitigation based on the multi-epoch partial parameterization

combined test consisting of the ADF test, JB test and $t$-test is applied. It is worth noting that these unmodeled errors may need to be parameterized afterwards.

3. Selection of satellite (pair).

According to the certain indicator (e.g., elevation, SNR, temporal correlations of observations or other statistics of observation residuals), the unmodeled errors of certain satellite (pair) that are most in need of parameterization are determined. The basic principle of the selection of satellite (pair) is that the observations are contaminated by the unmodeled errors are always highly correlated with certain indicators such as elevation, SNR, temporal correlations of observations or other statistics of observation residuals.

4. Multi-epoch partial parameterization and corrections.

The method of parameterization in the section on multi-epoch partial parameterization is applied. That is, the equations from (7.31) to (7.46) is used iteratively epoch-by-epoch. After the error equation is reconstructed, the corrections, new positioning results and other relevant quantities are obtained.

5. Result comparison.

The new positioning solutions are determined whether they are improved or not (e.g., by applying the state equation to obtain the prediction solution). The parameterization is stopped until the precision of the positioning results cannot be improved.

## 7.4  Results and Discussion

### *7.4.1  Results and Discussion of Unmodeled Error Detection*

For the experiment of simulated data, a simulated example is implemented to validate the Li's procedure. According to the former analysis, without loss of generality, the trend term, periodic term, colored noise and white noise are all simulated. The simulated time series is generated with Eq. (7.9) where the sample size is taken as 1000 epochs. The white noises are simulated with three different variances, namely, $W(0.2) \sim N(0, 0.2^2)$, $W(0.05) \sim N(0, 0.05^2)$, $W(0.01) \sim N(0, 0.01^2)$, for specifying the strength of white noise in time series. The colored noise is simulated as a GM process realized by an autoregressive moving average (ARMA) model of ARMA(7,7). Here, the parameters (7,7) are used to specify the pink noise that is one of the most common colored noises in real applications. Then the colored noise $s$ is generated as follows

$$
\begin{aligned}
s_t = {} & 0.5s_{t-1} + 0.125s_{t-2} + 0.063s_{t-3} + 0.036s_{t-4} - 0.026s_{t-5} \\
& - 0.007s_{t-6} - 0.005s_{t-7} + e_t + 0.5e_{t-1} - 0.125e_{t-2} + 0.063e_{t-3} \\
& - 0.036e_{t-4} - 0.026e_{t-5} + 0.007e_{t-6} - 0.005e_{t-7}
\end{aligned}
\tag{7.47}
$$

where all white noises $e$ follow normal distribution of N(0,0.04). The deterministic signals consist of trend and periodic terms. The trend term is simulated by $T_t = 0.0035t$, while the periodic term is simulated by $P_t = 0.4 \times \sin(2\pi t/300) + 0.4 \times \cos(2\pi t/300) + 0.3 \times \sin(2\pi t/60) + 0.3 \times \cos(2\pi t/60)$ with two periods of 300 and 60 epochs, respectively.

To specify the varying situations, we set up a set of simulation options. Firstly, only the white noises with different variances, and the lumped white noise and colored noise are simulated respectively. Secondly, the trend and periodic terms are added to the simulated data. In order to illustrate the advantages of the advised testing procedure, the alternative tests are also examined. We repeat here that our testing procedure includes the ADF-test for identifying the deterministic signals, and the combined JB-test and $t$-test for identifying the colored noises. The alternative tests we examine here include the KPSS-test for identifying the deterministic signals, and the combined LF-test and $t$ test for identifying the colored noises. A significance level of 5% is applied for all tests. For each simulation option, we run the simulations by 10,000 times. It is noted that for our simulations, the null hypotheses of these tests are that the time series is stationary and normally distributed with zero mean. Out of total 10,000 simulations, the testing results (empirical acceptance percentages) for simulated time series without and with deterministic signals are in Tables 7.1 and 7.2, respectively.

The results indicate that both the proposed and alternative procedures can effectively identify the stationarity and zero-mean normality of varying time series, and our proposed procedure performs better. Firstly, the ADF-test can always confirm the stationarity of time series with acceptance rates of at least 90% or even nearly 100% as shown in the second rows of both tables. However, as shown in the third row of Table 7.1, the KPSS-test cannot obtain the desirable results especially when the colored noises are lumped with the white noises for which the acceptance rates are as small as 4–10%. It implies that the KPSS-test should not be used when the deterministic signals are absent in the time series. The underlying reason is that the KPSS-test is based on the MA unit roots, which is very sensitive to the colored noises.

Let us compare the testing results between the combined JB-test and $t$-test and the combined LF-test and $t$-test, as shown in the fourth rows of Tables 7.1 and 7.2. In case of only white noises (the third to fifth columns of Table 7.1), 90% acceptance

**Table 7.1** The acceptance percentages of the testing procedure for the time series without the deterministic signals (unit: %)

| Hypothesis | Statistic | W(0.2) | W(0.05) | W(0.01) | W(0.2) + C | W(0.05) + C | W(0.01) + C | C |
|---|---|---|---|---|---|---|---|---|
| Stationarity | ADF | 100 | 100 | 100 | 100 | 100 | 100 | 100 |
|  | KPSS | 94.7 | 95.1 | 95.2 | 10.8 | 4.4 | 4.1 | 4.1 |
| Zero-mean normality | JB + t | 90.2 | 90.5 | 90.1 | 49.3 | 35.1 | 33.4 | 33.2 |
|  | LF + t | 90.6 | 90.6 | 90.1 | 55.8 | 47.2 | 46.5 | 46.4 |

The terms 'W' and 'C' denote the white noise and colored noise, respectively

**Table 7.2** The acceptance percentages of the testing procedure for the time series with the deterministic signals (unit: %)

| Hypothesis | Statistic | W(0.2) + C + T | W(0.05) + C + T | W(0.01) + C + T | W(0.2) + C + P | W(0.05) + C + P | W(0.01) + C + P | W(0.2) + C + T + P | W(0.05) + C + T + P | W(0.01) + C + T + P |
|---|---|---|---|---|---|---|---|---|---|---|
| Stationarity | ADF | 9.6 | 3.9 | 3.5 | 0 | 0 | 0 | 0 | 0 | 0 |
|  | KPSS | 10.4 | 3.9 | 3.7 | 0 | 0 | 0 | 0 | 0 | 0 |
| Zero-mean normality | JB + t | 0 | 0 | 0 | 13.6 | 6.6 | 6.3 | 0 | 0 | 0 |
|  | LF + t | 0 | 0 | 0 | 25.0 | 16.5 | 16.0 | 0 | 0 | 0 |

The terms of 'W', 'C', 'T' and 'P' denote the white noise, colored noise, trend term and periodic term, respectively rates are approximately obtained for confirming the zero-mean normality regardless of the white-noise variance. These results are consistent to the theoretical confidence 90.25%, i.e., the product of two theoretical confidences (95% × 95% = 90.25%). In this case, the combined LF-test and $t$-test obtain quite similar results, seeing the fifth row of Table 7.1. However, in case of the lumped white noise and colored noise or the pure colored noise. The combined JB-test and $t$-test slightly outperforms the combined LF-test and $t$-test, referring to the results from the sixth to ninth columns of Table 7.1. The null hypothesis that only white noise exists is accepted by the combined JB-test and $t$-test with 33–49%, while by the combined LF-test and $t$-test with 46–55%. The results of testing zero-mean normality also indicate that the acceptance rates are positively proportional to the strength (variance) of white noises. For the situation with deterministic signals, the acceptance rates of testing zero-mean normality in Table 7.2 are exactly 0 for both the combined JB-test and $t$-test and the combined LF-test and $t$-test for most of cases. For the cases from the sixth to eighth columns, the combined JB-test and $t$-test is much better than the combined LF-test and $t$-test with nearly half acceptance rates. Such results further confirm the validity of the combined JB-test and $t$-test. To sum up, based on the results of simulated experiments, the proposed procedure is overall better for the majority of circumstances.

Next, the experiment of real data is given. In order to further demonstrate the validity of the proposed procedure for significance testing of unmodeled errors, ten datasets were analyzed and for each dataset, two types of receivers were used to collect dual-frequency BDS data for 1 h with sampling interval of 1 s at the same time. The baseline lengths are from 0 to 50 km. Two types of high-end receivers were applied in these ten baselines, seeing the detailed information in Table 7.3. It is noted that all the datasets were collected in the same area, which means that the observation environment is quite similar.

All the datasets were processed by using our self-developed single-baseline RTK software (named by SRTK). To better study the properties of unmodeled errors, we

**Table 7.3** Data description of the ten datasets

| Dataset | Stations | Receiver type | Antenna type | Length |
| --- | --- | --- | --- | --- |
| No. 1 | T001-T002 | Trimble NetR9 | TRM59800.00 | 0.0 m |
| No. 2 | HKLM-HKQT | Trimble NetR9 | TRM59800.00 | 12.5 km |
| No. 3 | HKLM-HKCL | Trimble NetR9 | TRM59800.00 | 23.5 km |
| No. 4 | HKQT-HKCL | Trimble NetR9 | TRM59800.00 | 31.5 km |
| No. 5 | HKTK-HKCL | Trimble NetR9 | TRM59800.00 | 42.7 km |
| No. 6 | HKFN-T430 | Leica GR25 | LEIAR25.R4 | 5.0 m |
| No. 7 | HKST-HKKT | Leica GR25 | LEIAR25.R4 | 13.3 km |
| No. 8 | HKSS-HKKT | Leica GR25 | LEIAR25.R4 | 20.9 km |
| No. 9 | HKOH-HKNP | Leica GR25 | LEIAR25.R4 | 34.5 km |
| No. 10 | HKWS-HKNP | Leica GR25 | LEIAR25.R4 | 49.9 km |

should make sure that the same visible satellites are processed from different baselines. The cut-off angle of $15°$ is the lowest elevation that can meet this demand. The coordinates of all the stations are precisely known, serving as ground truths for validating results later. All phase ambiguities were correctly fixed in advance by using all observations of each baseline dataset with the least-squares ambiguity decorrelation adjustment (LAMBDA) method, where the elevation-dependent weighting scheme was applied and the tropospheric effects were corrected with the Hopfield model. For the ionospheric effects, we set up two strategies corresponding to two mathematical models: (A) Ignoring the ionospheric effects, referring to as the ionosphere-fixed model: (B) Eliminating the first-order ionospheric effects by using the IF model. Total of 8 BDS satellites were tracked for the whole period, and there are 28 time series of DD residuals for dual-frequency observations in each dataset. Since only 3 coordinates are assumed as unknowns, the number of observation redundancies is sufficiently large. Then the DD residuals can be used to investigate the properties of unmodeled errors. It is also worth noting that the code observations are both proved to be preprocessed by these two types of receivers based on our study. Specifically, the filter and carrier-smoothed code techniques are applied by the Trimble and Leica receivers, respectively. The similar conclusions can also be found in other literatures. Hence, we are not able to compare the raw code observations. In fact, the raw code observations always have the unmodeled errors due to their limited precision, and the high-precision positioning mainly depends on the phase observations. Therefore, there has not much need to test the significance of unmodeled errors in code observations and only the phase observations are discussed here.

The DD residuals derived from the models A and B on baselines No. 2–5 with Trimble receivers and the similar results but derived from the baselines No. 7–10 with Leica receivers are all computed. It can be intuitively seen that some of DD residuals are not stationary, and the deterministic (certain systematic) signals and/or colored noise may be in the unmodeled errors. The magnitudes of unmodeled errors are roughly positively proportional to the baseline length, especially for the B1 and B2 data. Besides, the patterns of unmodeled errors are similar with each other in each dataset from different frequencies, but there exist significant discrepancies for the different receiver types. In addition, compared with the model A, the unmodeled errors of model B reduced dramatically. That is, the ionospheric effects of unmodeled errors are significantly mitigated by model B. In conclusion, the unmodeled errors are correlated with the baseline length, the mathematical model and the receiver type.

The proposed procedure is then applied to identify the components of unmodeled errors with a significance level of 5%. The null hypotheses of ADF-test, JB-test and $t$-test are that the residual time series is stationary, normally distributed and zero-mean, respectively. For the reisudal time series of each baseline, we set up a moving window of 300 epochs. For each window, we apply the testing procedure. As usual, there is no need to detect the zero-mean normality when the time series is nonstationary. However, to obtain more detailed results, these three hypothesis tests are all applied no matter whether the tested time series is stationary or not. All the rejection rates of the null hypotheses are computed for two mathematical models. The results of the hypothesis tests are listed in Table 7.4. In general, the rejection rates of all statistic tests become larger as the baseline length increasing. Comparing the models A and B, the rejection rates of IF residuals (model B) are smaller than those of B1 and B2 residuals (model A). A closer look at the test results from the two receiver types, the rejection rates of ADF-test and JB-test for Trimble receivers are smaller than those for Leica receivers. Therefore, the rejection rates depend on the baseline length, the mathematical model and the receiver type, which agrees with the intuitive behaviors of GNSS residuals in related study by the authors.

Since the results have not significant frequency-dependence as shown in Table 7.4, the testing results of the proposed procedure can be further summarized in Table 7.5. It can be easily found that the DD residuals are all stationary on zero baseline (No. 1). Besides, the DD residuals can even be regarded as zero-mean normal distribution by approximately 90% for both the ionosphere-fixed model (A) and IF model (B). Therefore, the main error component on zero baseline is the white noise. For the ultra-short baseline (No. 6), the DD residuals are exactly stationary with acceptance rate of 100%, whereas they all do not obey the zero-mean normal distribution with small acceptance rates, i.e., 6.6% and 10.7% for models A and B, respectively. These testing results indicate that the unmodeled errors are stationary but include the colored noises. For the longer baselines (from No. 2–5 and No. 7–10), the acceptance rates of stationarity and zero-mean normality are almost negatively proportional to the baseline length. These indicate that the unmodeled error components on longer baselines are more complicated comparing to those on ultra-short baseline. Specifically, apart from the colored noises, the deterministic signals would become significant and they are positively correlated with the baseline length. It is also the evidence why there is no need to detect the zero-mean normality when the time series is nonstationary. In summary, our testing procedure can identify the unmodeled error components that are highly consistent with the former analysis in this section.

As stated before, the unmodeled errors mainly contain two types of components, namely, the deterministic signals and the colored noises. When the deterministic

**Table 7.4** Rejection rates of the null hypotheses from baselines No. 1 to 10 (unit: %)

| | | Trimble | | | | | Leica | | | | |
|---|---|---|---|---|---|---|---|---|---|---|---|
| | | No. 1 | No. 2 | No. 3 | No. 4 | No. 5 | No. 6 | No. 7 | No. 8 | No. 9 | No. 10 |
| ADF | B1 | 0 | 0 | 6.0 | 14.3 | 35.7 | 0 | 3.3 | 16.7 | 41.7 | 61.7 |
| | B2 | 0 | 1.2 | 9.5 | 16.7 | 29.8 | 0 | 3.3 | 23.3 | 48.3 | 68.3 |
| | IF | 0 | 0 | 0 | 0 | 0 | 0 | 0 | 0 | 0 | 0 |
| JB | B1 | 3.6 | 9.5 | 21.4 | 25.0 | 33.3 | 7.1 | 28.3 | 60.0 | 76.7 | 83.3 |
| | B2 | 1.2 | 13.1 | 33.3 | 40.5 | 46.4 | 10.7 | 63.3 | 76.7 | 83.3 | 88.3 |
| | IF | 7.1 | 7.1 | 8.3 | 11.9 | 11.9 | 4.8 | 13.3 | 15.0 | 16.7 | 26.7 |
| $t$ | B1 | 4.8 | 96.4 | 96.4 | 97.6 | 98.8 | 95.2 | 95.0 | 98.3 | 98.3 | 100 |
| | B2 | 10.7 | 90.5 | 96.4 | 96.4 | 97.6 | 91.7 | 91.7 | 95.0 | 95.0 | 100 |
| | IF | 10.7 | 88.1 | 90.5 | 96.4 | 96.4 | 89.3 | 83.3 | 90.0 | 93.3 | 95.0 |

**Table 7.5** Test results of the proposed procedure from baselines No. 1–10 (unit: %)

| | | Trimble | | | | | Leica | | | | |
|---|---|---|---|---|---|---|---|---|---|---|---|
| | | No. 1 | No. 2 | No. 3 | No. 4 | No. 5 | No. 6 | No. 7 | No. 8 | No. 9 | No. 10 |
| Stationarity | A | 100 | 99.4 | 92.3 | 84.5 | 67.3 | 100 | 96.7 | 80.0 | 55.0 | 35.0 |
| | B | 100 | 100 | 100 | 100 | 100 | 100 | 100 | 100 | 100 | 100 |
| Zero-mean normality | A | 91.7 | 5.4 | 2.4 | 1.8 | 1.8 | 6.6 | 2.5 | 1.7 | 0.8 | 0 |
| | B | 88.1 | 11.9 | 8.3 | 3.6 | 3.6 | 10.7 | 16.7 | 8.3 | 6.7 | 5.0 |

signals are significant, it indicates that there is misspecification between measurements and functional model. Therefore, the functional model compensation should be applied, i.e., modeling or parameterization the observations with significant deterministic signals. On the other hand, when the colored noises are significant, it means that the common-used stochastic model with ignoring the physical correlations does not match the reality. Hence the physical correlations should be considered into the stochastic model. In conclusion, if the proposed procedure is applied, one can know when and how the functional and stochastic models should be improved.

Then the effectiveness of the proposed procedure is validated by analyzing the properties of unmodeled errors, where the impacts of receiver, multipath and atmosphere are all discussed in detail. We divide the baselines into three types: zero baseline, ultra-short baseline and longer baseline, and analyze the results in detail.

For the results of zero baseline. We first analyze the zero-baseline data (No. 1), where the DD residuals with model A are shown in Fig. 7.3. The results of model B are not shown here because the ionospheric delays completely vanish on the zero baseline. Each color denotes one DD satellite pair. It can be easily seen that the DD residuals are highly random. This is because the external errors can be completely eliminated for zero baseline, only the receiver-specific white noises are remained [37, 38].

To further confirm that the zero-baseline residuals are of zero-mean normality and only receiver-dependent, the graphical and statistical analyses are carried out. Unlike the other GNSS satellite systems, BDS has three types of orbiting satellites, i.e., the geostationary earth orbit (GEO) satellites, the inclined geosynchronous satellite orbit

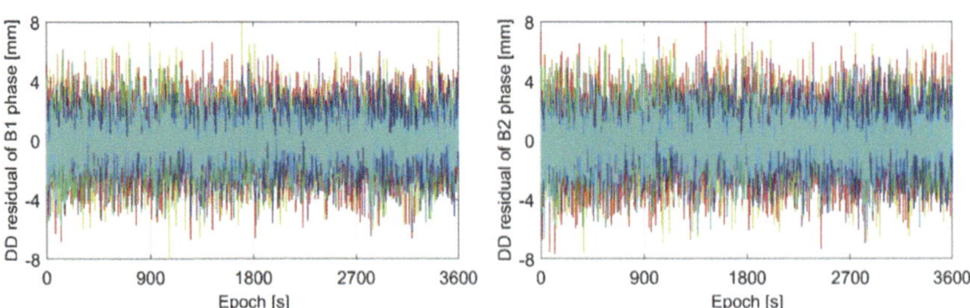

**Fig. 7.3** The DD residuals of zero baseline with model A

(IGSO) satellites, and the medium earth orbit (MEO) satellites. The PRN 7 is selected as the reference satellite, then the B1 observations of GEO PRN 3, IGSO PRN 10 and MEO PRN 11 are used as example. Figure 7.4 presents the histograms of their DD residuals and the corresponding fitted curves with zero-mean normal distributions. The graphical results obtained from B2 observations are quite similar to those from B1 observations although they are not shown here. It is confirmed from Fig. 7.4 that the distributions of zero-baseline DD residuals are rather close to the zero-mean normal distributions.

The sample moments are applied to quantitatively demonstrate the zero-mean normality of zero-baseline DD residuals. Table 7.6 lists the statistics of mean, STD, sekeness and kurtosis based on the DD residuals of all satellites. Typically, for a normal distribution, the theoretical values of skewness and kurtosis are 0 and 3, respectively. It is clear that for all three types of orbiting satellites, the computed statistics are all very close to the theoretical values of zero-mean normal distribution. It is therefore concluded that there are only the white noises on zero baseline. This result is highly consistent to the test results with the proposed procedure, thus certifying the validity of the testing procedure.

The unit of mean and STD is mm, while the unit of skewness and kurtosis is scale.

For the results of ultra-short baseline. The data of ultra-short baseline (No. 6) is analyzed. Only the results of model A are analyzed because the atmospheric effects can be basically eliminated, and the multipath would be the dominant error source on the ultra-short baseline of 5 m in our study [38]. The code multipath can be estimated by using the multipath combination function. It can be extracted by using one frequency code and two frequency phase observations to form the geometry-free and IF combination. The code multipath function $M$ on frequency $i$ reads

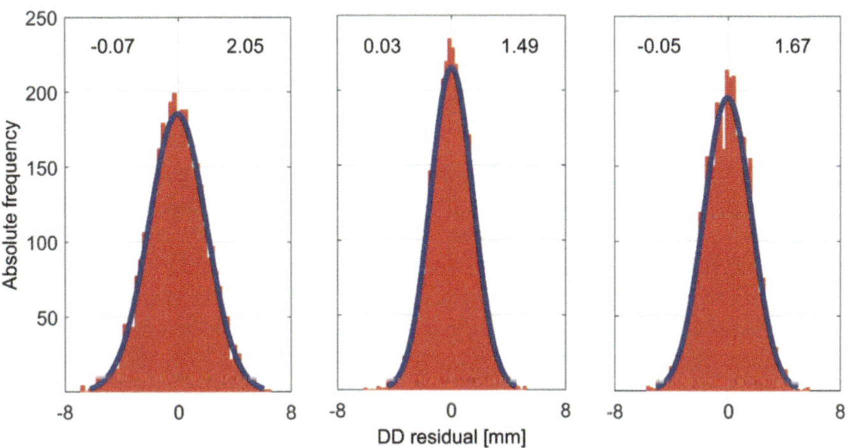



**Fig. 7.4** The histograms of zero-baseline DD residuals of GEO satellite PRN 3 (left), IGSO satellite PRN 10 (middle) and MEO satellite PRN 11 (right), and their corresponding fitted curves with zero-mean normal distributions (red lines). The numbers in the top left and top right corners indicate the corresponding means and STDs in unit of mm, respectively

**Table 7.6** The statistics of mean, STD, skewness and kurtosis of all DD residuals on zero baseline

| | GEO satellites | | IGSO satellites | | MEO satellites | |
|---|---|---|---|---|---|---|
| | B1 | B2 | B1 | B2 | B1 | B2 |
| Mean | − 0.04 | 0.00 | − 0.02 | − 0.02 | − 0.05 | − 0.03 |
| STD | 1.86 | 1.99 | 1.60 | 1.74 | 1.66 | 1.81 |
| Skewness | 0.01 | 0.02 | 0.00 | 0.02 | − 0.04 | − 0.03 |
| Kurtosis | 3.03 | 3.00 | 3.10 | 3.03 | 3.00 | 3.04 |

$$M_i = P_i - \frac{f_i^2 + f_j^2}{f_i^2 - f_j^2}\Phi_i + \frac{2f_j^2}{f_i^2 - f_j^2}\Phi_j \qquad (7.48)$$

where the subscripts $i$, $j$ ($i \neq j$) denote two frequencies; $f$ is the carrier phase frequency; $P$ and $\Phi$ are the code and phase measurements in unit of length, respectively. $M_i$ contains not only multipath on $P_i$, but also the terms of phase ambiguities and the relevant hardware delays. Then the code multipath $MP_i$ is derived as

$$MP_i = M_i - \overline{M}_i \qquad (7.49)$$

where $\overline{M}_i$ is the mean value of $M_i$ over a certain period. Without cycle slips, $MP_i$ is dominated by the code multipath. Hence, the multipath combination function can be applied to assess the multipath effects of single satellite. Note the BDS may suffer from the elevation-dependent code hardware variations. These satellite-induced code biases that are dependent on the elevations in this study have been tested to be insignificant on multipath estimation mainly due to the small elevation variation in duration of 1 h. Thus, the satellite-induced variations are ignored here. For saving the space, we illustrate the multipaths of satellites PRN 01 and PRN 10, and their corresponding DD residuals in Fig. 7.5, representing the weak and strong multipath, respectively.

Obviously, the DD residuals in Fig. 7.5 are much different from those in Fig. 7.3. The DD residuals with multipath effects are all not purely white nosies, and the systematic errors exist. The stronger the multipath is, the more systematic the DD residuals are. Such results are are again consistent with the analysis, and the unmodeled errors in such ultra-short baseline are indeed induced by the multipath.

To further validate the former conclusions in this section, the time-domain Allan variance is applied to study the noise characteristics of unmodeled errors. The Allan variance has been widely used for identifying the noise types. The non-overlapped Allan variance $\sigma_{\overline{Y}}^2$ is defined as

$$\sigma_{\overline{Y}}^2(T) = \frac{1}{2(N-1)} \sum_{i=1}^{N-1} \left(\overline{Y}_{i+1} - \overline{Y}_i\right)^2 \qquad (7.50)$$

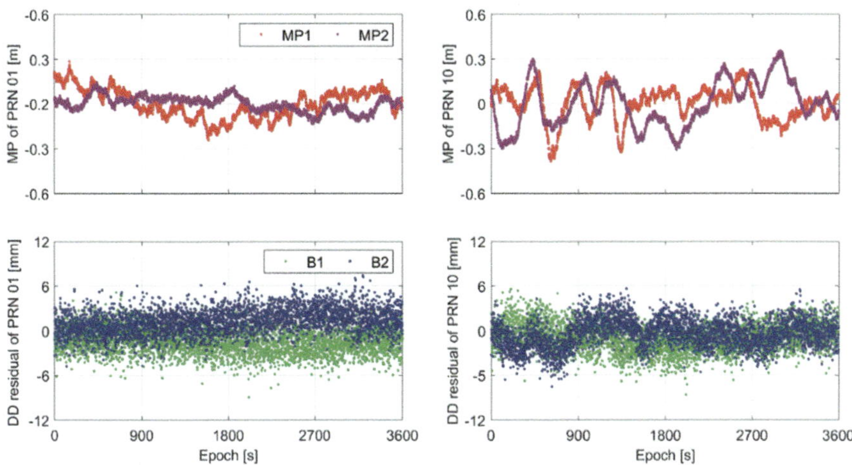

**Fig. 7.5** The weak (left) and strong (right) code multipaths and corresponding DD residuals for satellites PRN 01 and PRN 10 on ultra-short baseline, respectively

where $\overline{Y}_i$ is the $i$th mean value of $N$ fractional frequency values averaged over the cluster time $T$. According to a log-log plot of Allan STD, the noise types with their magnitudes can be determined. In GNSS applications, the Gaussian white noise, random walk noise, flicker noise and the first order GM process are four types of important stochastic processes. They are characterized by the regions of $-1/2, +1/2, 0$ and $\pm 1/2$ slopes, respectively [39, 40]. The strength of Gaussian white noise can be also estimated by $\sigma(1)$.

The Allan plots of DD residuals with weak and strong multipaths are shown in Fig. 7.6. They exhibit the similar patterns between B1 and B2 frequencies. It means that the noise characteristics are highly similar between two frequency observations. Due to the straight lines with the slopes of approximately $-1/2$ at the beginning, the white noise can be easily identified as the dominant error component with strength of $1.5\,\mathrm{mm}/\sqrt{\mathrm{Hz}}$ for cluster time $T < 128\,\mathrm{s}$ with weak multipath and $T < 16\,\mathrm{s}$ with strong multipath, respectively. For the rest cluster time with these two types of multipath impacts, the first order GM process is identified to some extent. Consequently, the white noise and the first order GM process are both existent in unmodeled errors, and the first order GM process may play a more important role in case of strong multipath since the cluster time of white noise in case of stong multipath is shorter. It is concluded that the noise characteristics with multipath are more complicated than the zero baseline results, which is also consistent with the analysis of Fig. 7.5. Besides, the strength of the white noise are almost the same under different conditions of multipath, thus indicating the influence of the white noise mainly comes from the receivers, which provides substantial evidence for the former results of the zero baseline.

The frequency-domain FFT is applied to identify the components of deterministic signals in unmodeled errors. FFT can reveal important frequency components of a

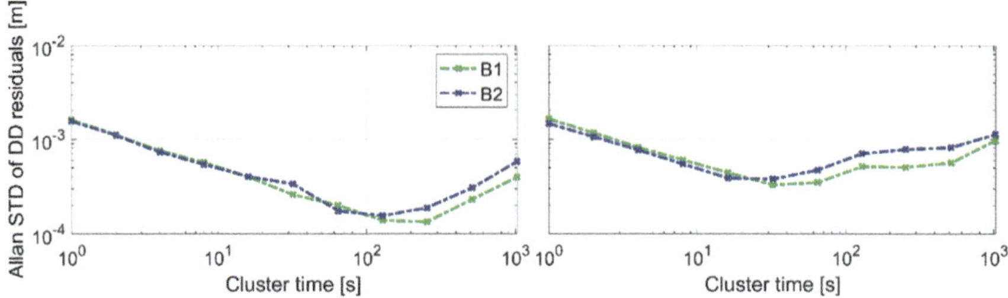

**Fig. 7.6** The Allan plots of DD residuals with weak multipath (left) and strong multipath (right) on ultra-short baseline

signal. The discrete form of the FFT is defined as follows [41]

$$p_k = \sum_{j=1}^{N} q_j w_N^{(j-1)(k-1)} \tag{7.51}$$

where the vector $p$ is the Fourier transform of a vector $q$ with length $N$. $j$ and $k$ are the indices that run from 1 to $N$; $w_N = e^{(-2\pi i)/N}$ is one of $N$ roots of unity with $i$ the imaginary unit. According to the FFT plot, the frequency range varies from the fundamental frequency (i.e., the $1/T$, with $T$ the observation length) to the Nyquist frequency that is the half of sample rate [42]. Then the interested sections with their corresponding amplitude spectra can be identified. For the deterministic signals, the frequency can be identified by the peaks of the FFT plots.

The FFT plots of DD residuals with weak and strong multipaths are illustrated in Fig. 7.7. Since 1-h GNSS unmodeled error series is employed with the sample rate of 1 Hz, the spectrum falls between $2.8 \times 10^{-4}$ and $5.0 \times 10^{-1}$ Hz. A significant discrepancy is found between weak and strong multipaths. Specifically, the amplitudes of the intermediate segment $5.0 \times 10^{-4} < f < 3.0 \times 10^{-3}$ with strong multipath are frequently close to or even higher than 0.5 mm. The results suggest that there will exhibit some insignificant deterministic signals (since the highest amplitude spectra still lie in the areas of fundamental frequency) in case of strong multipath.

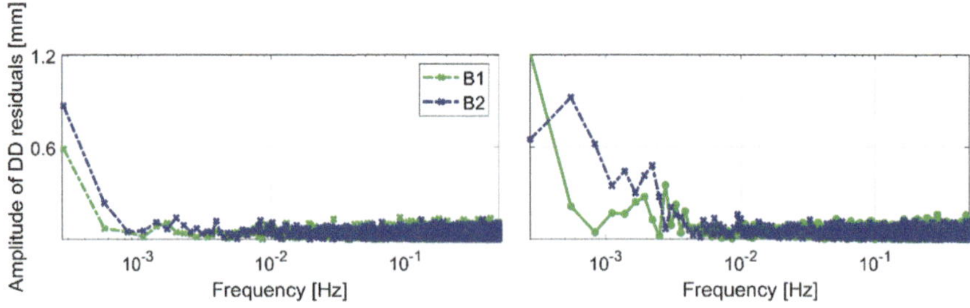

**Fig. 7.7** The FFT plots of DD residuals with weak multipath (left) and strong multipath (right) on ultra-short baseline

To summarize, as expected, the multipath is closely related to the unmodeled errors. When the multipath is weak, the colored noise is the dominant component; whereas the more complicated colored noise and deterministic signals may be both exsistent in case of strong multipath although the deterministic signals could be not as significant as colored noises. Besides, all the results certificate that suffering from the multipath effects, the proposed procedure can accurately and efficiently identify the components of the unmodeled errors.

For the results of longer baselines, finally, the longer baselines (from No. 2–5 and No. 7–10) with baseline lengths from 10 to 50 km are analyzed, where the atmospheric effets will become the dominant error source. Therefore, the models A and B are both studied. Since the baseline data are collected at the same time and small area, they should suffer from the similar atmospheric effects. The atmospheric effects consist of tropospheric and ionospheric effects. Analogously, to validate the results of the proposed procedure, the Allan plots of DD residuals of PRN 6 computed from baselines No. 7–10 are shown in Fig. 7.8. The white noise, flicker noise, random walk and the first order GM process can all be identified at the beginning and the intermediate segments, especially for the B1 and B2 data. The results are more complicated than those with purely multipath on ultra-short baseline No. 6. For the ending part, it is hard to identify the noise type due to the large uncertainty of the Allan variances. Besides, comparing the Allan plots between B1, B2 and IF modes, their shapes have no apparent difference, which agrees with the former analysis. For the same model, the shapes of Allan plots are similar for different baselines. It implies the noise characteristics are not necessrily dependent on the magnitudes of atmospheric effects. In fact, this point can be used to explain why the temporal correlations caused by unmodeled errors can be fitted by some empirical functions that are free from the baseline length [1]. As a result, compared to the situation with pure multipath, the components of unmodeled errors become more complicated in the situation of atmospheric effects, whereas the colored noise has weak dependence on the magnitute of atmospheric effects.

The FFT plots of DD residuals of PRN 6 computed from baselines No. 7–10 are shown in Fig. 7.9, where the deterministic terms can be easily identified. Unlike the results in case of pure multipath on ultra-short baseline, for the B1 and B2 data, the deterministic signals seem much more serious due to their larger amplitude spectra. Judging from Fig. 7.9, the deterministic signals with model A are much more significant in comparison with model B. These results agree again with the former analysis. In addition, the deterministic signals seem more significant with the increasing of baseline length. In summary, under the strong atmospheric conditions, the DD residuals are not stationary anymore. Furthermore, the deterministic signals will be influenced more easily than the colored noise when the strength of atmosphere effects changes.

To summarize, the atmospheric effect is the main attribution of unmodeled errors. At this time, the unmodeled errors will become not stationary. Therefore, it proves that the proposed procedure is indeed effective with high efficiency.

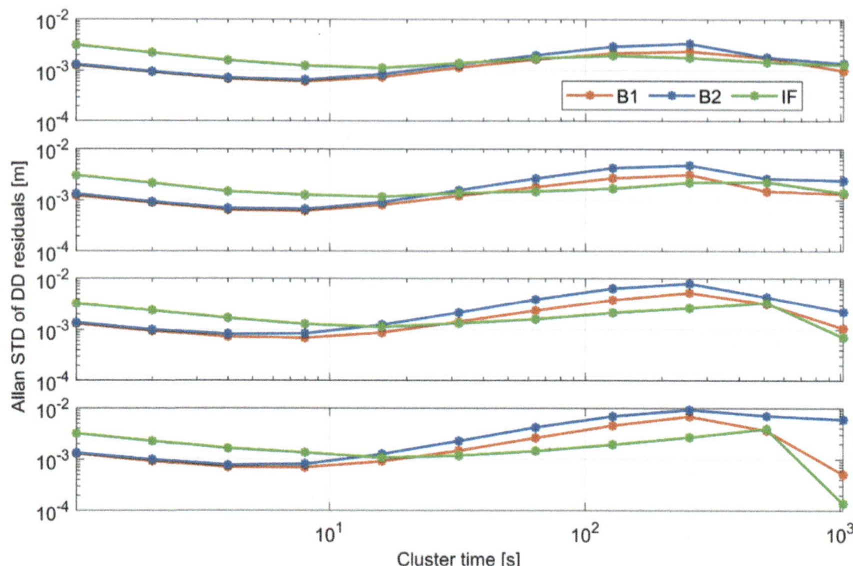

**Fig. 7.8** The Allan plots of DD residuals with models A (for which both B1 and B2 Allan STDs are analyzed) and B (for which IF Allan STDs are analyzed) for satellite PRN 6 on baselines No. 7–10 (from top to bottom)

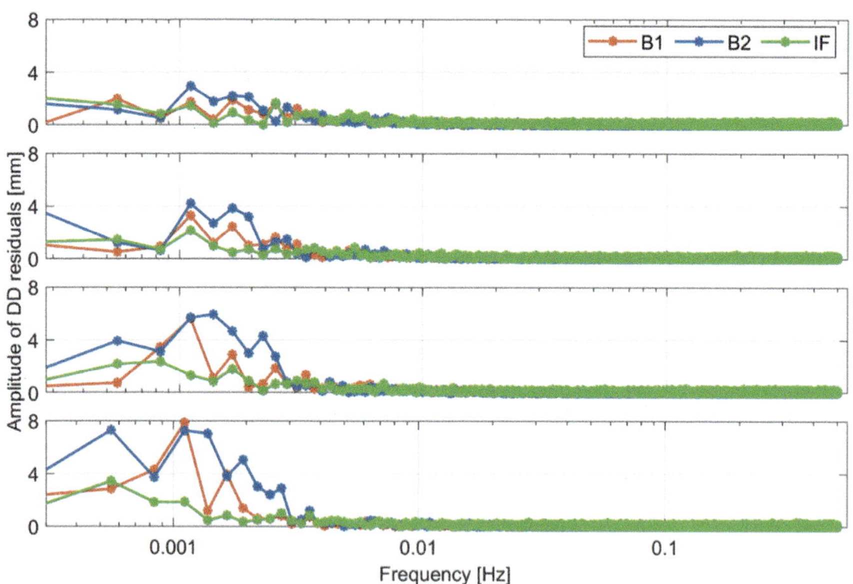

**Fig. 7.9** The FFT plots of DD residuals with model A (for which both B1 and B2 results are analyzed) and B (for which IF results are analyzed) for satellite PRN 6 on baselines No. 7–10 (from top to bottom)

## 7.4.2 Results and Discussion of Unmodeled Error Compensation

In order to validate the performance of the proposed multi-epoch partial parameterization method, 2-h single-frequency GPS/BDS observations were collected using Leica GR25 receivers with a sampling interval of 1 s at a network consisting four stations in the same area, and the observations are obtained at the same time on 2 April, 2016. The RTK is taken as an example, and six baselines were formed with lengths from approximately 5–25 km from these four stations A, B, C and D. The coordinates of all the stations are precisely known and served as ground truths. The traditional single-epoch RTK and the proposed RTK methods were implemented in self-developed RTK software for the precise multi-frequency and multi-GNSS RTK processing. The processing strategies of the traditional and proposed RTK methods are presented in Table 7.7, where the ambiguities were fixed by the LAMBDA method. It can be seen that the processing strategies are quite common and the same for these two different methods, thus ensuring the reliability of the analysis later.

When the proposed method is used, the elevation is chosen as the indicator that determines the order of parameterization where the unmodeled errors are significant. The main reason is that, as usual, when the elevation is lower, the unmodeled errors are more easily significant mainly because the atmospheric delays have the longer propagation path. Additionally, the elevation can be computed by real time which is also easy-to-implement. For the issue that how many consecutive epochs of unmodeled errors can be regarded highly correlated during the partial parameterization, we set 5 epochs as the width of moving window if the sampling rate is 1 Hz. The reason is that, in this situation, the temporal correlations between two consecutive epochs are usually higher than 0.98 or even 0.99 [1]. Hence, when the width of moving window is 5 epochs, the temporal correlation between the first and fifth observations are still higher than 0.9. Therefore, the unmodeled errors can be regarded the same within a short time (i.e., 5 s in this study).

**Table 7.7** Common processing strategies of the traditional and proposed RTK methods

|  | Processing strategies |
| --- | --- |
| Used observations | DD pseudorange observations |
|  | DD carrier phase observations |
| Cut-off elevation | $8°$ |
| Strategy of ambiguity resolution | LAMBDA |
| Troposphere correction | Modified Hopfield model |
| Ionosphere correction | Ionosphere-fixed model |
| Weighting function | Elevation model |
| Observation variances | UD pseudorange observations: 0.2 m |
|  | UD carrier phase observations: 2 mm |

Figure 7.10 illustrates the positioning errors of the traditional and proposed methods for six baselines, where the blue, green and red dots denote the east, north and up directions, respectively. The results clearly show that the positioning precisions are indeed improved by the proposed method for all these six baselines. Hence, it demonstrates the effectiveness of the multi-epoch partial parameterization. A closer look at the positioning results, the precisions of up direction are improved more significantly than the other directions. The reason may be that the unmodeled errors have high correlations with the positioning solutions of the up direction. Therefore, when the unmodeled errors are mitigated, the positioning solutions of the up direction are improved significantly.

Figures 7.11 and 7.12 illustrate the DD GPS/BDS code and phase residuals of traditional and proposed methods for six baselines, respectively. It is worth noting that each color denotes one satellite pair. Compared with the observation residuals from the traditional method, the ones from the proposed method are more stationary, especially for the phase ones. It can be further seen that, for some certain satellite pairs, the observation residuals with large fluctuations are mitigated to a great extent after using the proposed method. Hence, the significant unmodeled errors are proved to be mitigated at this time.

Figure 7.13 illustrates the code and phase unmodeled error corrections based on the proposed method for six baselines, where each color denotes one satellite pair. It can be clearly seen that these code and phase unmodeled errors can be up to 4 m and 60 mm or larger, respectively. Hence, these unmodeled effects cannot be simply ignored, otherwise they will have adverse impacts on GNSS positioning. Judging from the behaviors of these unmodeled error parameters, they are highly correlated

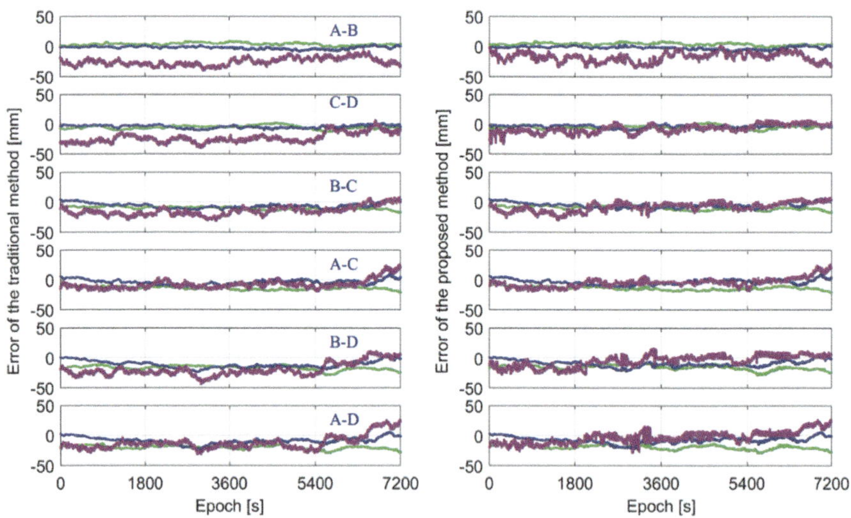

**Fig. 7.10** Positioning errors of the traditional (left) and proposed (right) methods for six baselines. The blue, green and red dots denote the east, north and up directions, respectively

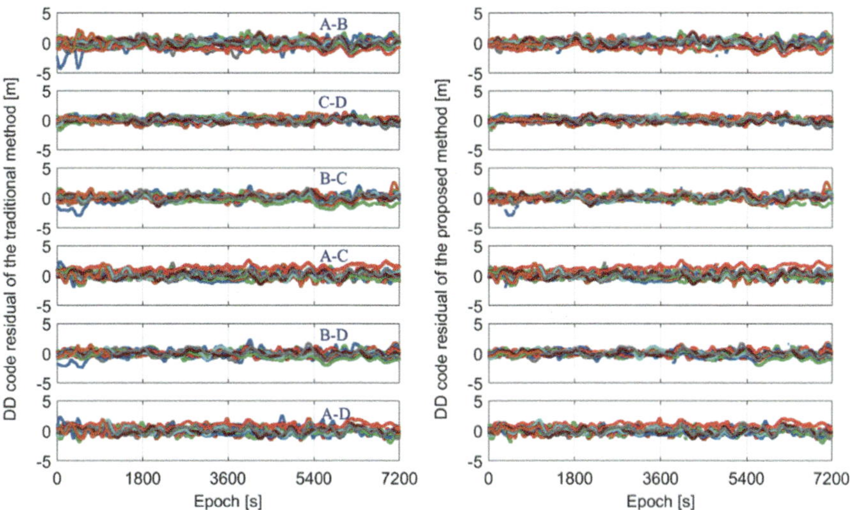

**Fig. 7.11** DD GPS/BDS code residuals of the traditional (left) and proposed (right) methods for six baselines. Each color denotes one satellite pair

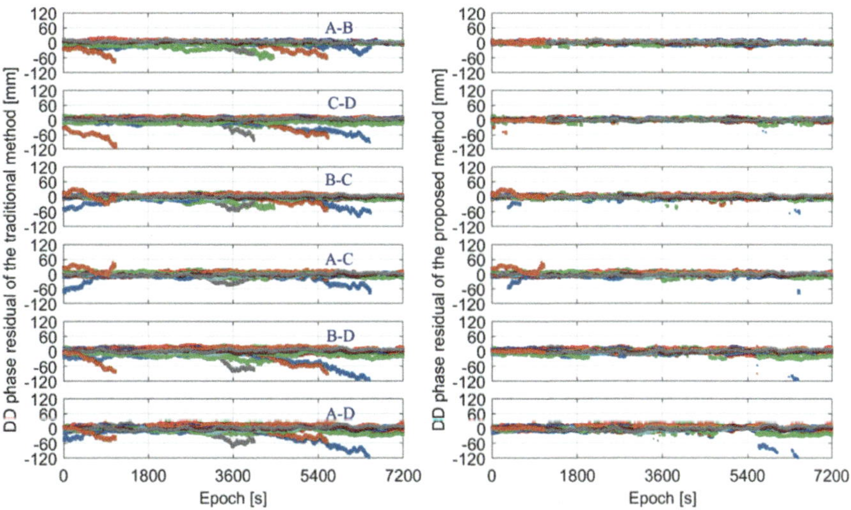

**Fig. 7.12** DD GPS/BDS phase residuals of the traditional (left) and proposed (right) methods for six baselines. Each color denotes one satellite pair

within a short time, just like a deterministic signal as previously analyzed. That is, the multi-epoch partial parameterization can indeed work for capturing these significant unmodeled effects.

Table 7.8 shows the three-dimensional (3D) bias and 3D root mean square (RMS) of the traditional and proposed methods for six baselines. It can be clearly seen that,

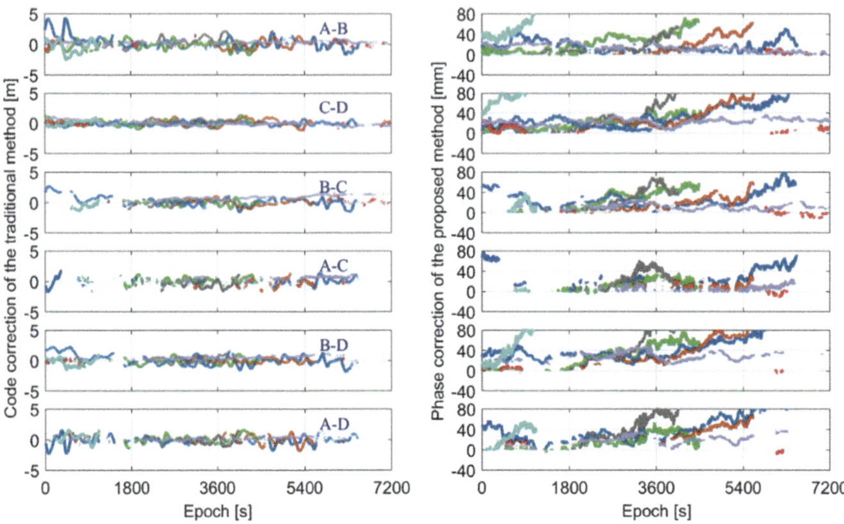

**Fig. 7.13** Unmodeled error corrections based on the proposed method for six baselines. Each color denotes one satellite pair

compared with the traditional method, the 3D biases of the proposed method are all smaller. Similarly, the 3D RMSs of the proposed method is significantly smaller than the ones of the traditional method. Specifically, the mean improvements of the 3D bias and 3D RMS are 28.79% and 24.22%, respectively. It can be found that the precision of up direction is improved to a great extent, where the precision can be increased by up to 56.90%. It indicates that the unmodeled effects are mitigated and have large dependence on the up direction, which is consistent with the aforementioned analysis. In conclusion, according to the experiment results and corresponding analysis in this section, the proposed method can indeed mitigate the significant unmodeled errors in real time.

**Table 7.8**  3D bias and 3D RMS of the traditional and proposed methods for six baselines (mm)

| Error | Method | A-B | C-D | B-C | A-C | B-D | A-D |
|---|---|---|---|---|---|---|---|
| 3D bias | Traditional method | 25.94 | 24.74 | 19.30 | 16.43 | 27.95 | 25.18 |
|  | Proposed method | 18.41 | 10.30 | 13.77 | 15.05 | 18.86 | 21.21 |
| 3D RMS | Traditional method | 26.93 | 26.42 | 21.35 | 18.95 | 30.85 | 28.38 |
|  | Proposed method | 20.28 | 13.13 | 16.57 | 17.75 | 22.34 | 24.41 |

## 7.5 Conclusion

The GNSS unmodeled errors are studied systematically and this research is a solid foundation to further improve the accuracy and reliability of GNSS applications. A testing procedure has been proposed to test the significance of unmodeled errors in GNSS observations. It is proved to be effective and feasible with the various simulated and real data as well as time-domain Allan variance analysis and frequency-domain FFT. Besides, in GNSS positioning, the unmodeled errors are objectively existent and cannot be easily eliminated since they are time- and space-variable. The attributions of these unmodeled effects mainly come from the ionosphere, troposphere and multipath. Besides, the receiver may also have an impact on the behaviors of the unmodeled errors. Three dominant components are identified in GNSS unmodeled error: nonstationary signal, stationary signal and white noise. They can be largely understood as the deterministic signal (including trend and periodic terms), colored noise (including random walk noise, flicker noise and the first order GM process) and Gaussian white noise, respectively. The magnitudes of unmodeled errors are positively correlated with the multipath and atmospheric effects. Then the unmodeled errors with different frequencies share similar patterns. Under the conditions of multipath, the colored noise is the dominant error source for unmodeled errors. Then the ionospheric and tropospheric effects are the main attributions of unmodeled errors since these effects will become not stationary. Besides, unlike the multipath, the atmospheric effects have marginal effects on noise characteristics and will influence considerably the deterministic signals. As future work, under different application modes (e.g., RTK, PPP) and environments, once the error types are identified with the Li's procedure, an open problem raises for developing their own specific compensation methods.

Also, we present a method for real-time mitigating the unmodeled errors in GNSS precise positioning, which is especially suitable when there are enough redundant observations. In single-frequency and multi-GNSS scenario, since the ionospheric delays cannot be mitigated by the IF combination in terms of two or more carrier phases with different frequencies, the unmodeled effects certainly will be significant more frequently. Therefore, the corresponding unmodeled error mitigation is urgently needed in this situation. In essence, the proposed method is mainly based on the multi-epoch partial parameterization, where only the significant unmodeled errors are captured. It is worth noting that the proposed can also be used in post time, hence it can be introduced to RTK, PPP or any other positioning modes. According to the experiment and analysis, the results show that this proposed method is effective and can improve the positioning precision significantly. In addition, because the unmodeled errors have high dependence on the positioning results of the up direction, the proposed method can improve the precision of the up direction to a great extent.

# References


1. Zhang Z, Li B, Shen Y (2017) Comparison and analysis of unmodelled errors in GPS and BeiDou signals. Geod Geodyn 8:41–48
2. Schön S, Wieser A, Macheiner K (2005) Accurate tropospheric correction for local GPS monitoring networks with large height differences. Proceedings of the ION GNSS 2005, pp 250–260
3. Hoque MM, Jakowski N (2007) Higher order ionospheric effects in precise GNSS positioning. J Geod 81:259–268
4. Schüler T (2006) Impact of systematic errors on precise long-baseline kinematic GPS positioning. GPS Solut 10:108–125
5. Bock Y, Nikolaidis RM, de Jonge PJ, Bevis M (2000) Instantaneous geodetic positioning at medium distances with the global positioning system. J Geophys Res Solid Earth 105:28223–28253
6. Ragheb AE, Clarke PJ, Edwards SJ (2007) GPS sidereal filtering: coordinate- and carrier-phase-level strategies. J Geod 81:325–335
7. Zhong P, Ding X, Yuan L, Xu Y, Kwok K, Chen Y (2010) Sidereal filtering based on single differences for mitigating GPS multipath effects on short baselines. J Geod 84:145–158
8. Moore M, Watson C, King M, McClusky S, Tregoning P (2014) Empirical modelling of site-specific errors in continuous GPS data. J Geod 88:887–900
9. Fuhrmann T, Luo X, Knöpfler A, Mayer M (2015) Generating statistically robust multipath stacking maps using congruent cells. GPS Solut 19:83–92
10. Dong D, Wang M, Chen W, Zeng Z, Song L, Zhang Q, Cai M, Cheng Y, Lv J (2016) Mitigation of multipath effect in GNSS short baseline positioning by the multipath hemispherical map. J Geod 90:255–262
11. Bilich A, Larson KM, Axelrad P (2008) Modeling GPS phase multipath with SNR: case study from the Salar de Uyuni, Boliva. J Geophys Res Solid Earth 113.
12. Zhong P, Ding XL, Zheng DW, Chen W, Huang DF (2008) Adaptive wavelet transform based on cross-validation method and its application to GPS multipath mitigation. GPS Solut 12:109–117
13. Phan QH, Tan SL, McLoughlin I (2013) GPS multipath mitigation: a nonlinear regression approach. GPS Solut 17:371–380
14. Lau L, Cross P (2007) Development and testing of a new ray-tracing approach to GNSS carrier-phase multipath modelling. J Geod 81:713–732
15. Hopfield HS (1969) Two-quartic tropospheric refractivity profile for correcting satellite data. J Geophys Res 74:4487–4499
16. Saastamoinen J (1973) Contributions to the theory of atmospheric refraction. Bull Geod 107:13–34
17. Leandro RF, Langley RB, Santos MC (2008) UNB3m_pack: a neutral atmosphere delay package for radiometric space techniques. GPS Solut 12:65–70
18. Li B, Feng Y, Shen Y, Wang C (2010) Geometry-specified troposphere decorrelation for subcentimeter real-time kinematic solutions over long baselines. J Geophys Res Solid Earth 115
19. Schüler T, Diessongo H, Poku-Gyamfi Y (2011) Precise ionosphere-free single-frequency GNSS positioning. GPS Solut 15:139–147
20. Weng D, Ji S, Chen W, Li Z, Xu Y, Ye L (2015) Assessing and mitigating the effects of the ionospheric variability on DGPS. GPS Solut 19:107–116
21. Tinin MV (2015) Eliminating diffraction effects during multi-frequency correction in global navigation satellite systems. J Geod 89:491–503
22. Tiberius C, Borre K (1999) Probability distribution of GPS code and phase data. Z Vermess 124:264–273
23. Luo X, Mayer M, Heck B (2011) On the probability distribution of GNSS carrier phase observations. GPS Solut 15:369–379
24. Zhang Z, Li B, Shen Y (2018) Efficient approximation for a fully populated variance-covariance matrix in RTK positioning. J Surv Eng 144:04018005

25. Paziewski J, Wielgosz P (2015) Accounting for Galileo–GPS inter-system biases in precise satellite positioning. J Geod 89:81–93
26. Teunissen PJG (2007) Best prediction in linear models with mixed integer/real unknowns: theory and application. J Geod 81:759–780
27. Yu W, Ding X, Dai W, Chen W (2017) Systematic error mitigation in multi-GNSS positioning based on semiparametric estimation. J Geod 91:1491–1502
28. Bischoff W, Heck B, Howind J, Teusch A (2005) A procedure for testing the assumption of homoscedasticity in least squares residuals: a case study of GPS carrier-phase observations. J Geod 78:397–404
29. Hamilton JD (1994) Time series analysis. Princeton University Press, New Jersey
30. Didova O, Gunter B, Riva R, Klees R, Roese-Koerner L (2016) An approach for estimating time-variable rates from geodetic time series. J Geod 90:1207–1221
31. Said SE, Dickey DA (1984) Testing for unit roots in autoregressive-moving average models of unknown order. Biometrika 71:599–607
32. Dickey D, Fuller W (1979) Distribution of the estimators for autoregressive time series with a unit root. J Am Stat Assoc 74:427-431
33. Jarque CM, Bera AK (1987) A test for normality of observations and regression residuals. Int Stat Rev 55:163–172
34. Luo X (2013) GPS stochastic modelling: signal quality measures and ARMA processes. Springer, Berlin
35. Lilliefors HW (1967) On the Kolmogorov-Smirnov test for normality with mean and variance unknown. J Am Stat Assoc 62:399–402
36. Teunissen P, Khodabandeh A (2013) BLUE, BLUP and the Kalman filter: some new results. J Geod 87(5):461–473
37. Li B (2016) Stochastic modeling of triple-frequency BeiDou signals: estimation, assessment and impact analysis. J Geod 90:593–610
38. Zhang Z, Li B, Shen Y, Yang L (2017) A noise analysis method for GNSS signals of a standalone receiver. Acta Geod Geophys 52:301–316
39. IEEE (2008) Standard specification format guide and test procedure for single-axis laser gyros, IEEE Std
40. IEEE (2008) Standard specification format guide and test procedure for linear, single-axis, non-gyroscopic accelerometers, IEEE Std
41. Frigo M, Johnson SG (1998) FFTW: an adaptive software architecture for the FFT. Proceedings of the 1998 IEEE international conference on acoustics, speech and signal processing, pp 1381–1384
42. Jenkins WK (1999) Fourier series, Fourier transforms and the DFT. CRC Press, Florida

# Chapter 8
# Data Quality Control

## 8.1 Introduction

The proliferation of Global Navigation Satellite Systems (GNSSs) has driven widespread adoption of Real-Time Kinematic (RTK). Nevertheless, the presence of outliers in GNSS measurements poses a significant threat to the achievable accuracy and reliability of RTK solutions. Consequently, effective identification and exclusion of these erroneous measurements are critical for robust positioning.

Two principal methodological approaches exist for handling outliers: detection, identification, and adaptation (DIA) and robust estimation. DIA procedure primarily employs hypothesis testing theory to distinguish between a null hypothesis (no outliers) and one or more alternative hypotheses (specifying potential outliers). These methods are explicitly designed within the functional model framework, requiring assumptions about the nature, quantity, and location of potential outliers. Despite its widespread use, the DIA method faces inherent limitations in practical implementation. It cannot guarantee infallible testing decisions or unbiased parameter estimates. Challenges such as missed detections, false alarms, and incorrect exclusions inevitably arise due to factors including the underlying measurement geometry, the distinguishability (separability) between competing hypotheses, the chosen test statistics, and the predetermined critical values governing the tests.

Robust estimation, conversely, operates by modifying the stochastic model through equivalent weight functions. These functions aim to mitigate or eliminate the prejudicial influence of suspect observations on the final parameter estimates. Such robust techniques are extensively applied across various GNSS data processing domains, including deformation analysis, least-squares collocation, and Kalman filtering. Numerous enhanced robust estimators have been developed to achieve greater robustness and higher breakdown points, such as the median method, least trimmed squares, and sign-constrained robust least-squares. Among these, the robust M-estimator stands out due to its practical advantages: ease of implementation and computational efficiency. M-estimation is versatile, applicable to both independent



B. Li et al., *GNSS Real-Time Kinematic Positioning*, Navigation: Science and Technology 17, https://doi.org/10.1007/978-981-96-9116-6_8

and correlated measurements using appropriate reweighting schemes. For handling correlated GNSS data in RTK, the bifactor equivalent weight function, particularly following the Institute of Geodesy and Geophysics 3 (IGG3) scheme, is a standard choice. The IGG3 approach classifies potentially problematic measurements into three categories: 1) complete exclusion, 2) downweighting, or 3) retention with full weight.

This chapter first introduces two categories of outlier detection methods. The first category focuses on the stochastic model, aiming to mitigate the impact of outliers on parameter estimation. The second category is based on the functional model, targeting the elimination of outliers influence on parameter estimation. Specifically, robust estimation methods are commonly applied in the posterior stage to process outliers in the stochastic model. For the functional model, the DIA method is utilized to handle outliers.

## 8.2  Mitigation and Elimination of GNSS Outliers

GNSS observations often inevitably contain outliers in harsh environments. Outliers or misspecifications in the functional model generally result in biased least-squares (LS) estimators.

When the observations contain outliers, the outlier model is usually denoted as follows

$$y = Ax + \Delta + \epsilon \tag{8.1}$$

where $A$ and $y$ denote the design matrix of the functional model and observation vector, respectively. The $x$ is the estimated vector of parameters. Besides, $\epsilon$ and $\Delta$ are the measurement noise and the outlier, respectively.

The outlier should be processed and removed and it relates to the reliability theory in the processing. The reliability theory includes internal reliability and external reliability, which exhibit the ability to detect the outliers and resist the influence of undiscoverable outliers, respectively. Besides, there are essentially two main ideas for outlier detection. Firstly, when the outliers are non-stochastic, the mean shift model is researched and eliminated as the error model. It is a perspective that the outliers can affect the mean of observations. Secondly, when the outliers are stochastic, it is another perspective that the outliers can affect the variance of observations. Therefore, the variance inflation model is introduced as the error model. Based on the above descriptions, the outlier detection correspondingly has two major classes of methods. The first one is data snooping, which classifies the outlier to detect and eliminate based on the functional model. At this time, relatively pure observations are obtained and they meet the conditions of LS estimation. The second one is the robust estimation method. The outlier is introduced into the stochastic model. When there exists an outlier, the corresponding weight of this observation is set as zero to exclude the effect of the outlier. The adjustment results are successively iterated and the method requires an adjustment factor $R$, which is as follows

$$v = Ry \tag{8.2}$$

$$R = \left(P^{-1} - A\left(A^{\mathrm{T}}PA\right)^{-1}A^{\mathrm{T}}\right)P \tag{8.3}$$

where $P$ donates the weight matrix. The adjustment factor is the geometric condition of the adjustment, and it reflects the effect of the observation on the residuals.

Missed detection, false alarm, and wrong identification, usually cannot be avoided due to the geometry of the observation model in the DIA method [1, 2]. The robust estimation method restrains the impacts of outliers on the final parameter solutions by minimizing the score function with higher robustness and breakdown point when it excludes the doubtful observations [3]. The robust estimation methods include M estimation, L estimation, and R estimation. The M estimation is a generalized maximum likelihood estimation and can be further divided into two categories: the iterative method with variable weights and the P minimum norm.

The iterative method with variable weights is briefly introduced and divided into the following steps:

The mathematical model is as follows

$$v = A\hat{x} - y, \quad P \tag{8.4}$$

The first parameter estimation and their residuals are then solved as follows

$$\hat{x}^{[1]} = \left(A^{\mathrm{T}}PA\right)^{-1}A^{\mathrm{T}}Py \tag{8.5}$$

$$v^{[1]} = A\hat{x}^{[1]} - y \tag{8.6}$$

And then iterative calculation continues. The iteration is stopped until the difference between the parameter estimates obtained at the $k$th iteration and the $k - 1$th iteration meets

$$\left|\hat{x}^{[k]} - \hat{x}^{[k-1]}\right| \leq \omega \tag{8.7}$$

The $\omega$ is a tiny amount as constant threshold. Finally, the parameter estimation and their residuals are solved.

$$\hat{x}^{[k]} = \left(A^{\mathrm{T}}\overline{P}^{[k-1]}A\right)^{-1}A^{\mathrm{T}}\overline{P}^{[k-1]}y \tag{8.8}$$

$$v^{[k]} = A\hat{x}^{[k]} - y \tag{8.9}$$

The key to the iterative method with variable weights is to choose the $\rho$ function. The $\rho$ function is basically selected by experience, and it is a function of residual. The weight function $p_i$ can be calculated from the $\rho$ function, with $\varphi(v_i) = \frac{\partial \rho(v_i)}{\partial v_i}$ and $p_i = \frac{\varphi(v_i)}{v_i}$. And the most classic scheme is the IGG3 method. The weight function $p_i$ of the IGG3 method is determined as follows

$$
p_i = \begin{cases} 1 & |\tilde{v}_i| \le k_0 \\ \frac{k_0}{|\tilde{v}_i|}\left(\frac{k_1-|\tilde{v}_i|}{k_1-k_0}\right)^2 & k_0 < |\tilde{v}_i| \le k_1 \\ 0 & |\tilde{v}_i| > k_1 \end{cases} \tag{8.10}
$$

where the $k_0$ and $k_1$ are two constant thresholds. Without loss of generality, $k_0 = 1.0$ and $k_1 = 3.0$ in this study. The $\tilde{v}_i$ is the standardized residual as follows

$$
\tilde{v}_i = \frac{v_i}{\hat{\sigma}_0 \sqrt{q_{v_i v_i}}} \tag{8.11}
$$

where $q_{v_i v_i}$ is the $i$th diagonal element of the cofactor matrix $\boldsymbol{Q}_{vv}$, $\hat{\sigma}_0$ is the posterior variance factor as follows

$$
\hat{\sigma}_0 = \sqrt{\frac{\boldsymbol{v}^{\mathrm{T}} \boldsymbol{P} \boldsymbol{v}}{r}} \tag{8.12}
$$

where $r$ is the number of the redundant observations. In robust estimation, other types of weight functions can also be adopted according to the real applications. For instance, one can use the Huber function, which is as follows

$$
p_i = \begin{cases} \tilde{v}_i^2 & |\tilde{v}_i| \le 2\sigma \\ 4\sigma(\tilde{v}_i) - 4\sigma^2 & |\tilde{v}_i| > 2\sigma \end{cases} \tag{8.13}
$$

where the $\sigma$ is the standard deviation of the residuals. The Hampel function can also be used as follows

$$
p_i = \begin{cases} |\tilde{v}_i| & 0 \le |\tilde{v}_i| < k_0 \\ k_0 & k_0 \le |\tilde{v}_i| < k_1 \\ \frac{k_2-(\tilde{v}_i)}{k_2-k_0} - k_0 & k_1 \le |\tilde{v}_i| < k_2 \\ 0 & |\tilde{v}_i| \ge k_2 \end{cases} \tag{8.14}
$$

where $k_2$ is a constant threshold.

GNSS outliers elimination includes three steps: detection, identification, and adaptation. The DIA method relies on making a decision between a null and a set of alternative hypotheses, $H_0$ and $H_1$. The null hypothesis and alternative hypothesis are usually formed as

$$
H_0 : (v_i) = 0 \tag{8.15}
$$

$$\mathrm{H}_1 \colon (v_i) \neq 0 \tag{8.16}$$

The $v_i$ is the $i$th element of the residual vector $v$. The null hypothesis provides an unbiased estimation of when the mathematical model is correct. Instead, the alternative hypotheses are used to know and successfully eliminated the suspected outliers and the subset of affected faulty observation.

When the null hypothesis does not hold, the alternative hypothesis is tested. The classical DIA method is used in an alternative hypothesis test. It is briefly introduced and divided into the following three steps:

## (1) Detection

In the detection step, conventionally, the residual $\hat{\epsilon}$ and global test statistic $\mathrm{T}_q$ are formed as

$$\hat{\epsilon} = y - A\hat{x} \tag{8.17}$$

$$\mathrm{T}_q = \hat{\epsilon}^{\mathrm{T}} Q_{yy}^{-1} \hat{\epsilon} \tag{8.18}$$

where $y$ is the remaining term of the observation equation. The $q$ and $Q_{yy}$ are the number of redundant observations and the variance matrix of the observations, respectively. According to the critical value with the significance level $\alpha_1$, it indicates that there are outliers in the observations if $\mathrm{T}_q > \chi^2_{\alpha_1}(q, 0)$. Therefore, an identification procedure should be conducted. On the contrary, the observation does not contain outliers and thus the solutions can be accepted.

## (2) Identification

Once the above detection procedure detects outliers, an identification procedure should be conducted by searching among the alternative hypotheses. At this time, the local test statistic $w_i$ is constructed and used in the identification step for the most likely model misspecification. The detailed formula is as follows

$$Q_{\hat{\epsilon}\hat{\epsilon}} = Q_{yy} - A \left( A^{\mathrm{T}} Q_{yy}^{-1} A \right)^{-1} A^{\mathrm{T}} \tag{8.19}$$

$$|w_i| = \left| \frac{c_i^{\mathrm{T}} Q_{yy}^{-1} \hat{\epsilon}}{\sqrt{c_i^{\mathrm{T}} Q_{yy}^{-1} Q_{\hat{\epsilon}\hat{\epsilon}} Q_{yy}^{-1} c_i}} \right| \tag{8.20}$$

where $c_i = [0, \ldots, 0, 1, 0, \ldots, 0]^{\mathrm{T}}$; $Q_{\hat{\epsilon}\hat{\epsilon}}$ represent the cofactor matrix of residuals. The $N_{\frac{1}{2}\alpha_2}(0, 1)$ is the critical value of a standard normal distribution with a significance level $\alpha_2$. When $|w_i| > N_{\frac{1}{2}\alpha_2}(0, 1)$, the outlier is existed in the $i$th observation and the corresponding observation needs to be removed. The data snooping is stopped until the global test is accepted, and then an adaptation procedure should be conducted.

(3) Adaption

After the identification of the suspected misspecification, the remaining observations are used to conduct the LS estimation again. Then the relatively reliable parameter solutions can be obtained. The bias in the unknown parameters can only be completely removed after the hypothesis testing.

The above-mentioned DIA method uses the normality ($N$) test. And the joint ($F$) test, student ($t$) test, and chi-square ($\chi^2$) test can also be used to construct test statistics.

## 8.3 Importance About Data Quality Control

In a data adjustment system, the functional model describes the relationship between observations and parameters, while the stochastic model describes the observation precisions and their correlations to each other. The stochastic model can be specified by a covariance matrix, being the second-order central moments of the random observation errors. Despite the principle that an arbitrarily positive-definite covariance matrix can be used to compute the unbiased estimator in LS adjustment, one can never achieve the optimal estimate with the minimal variance unless the correct stochastic model is applied [4–6].

In GNSS applications, the stochastic model is very important for reliable integer ambiguity resolution [7, 8] and for precise positioning [9–11]. Compared with the correct stochastic model, any approximate stochastic model will result in the smaller success rate of both integer least squares and integer bootstrapped ambiguity resolution [12]. Hence, refining the GNSS stochastic model is a worthy aspiration and significant research efforts have been done in the past two decades. The earlier studies were based on the elevation dependence of random observation errors [13], and later took into account the physical correlations, typically the between-frequency cross correlation and time correlation [14]. Based on these studies, it is concluded that in general the observation precision is elevation-dependent and the cross correlation and time correlation may exist. Moreover, these stochastic characteristics vary with both receiver and observation types.

Besides achieving the precise parameter estimator, the correct stochastic model is also required to retrieve the objective precision measures and the covariance matrix of the estimator. For short baselines and short observation sessions, the physical correlations have no significant effects on the baseline solutions, but significant effects on the covariance matrix of the baselines, as numerically shown in [15]. Existing studies of refining GNSS stochastic models almost all focus on the improvement of positioning. In fact, the stochastic model is even more important for the reliability of quality control, where the covariance matrix is involved in testing statistics, for instance, the overall statistic for model specification and the $w$ statistic for outlier detection. Such statistics are known to be sensitive to the stochastic model [16].

However, in the GNSS community, the influence of the stochastic model on the statistical reliability tests has been rarely studied. Teunissen [17] derived the analytical formulae of minimal detectable biase (MDB) for canonical forms of different GNSS application models. Li et al. [18] numerically demonstrated the impact of the elevation-dependent model on the overall and $w$ statistic tests as an initial study. In this chapter, we will synthetically study the influence of the GNSS stochastic model on the statistic tests involved in reliability with triple frequency BeiDou Global Navigation Satellite System (BDS) observations as an example. We first apply the variance component estimation (VCE) method to achieve the realistic elevation-dependent precisions, cross correlations and time correlations. Compared with the empirical stochastic models, we numerically demonstrate the influence of these realistic stochastic properties on the overall and $w$ statistic tests. In addition, the MDBs together with separability defined by the correlation coefficient of two $w$ statistics are examined. To the best of our knowledge, this chapter is the first comprehensive study on the reliability influence of BDS stochastic modeling. The achieved results will be very helpful for users to do quality control in real applications.

As well known in the Gauss-Markov model, the LS solution is optimal only when no outlier exists neither any other misspecifications of the functional and stochastic model [4]. It is therefore important to validate this pre-condition by using some proper statistical testing. Often, two test statistics, overall test and $w$-test, are popularly applied to check the specification of the mathematic model. The overall test is to test the overall discrepancy between the underlying observation model and the real observations, while the $w$-test is to test whether outliers in individual observations are present. In GNSS applications, one can apply these two statistical tests to both float and fixed solutions.

Once the float solution is obtained in the first step of solving the mixed GNSS model, one can apply the overall test to check the compatibility of the mathematic model. The overall test statistic is [19]

$$
T_q = \frac{\hat{\epsilon}_y^{\mathrm{T}} Q_{yy}^{-1} \hat{\epsilon}_y}{q} \tag{8.21}
$$

For the null hypothesis that no misspecification exists, the overall statistic has a Fisher distribution with $q = m - n - p = f(s - 1) - 3$ and $\infty$ degrees of freedom, i.e., $T_q \sim F(q, \infty)$. Given the correct stochastic model $Q_{yy}$, it is emphasized that the expectation of $T_q$ is equal to 1 if the function model is overall well-specified. Therefore, given a significance level $\alpha$, if $T_q < F_{1-\alpha}(q, \infty)$, we accept the null hypothesis that there is no misspecification in the functional and/or stochastic model; otherwise, we accept the alternative hypothesis that the misspecification exists in the functional and/or stochastic model.

If the null hypothesis is rejected, one may then need to further identify the cause of the misspecification between model and data. Usually, one starts with testing for outliers in individual observations by using the $w$-test. The $w$-test statistic of the $i$th observation reads [19]

$$w_i = \frac{c_i^{\mathrm{T}} Q_{yy}^{-1} \hat{\epsilon}_y}{\sqrt{c_i^{\mathrm{T}} Q_{yy}^{-1} Q_{\hat{\epsilon}\hat{\epsilon}} Q_{yy}^{-1} c_i}} \qquad (8.22)$$

where $c_i$ is $m$-column vector with all elements of 0 except the $i$th element of 1. The $w_i$ is standard normally distributed with zero mean [i.e., $w_i \sim N(0, 1)$] for null hypothesis $H_0$ and with non-zero mean (i.e., non-centrality parameter $\sqrt{c_i^{\mathrm{T}} Q_{yy}^{-1} Q_{\hat{\epsilon}\hat{\epsilon}} Q_{yy}^{-1} c_i}\,\nabla$ with $\nabla$ an unknown scalar as expectation of bias) for alternative hypothesis $H_a$. In any statistical hypothesis test, one has to encounter the type I error of false alarm and the type II error of wrong detection, namely, the error of rejecting a correct hypothesis and the error of accepting a wrong hypothesis [20]. In the $w$-test, with a significance level $\alpha$, the null hypothesis will be accepted that the $i$th observation is not an outlier if $|w_i| < N_{1-\alpha/2}$; otherwise, the corresponding alternative hypothesis will be accepted if it has the largest $|w_i|$ of all $m$ alternatives. In such case, the corresponding detection power, $\gamma = 1 - \beta$ with $\beta$ the probability of the type II error, can be computed under $H_a$.

In theory, with a significance level $\alpha_0$, the larger $|\nabla|$ will receive the larger detection power $\gamma$. If the detection power is further controlled to a level $\gamma_0$, the absolute non-centrality parameter $\sqrt{\lambda_0}$ as a function of $\alpha_0$ and $\gamma_0$ can be obtained. For instance, for $\alpha_0 = 0.001$ and $\gamma_0 = 0.8$, it follows that $\lambda_0 = 17$. Once the non-centrality parameter is known, the corresponding size of the bias is [19]

$$|\nabla| = \sqrt{\frac{\lambda_0}{c_i^{\mathrm{T}} Q_{yy}^{-1} Q_{\hat{\epsilon}\hat{\epsilon}} Q_{yy}^{-1} c_i}} \qquad (8.23)$$

If the outlier is smaller than this size, the testing power will be smaller than $\gamma_0$. Hence, this size is defined as the MDB related to the probabilities of $\alpha_0$ and $\gamma_0$.

For the fixed solutions, one can apply the overall test, $w$-test and compute the MDB exactly following (8.21), (8.22) and (8.23), respectively; But now $\check{\epsilon}_y$ and $Q_{\check{\epsilon}\check{\epsilon}}$ must be used instead of their float counterparts $\hat{\epsilon}_y$ and $Q_{\hat{\epsilon}\hat{\epsilon}}$. Note in the overall test the degree of freedom becomes $q = 2f(s-1) - 3$ since the $f(s-1)$ double-differenced (DD) ambiguities are fixed.

To intuitively get some insight on how the stochastic model (covariance matrix $Q_{yy}$) affects the LS solutions and the hypothesis testing statistics, we assume simply that the structure of $Q_{yy}$ is correct but scaled by a factor $\kappa$, i.e., $Q_{yy} \to \kappa Q_{yy}$. Then the LS float estimate, $\hat{b}$, is invariant but its covariance matrix $Q_{\hat{b}\hat{b}} \to \kappa Q_{\hat{b}\hat{b}}$, which is the case also for the fixed solution $\check{b}$. The overall and $w$-statistics as well as MDB become $T_q \to T_q/\kappa$, $w_i \to w_i/\sqrt{\kappa}$ and $|\nabla| \to \sqrt{\kappa}|\nabla|$, respectively. It is obvious that the scaled stochastic model has immediate effect on both the overall and $w$-statistics and MDB although it does not affect the parameter estimate $\hat{b}$. In the following, we will numerically demonstrate how the elevation-dependent precisions, the cross correlations and time correlations in the stochastic model affect the hypothesis tests by comparing between the realistic and empirical stochastic models.

## 8.4  Results and Discussion

This section will give comprehensive analysis of the outlier handling method and the impacts on the the overall test, $w$-test, and MDB.

### 8.4.1  Analysis of the Outlier Processing Method

In order to validate the effectiveness of the outlier handling method and evaluate positioning accuracy, this section conducts experiments on five sets of landslide monitoring data from the southwestern region, collected on the 288th day of the year 2021 over a 24-h period. These datasets are named Test1, Test2, Test3, Test4, and Test5, with baseline lengths approximately 92 m, 143 m, 53 m, 133 m, and 75 m respectively. The primary processing involves Global Positioning System (GPS) L1 + L2 observations and BDS B1 + B2 observations, with ionospheric delays corrected using the Klobuchar model and tropospheric delays corrected using the Saastamoinen model. The ambiguity resolution is achieved through the least-squares ambiguity decorrelation adjustment (LAMBDA) method. Specific processing strategies are detailed in Table 8.1.

To verify the efficacy of two categories of outlier treatment methods and identify the optimal approach under complex conditions, ten different schemes were designed based on these methods, differing in threshold settings. Schemes B, C, and D represent three data snooping approaches with varying thresholds, while E,

**Table 8.1**  Processing strategies

| Content | Processing strategy |
| --- | --- |
| Used signals | GPS L1 + L2, BDS B1 + B2 |
| Data processing | Carrier phase differential technique |
| Solution method | Real-time single epoch solution |
| Ambiguity resolution strategy | LAMBDA |
| Ambiguity fixing threshold | 2.0 |
| Sampling interval (s) | 5 |
| Ionospheric delay correction | Klobuchar model |
| Tropospheric delay correction | Saastamoinen model |
| Cut-off elevation angle (°) | 15 |

F, G, E1, F1, and G1 denote six robust estimation schemes paired with two distinct threshold values for iteration control, set at 0.01 m and 9 m. The specific outlier handling measures for each scheme are outlined in Table 8.2. To comprehensively assess the performance of these schemes, analysis includes ambiguity fixing rates and positioning errors.

Ambiguity fixing rates for the five datasets under different schemes are presented in Table 8.3. It is evident that the data snooping schemes successfully identify and remove contaminated observations. Among them, scheme B, which employs a lower threshold, achieved the highest ambiguity fixing rate, increasing from 60.8% to 63.2%. Robust estimation schemes perform better with a larger iteration threshold, demonstrating superior performance in obtaining more fixed solutions. Notably, the ambiguity fixing rates of schemes E and F are lower than that of scheme A, which highlights the importance of threshold selection in robust estimation. Scheme E1 shows the highest ambiguity fixing rate among the robust estimation schemes. This indicates that a combination of a small threshold and a large iteration threshold in robust estimation is more effective for ambiguity resolution. Both data snooping and robust estimation have proven effective in enhancing ambiguity fixing rates.

Due to space constraints, only detailed results for Test1 are provided. In satellite positioning, a higher number of visible satellites and a more stable satellite geometry are preferred. Figure 8.1 illustrates the average number of visible GPS + BDS satellites and the position dilution of precision (PDOP) in Test1. The average number of visible satellites is 21, but there is significant fluctuation, indicating unstable signal reception and poor observation quality. All PDOP values exceed 1, with an average of approximately 1.6, and some values exceeding 3, indicating that the satellite geometry layout is generally good but occasionally exhibits poor spatial structure. As the number of satellites decreases, PDOP values increase correspondingly. Figure 8.2 shows the sky plot of Test1. The figure shows that there are some GPS and BDS satellites with low elevation angles, which may cause signal reflection, diffraction,

**Table 8.2** Outlier processing scheme

| Scheme | Outlier handling method | Threshold for iteration control | Threshold value |
| --- | --- | --- | --- |
| A | No outlier treatment | – | – |
| B | Data snooping | – | $N_{\frac{1}{2}\alpha_2}(0, 1) = 1.960$ |
| C | Data snooping | – | $N_{\frac{1}{2}\alpha_2}(0, 1) = 2.576$ |
| D | Data snooping | – | $N_{\frac{1}{2}\alpha_2}(0, 1) = 3.291$ |
| E | Robust estimation | Small ($\omega = 0.01$ m) | $k_0 = 1.0, k_1 = 2.5$ |
| F | Robust estimation | Small ($\omega = 0.01$ m) | $k_0 = 1.5, k_1 = 3.5$ |
| G | Robust estimation | Small ($\omega = 0.01$ m) | $k_0 = 2.5, k_1 = 6.0$ |
| E1 | Robust estimation | Large ($\omega = 9$ m) | $k_0 = 1.0, k_1 = 2.5$ |
| F1 | Robust estimation | Large ($\omega = 9$ m) | $k_0 = 1.5, k_1 = 3.5$ |
| G1 | Robust estimation | Large ($\omega = 9$ m) | $k_0 = 2.5, k_1 = 6.0$ |

**Table 8.3**  Ambiguity fixing rates of observation data (%)

| Scheme | Test1 | Test2 | Test3 | Test4 | Test5 |
|---|---|---|---|---|---|
| A | 60.8 | 32.3 | 41.4 | 82.8 | 72.0 |
| B | 63.2 | 36.7 | 44.8 | 83.9 | 76.6 |
| C | 62.7 | 35.7 | 44.7 | 83.8 | 76.3 |
| D | 62.1 | 34.7 | 44.1 | 83.4 | 76.0 |
| E | 54.5 | 24.1 | 38.6 | 71.6 | 69.7 |
| F | 58.2 | 27.7 | 40.2 | 78.6 | 72.3 |
| G | 61.3 | 32.3 | 42.8 | 83.2 | 75.1 |
| E1 | 66.5 | 39.8 | 46.3 | 85.2 | 77.7 |
| F1 | 64.8 | 37.4 | 45.4 | 84.6 | 77.0 |
| G1 | 63.3 | 35.0 | 44.6 | 84.1 | 76.3 |

or obstruction. Therefore, a cutoff elevation angle of 15° was set to exclude unsuitable satellites that are not suitable for data processing.

We also analyzed the positioning errors and observation residuals of Scheme A, which are shown in Figs. 8.3, 8.4, and 8.5. Without outlier handling (scheme A), the positioning results, particularly in the U direction, exhibit noticeable instability issues. The data snooping scheme (Fig. 8.6) can effectively detect and remove outliers, with scheme B performing best in reducing U-direction errors compared to the other two snooping schemes. For robust estimation schemes (Fig. 8.7), those with a large iteration threshold show more pronounced convergence than those with a small iteration threshold at the same threshold values. Scheme E1, with a small threshold combined with a large iteration threshold, significantly reduces U-direction errors

**Fig. 8.1**  Satellite number and PDOP value of Test1

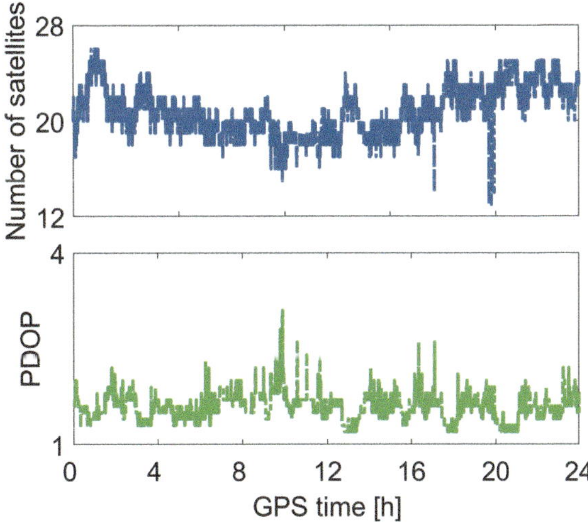

**Fig. 8.2** Skyplot of Test1

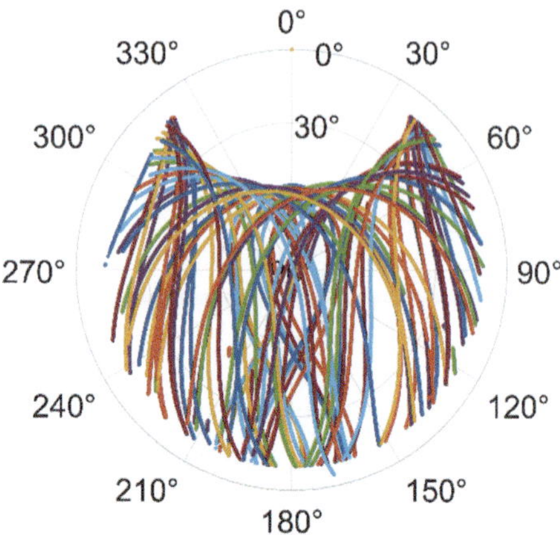

compared to scheme E, illustrating its superior performance in satellite positioning under complex conditions.

Furthermore, the root mean square error (RMSE) of the positioning results for each scheme is computed. Data snooping schemes, especially scheme B with a small threshold, can identify and remove more outliers. Compared to scheme A, the positioning results improved by 0.059 m, 0.017 m, and 0.062 m, respectively. Among various robust estimation schemes, those combining a small threshold with a large iteration threshold (like E1) perform better than scheme A. For example, E1 improves

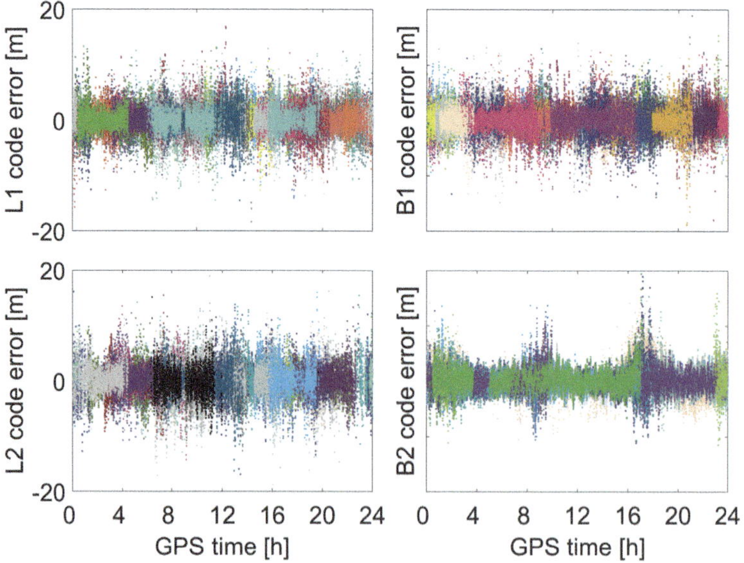

**Fig. 8.3** Double-differenced pseudorange residuals of scheme A

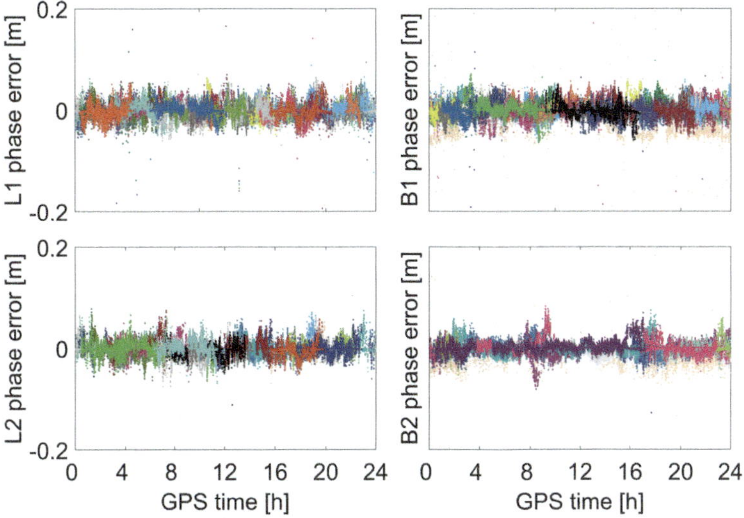

Fig. 8.4  Double-differenced phase residuals of scheme A

Fig. 8.5  Positioning errors
of scheme A

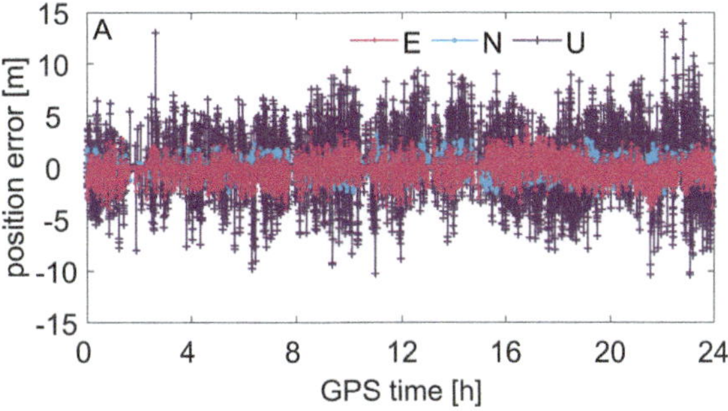

the positioning accuracy by 0.098 m, 0.055 m, and 0.209 m, respectively. This analysis confirms that both data snooping and robust estimation are effective in dealing with outliers under complex conditions.

We further validate the effectiveness of outlier treatment by listing the positioning availability for data snooping and robust estimation schemes. The availability for data snooping schemes surpasses that of scheme A in all horizontal components. While the availability of robust estimation schemes E and F is slightly worse than scheme A, other robust estimation schemes (including E1) show improvements over scheme A. Among them, scheme E1 achieves a positioning availability of 96.50% when the horizontal component is less than 2.0 m. This reinforces the notion that a small threshold combined with a large iteration threshold in robust estimation is optimal for accurate outlier detection and reasonable weighting of observations in complex scenarios.

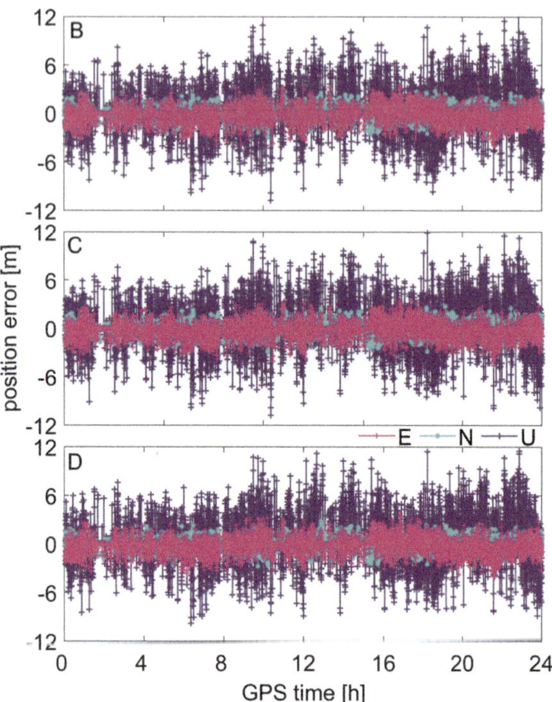

**Fig. 8.6** Positioning errors of data snooping

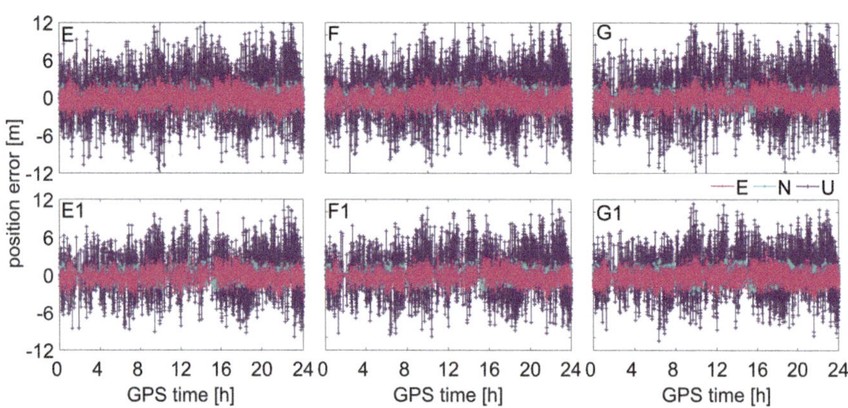

**Fig. 8.7** Positioning errors of robust estimation

## 8.4.2   Analysis of the Overall Test, w-test and MDB

In the GNSS community, the influence of the stochastic model on the statistical reliability tests has been rarely studied. In this section, we analyze the impacts of individual stochastic quantities, elevation-dependent precisions, cross correlations and time correlations, on the reliability, including the overall test, $w$-test and MDB based on the BDS stochastic model. It is important to verify the prerequisite conditions through appropriate statistical tests, so some statistical introductions and explanations will be provided.

The purpose of this section is to investigate the numerical impacts of a realistic stochastic model on the hypothesis tests compared with those with empirical stochastic model. In principle, the GNSS stochastic modelling should be purely based on the random noise. To completely avoid the influence of any remaining systematic biases on the numerical analysis, for instance, multipath, atmospheric bias etc., the zero baseline is employed in this study. Another benefit of zero baseline data is that we exactly know the baseline component which can serve for the latter analysis.

Two data sets of triple frequency BDS observations are collected on zero baselines by using ComNav and Trimble receivers, respectively. The total number of epochs is 13,582 and 86,400 for ComNav and Trimble baseline, respectively, both with sampling interval of 1 s. In the whole computations, the cut-off elevation is taken by $10°$.

The DD integer ambiguity resolution is the precondition to analyze the stochastic model. In this study, the data sets are collected on zero baselines. Such information can be applied to extremely enhance the model strength such that the ambiguity resolution can be reliably done epoch by epoch with the LAMBDA method.

Given the data window $K = 60$ epochs, we estimate the precision of each satellite per frequency and observation type. Then the precision estimates of all satellites for unique observation type are sorted in ascending order of elevations. For each elevation interval $0.5°$ from $10°$ to $90°$, we take the mean of the precision estimates in this elevation interval as the precision of this elevation. The results of elevation-dependent precisions are computed for all three-frequency phase and code observations.

In addition, two elevation-dependent functions are analyzed, denoted by model A and B, respectively. We choose model A as [18]

$$\sigma_\theta = f(\boldsymbol{c}|\theta) = c_1/(\sin\theta + c_2) \tag{8.24}$$

and its reduced version model B

$$\sigma_\theta = f(\boldsymbol{c}|\theta) = c/\sin\theta \tag{8.25}$$

It is noted that we choose these two elevation-dependent models just for a case study due to their simplicity.

Moreover, these two models are representative. That is, model (8.24) can fit the elevation-dependent observation precisions very well, while model (8.25) poorly, see the latter results. One can of course choose other elevation-dependent models, for instance, exponential function [13], which may result in the different numerical results but will not affect our conclusions.

The results show that the observation precisions are overall elevation-dependent for all triple frequency phase and code observations, although the dependence patterns differ from the observation and receiver types. For the ComNav receiver, the phase precisions of B3 are lower than those of B1 and B2, but the corresponding code precisions are much higher over all elevations. For the Trimble receiver, we cannot see the obvious precision difference for phase between frequencies, while the code precisions of B1 are relatively larger for low elevations. The model A can overall fit the precisions better than model B. For model B, the over-fitting problem exhibits. In other words, the precision values of low elevations are overly enlarged whilst those of high elevations are overly reduced.

The estimated cross-correlation coefficients are presented in Table 8.4. For each receiver, six cross-correlation coefficients are computed for phase and code observations amongst three frequencies. For the ComNav receiver, all cross-correlation coefficients deviate from 0 with values smaller than 0.2; especially for phase, which means that no cross correlation exists. However, for the Trimble receiver, very significant cross correlation with correlation coefficient 0.76 exists between B2 and B3 code observations.

The impact of elevation-dependent models on reliability takes precedence in our discussion. We demonstrate the impact of observation precisions on reliability by comparing two elevation-dependent models, A and B. Hereafter they are also called weighting models. As an example, the Trimble baseline data was processed with these two models, respectively.

We computed the statistics of overall test for single-epoch float and fixed solutions with two models. Given a significance level $\alpha = 0.05$, the critical values are computed by $F_{0.95}(q, \infty)$ for float and fixed solutions with $q = 3s - 6$ and $6s - 9$ for $f = 3$, respectively. The results of model A differ significantly from model B. Since the baseline data was collected in an ideal environment, very few outliers were found and excluded in our post-processing. In other words, there is no outlier in the observations used anymore. In such case, if the model specifies the observations very well, the expectation of the overall statistics in principle is equal to 1. The statistics of model A is indeed overall close to 1, but those of model B have significant deviations from 1. The mean of all epoch statistics can be deemed as an empirical approximation to expectation. Therefore, the smaller the difference of the computed mean from 1 is, the better the corresponding elevation-dependent model is. The means of overall statistics are calculated. The result indicates that the model A is best, follows by the model B. The deviations of means of overall statistics from 1 are only 0.03 and 0.02 for float and fixed solutions of model A, while they are 1.33 and 1.16 for model B.

In absence of outliers, the computed statistics should be smaller than the critical values statistically. If the statistic is larger than its associated critical value, it leads to

**Table 8.4** Estimated cross-correlation coefficients for all three-frequency phase and code observations of two types of receivers

|  | $\varrho^{[c]}_{\phi_1\phi_2}$ | $\varrho^{[c]}_{\phi_1\phi_3}$ | $\varrho^{[c]}_{\phi_2\phi_3}$ | $\varrho^{[c]}_{p_1p_2}$ | $\varrho^{[c]}_{p_1p_3}$ | $\varrho^{[c]}_{p_2p_3}$ |
|---|---|---|---|---|---|---|
| ComNav | 0.00 | − 0.00 | − 0.00 | 0.17 | − 0.01 | − 0.02 |
| Trimble | −0.01 | − 0.01 | − 0.07 | 0.12 | 0.13 | 0.76 |

a false alarm. The probabilities of false alarm are computed for both float and fixed solutions with three models. The results showed that model A is prominently better than the model B. The probabilities of false alarm for model A are smaller than 4 and 7% for float and fixed solutions, respectively. For model B, they are worst and even reach about 67 and 80% for float and fixed solutions, respectively. Roughly, the probabilities of false alarm of model A are smaller than those of model B by more than 10 times. Such performance reveals that if the elevation-dependent function is not properly specified, it will derive even worse results than the elevation-independent model.

In the single-epoch float solution model, the ambiguity parameters are to be estimated for all three frequency phase observations. Such model formation leads to that the denominators of (8.22) and (8.23) are zero, i.e., $c_i^T Q_{yy}^{-1} Q_{\hat\epsilon\hat\epsilon} Q_{yy}^{-1} c_i = 0$, for phase observations. It means that the statistics of $w$-test and MDB cannot be computed for phase observations with single-epoch float solutions. Therefore, we focus on analyzing the statistics of $w$-test and MDB for single-epoch fixed solutions.

The result shows the computed $w$-statistics as a function of elevations for all triple frequency code and phase observations with two elevation-dependent models, A and B. Recall the theoretical relation that $w_i \rightarrow w_i/\sqrt{\kappa}$ if $Q_{yy} \rightarrow \kappa Q_{yy}$. It means that the downscaling variance ($\kappa < 1$) derives the larger $w_i$ statistic, and vice versa. As a result, the $w$-statistics of low elevations are smaller than those of high elevations, especially for code of B2 and B3. The model A outperforms the model B, where its $w$-statistics are basically comparable for all elevation observations.

For a normal observation, $w_i$ is of standard normal distribution. Given the significance level $\alpha = 0.05$, the empirical probability of false alarm is computed as a ratio between the number of $w$-statistics outside the confident region $[N_{\alpha/2}, N_{1-\alpha/2}]$ and the total number of $w$-statistics. Given the elevation intervals of $10°$ from $10°$ to $90°$, there are total 8 elevation intervals. For each elevation interval, this empirical probability of false alarm can be computed.

Let us now analyze the MDB results with three weighting models. The MDBs are computed with single-epoch fixed solution following (8.23) for triple frequency code and phase observations. Again, recall the theoretical relation that $|\nabla| \rightarrow \sqrt{\kappa}|\nabla|$ if $Q_{yy} \rightarrow \kappa Q_{yy}$. It means that the MDB is positively proportional to the observation precision with arithmetic square root of a scalar. The more precise observation will receive a smaller MDB, and vice versa. In other words, with a given significance level $\alpha$ and detection power $\gamma$, the detectable outlier becomes smaller if the observation precision is improved. The model A receives the realistic MDBs since it can reflect the precisions of observations realistically. However, with model B unrealistic MDBs are obtained, which can be either too small or too large. Compared to the MDBs of model A, the model B obtains too large MDBs for low elevations while too small ones for high elevations. In other words, the outliers at low elevations that can be actually detected become non-detectable in terms of MDB with a certain reliability. More conservatively, some normal observations at high elevations may be wrongly excluded as outliers.

To investigate the impact of cross correlation on reliability, we use the B2 and B3 code observations of the Trimble baseline for the baseline resolution, where the

B2 and B3 code is strongly correlated with correlation coefficient 0.76. Here we do not incorporate B1 data just for simplicity due to its minor correlations with B2 and B3 data. Single-epoch sinle-differenced (SD) model with only B2 and B3 code observations reads

$$
\mathrm{E}\left(\begin{bmatrix} \boldsymbol{p}_2 \\ \boldsymbol{p}_3 \end{bmatrix}\right) = \begin{bmatrix} \boldsymbol{G}\ \boldsymbol{e}_s\ \boldsymbol{0} \\ \boldsymbol{G}\ \boldsymbol{e}_s\ \boldsymbol{e}_s \end{bmatrix} \begin{bmatrix} \boldsymbol{x} \\ \mathrm{d}t_2 \\ \mathrm{d}t_3 \end{bmatrix}, \begin{bmatrix} \boldsymbol{Q}_{p_2 p_2} & \varrho_c \boldsymbol{Q}_c \\ \varrho_c \boldsymbol{Q}_c & \boldsymbol{Q}_{p_3 p_3} \end{bmatrix}
\tag{8.26}
$$

where $\boldsymbol{Q}_{p_2 p_2} = \mathrm{diag}\left(\left[\left(\sigma_{p_2}^1\right)^2, \ldots, \left(\sigma_{p_2}^s\right)^2\right]\right)$ with $\sigma_{p_2}^s$ is the undifference observation precision of satellite $s$ computed by elevation-dependent model A (8.24). $\boldsymbol{Q}_{p_3 p_3}$ is similar to $\boldsymbol{Q}_{p_2 p_2}$ computed with its own elevation-dependent parameters. The matrix, $\boldsymbol{Q}_c = \boldsymbol{Q}_{p_2 p_2}^{\frac{1}{2}} \boldsymbol{Q}_{p_3 p_3}^{\frac{1}{2}}$, is scaled by a cross-correlation coefficient $\varrho_c$ between B2 and B3 code observations.

For the realistic stochastic model with $\varrho_c = 0.76$ and the empirical model with $\varrho_c = 0$, one can solve the model (8.26) to obtain the corresponding LS solutions epoch by epoch. Then the statistics of overall tests are computed with respect to two stochastic models. They are very close to each other and their means are 1.0756 and 0.9176, respectively. With significance level $\alpha = 0.01$, the probabilities of false alarm are 2.24 and 2.75% for stochastic models with $\varrho_c = 0.76$ and 0, respectively.

The $w$-test statistics of all observations are computed for these two stochastic models with and without cross correlations. For these two stochastic models, the means of $w$-test statistics are $5 \times 10^{-5}$ and 0.001 with standard deviations 1.0345 and 0.9544, respectively. Although the $w$-test statistics with stochastic model of $\varrho_c = 0.76$ are slightly closer to the standard normal distribution, they are practically very similar to those with stochastic model of $\varrho_c = 0$. For the significance level $\alpha = 0.01$, the probabilities of false alarm is 0.79 and 1.34%, respectively. In summary, the cross correlation in stochastic model has very minor effects on the overall and $w$-test.

Let us now analyze the impact of cross correlation on the MDBs. By considering and ignoring the cross correlations in stochastic model, the MDBs are computed for all B2 and B3 code observations of all satellites. Considering the cross correlation will decrease the MDBs, namely, the smaller outliers are detectable if the cross correlations are properly assimilated. In principle, the information content in the correlated B2 and B3 observations should be less than that in the B2 and B3 observations if they are independent. Hence, with less information contents for correlated B2 and B3 observations, the outlier detection should become difficult and the MDBs should be larger. Such contradiction attracts our further analysis. The correlation coefficient of two $w$-statistics of B2 and B3 code observations for a satellite is defined as [21]

$$
\varrho_{w_i w_j} = \frac{c_i^{\mathrm{T}} \boldsymbol{\Omega} c_j}{\sqrt{c_i^{\mathrm{T}} \boldsymbol{\Omega} c_i} \sqrt{c_j^{\mathrm{T}} \boldsymbol{\Omega} c_j}} = \frac{\boldsymbol{\Omega}(i, j)}{\sqrt{\boldsymbol{\Omega}(i, i)} \sqrt{\boldsymbol{\Omega}(j, j)}}
\tag{8.27}
$$

where $\boldsymbol{\Omega} = \boldsymbol{Q}_{yy}^{-1}\boldsymbol{Q}_{\hat{\epsilon}\hat{\epsilon}}\boldsymbol{Q}_{yy}^{-1}$. For a larger correlation coefficient $\varrho_{w_i w_j}$ between the $i$th and $j$th observations, it is more difficult to discriminate the outlier exactly on either the $i$th or $j$th observation. In other words, one may detect out the outlier, but wrongly position the outlier location, which will derive the type III error [21, 22]. In real applications, if the $w$-test statistics of two observations are highly correlated and an outlier is statistically detected on either one observation, an advisable strategy is to exclude these two observations simultaneously to control the type III error.

The correlation coefficients $\varrho_{w_i w_j}$ between two $w$-statistics of B2 and B3 code observations for individual satellites are computed. The mean correlation coefficients over the whole observation span for all satellites are computed as well. The correlation is increased from about 0.2 ($\varrho_c = 0$) to 0.8 ($\varrho_c = 0.76$). That makes sense since the cross correlation makes the B2 and B3 observations of one satellite correlated and then their $w$-statistics correlated. Therefore, if the outlier is detected for B2 or B3 observation, it is advisable to exclude both B2 and B3 observations of this satellite to control the type III error for high reliability of positioning solutions.

To demonstrate the impact of time correlation on reliability, we solve the baseline solutions based on the SD model with two-consecutive epoch observations of triple frequency code observations of ComNav receivers, where all triple frequency code observations are time correlated. The associate model reads

$$\mathrm{E}\left(\begin{bmatrix} \boldsymbol{p}_k \\ \boldsymbol{p}_{k+1} \end{bmatrix}\right) = \begin{bmatrix} \boldsymbol{e}_3 \otimes \boldsymbol{G}_k & \boldsymbol{I}_3 \otimes \boldsymbol{e}_s & \boldsymbol{0} \\ \boldsymbol{e}_3 \otimes \boldsymbol{G}_{k+1} & \boldsymbol{0} & \boldsymbol{I}_3 \otimes \boldsymbol{e}_s \end{bmatrix}\begin{bmatrix} \boldsymbol{x} \\ \mathrm{d}t_k \\ \mathrm{d}t_{k+1} \end{bmatrix} \tag{8.28}$$

with stochastic model

$$\begin{bmatrix} 1 & \varrho_t \\ \varrho_t & 1 \end{bmatrix} \otimes \boldsymbol{Q}_{pp} \tag{8.29}$$

where $\boldsymbol{Q}_{pp}$ is the covariance matrix of single-epoch SD code observations computed by elevation-dependent model A (8.24). $\varrho_t$ is the time correlation coefficient between two consecutive epochs, which is equal to 0.6 for all triple frequency code observations.

For two stochastic models specified by $\varrho_t = 0$ and $\varrho_t = 0.6$, i.e., ignoring and considering time correlation, the statistics of overall and $w$-test and MDBs are computed. The means of overall-test statistics are 0.9296 and 1.0098 for $\varrho_t = 0$ and 0.6, respectively. The corresponding probabilities of false alarm are 1.97 and 1.89% for significance level $\alpha = 0.01$. The means of $w$-test statistics are 0.0014 and 0.0007 with respect to $\varrho_t = 0$ and 0.6. The corresponding probabilities of false alarm is 0.79 and 1.16% for the significance level $\alpha = 0.01$. Therefore, in general, the time correlation has minor impact on the overall and $w$ tests.

The MDB results of all triple frequency code observations are computed. For each baseline solution, triple frequency code observations of two epochs are involved. The MDBs of B3 is smallest, followed by B2 and B1. This is due to the B3 code is most precise and then B2 and B1. Similar to the impact of cross correlation on MDB and correlation of $w$-statistics, the MDBs of all observations are reduced when taking into account the time correlation, while the correlations of $w$-statistics between two epochs are significantly increased. As a result, it is more difficult to discriminate exactly on which observation the outlier has occurred if the observations are time correlated. Again an advisable strategy is to exclude the observations of these two epochs simultaneously to control the type III error.

## 8.5   Conclusion

The importance of stochastic model on achieving optimal parameter estimator and realistic covariance matrix of the estimator has been well documented by GNSS researchers in past many years. That is, the ambiguity resolution and positioning can be improved by refining the stochastic model. However, the importance of stochastic model on the reliability of quality control has been rarely studied, where the covariance matrix is involved in statistical reliability tests. In this chapter, we have synthetically studied the influence of the stochastic model on the statistical tests with triple frequency BDS as an example. Compared with the empirical stochastic models, the influence of estimated realistic stochastic models on the overall and w statistical tests as well as the MDBs have been numerically investigated. Based on our studies, the conclusions are summarized as follows:

The GNSS observation precision is in general elevation-dependent and the cross and time correlation may exist. These stochastic characteristics differ from the receiver and observation types and frequencies, which should be taken into account for establishing a realistic stochastic model.

Comparison of elevation-dependent and -independent models in overall and $w$ tests reveals that a realistic elevation-dependent model can reduce the probabilities of both false alarm and wrong detection. Without proper elevation-dependent model, the probabilities of false alarm and wrong detection could be even worse than those of elevation-independent model.

The cross and time correlations have very marginal effects on the baseline (positioning) solutions [8]. However, they affect the covariance matrix of the baseline solutions and then the reliability test statistics significantly. In other words, one may not expect the improved baseline solutions by properly considering the physical correlations, but indeed the more realistic reliability results. That is, with taking into account the physical correlations, the probabilities of both false alarm and wrong detection will be reduced in statistical reliability tests; the MDBs become smaller with more difficulty of discriminating the outlier location. Hence, when the physical correlations exist amongst observations, an advisable strategy is to exclude these observations simultaneously to control the type III error for reliable positioning.

This chapter primarily focuses on a comprehensive study of the impact that stochastic models have on statistical reliability tests. Employing authentic random models enables the derivation of reasonable outcomes in reliability testing, which in turn facilitates users to make objective decisions concerning quality control in practical GNSS applications.

# References


1. Hekimoglu S, Berber M (2003) Effectiveness of robust methods in heterogeneous linear models. J Geod 76:706–713
2. Wieser A (2004) Reliability checking for GNSS baseline and network processing. GPS Solut 8:55–66
3. Yang L, Shen Y (2020) Robust M estimation for 3D correlated vector observations based on modified bifactor weight reduction model. J Geod 94:31
4. Koch K (1988) Parameter estimation and hypothesis testing in linear models. Springer, New York
5. Teunissen PJG, Amiri-Simkooei AR (2008) Least-squares variance component estimation. J Geod 82:65–82
6. Li B, Shen Y, Lou L (2011) Efficient estimation of variance and covariance components: a case study for GPS stochastic model evaluation. IEEE Trans Geosci Remote Sens 49:203–210
7. Amiri-Simkooei AR, Jazaeri S, Zangeneh-Nejad F, Asgari J (2016) Role of stochastic model on GPS integer ambiguity resolution success rate. GPS Solut 20:51–61
8. Li B (2016) Stochastic modeling of triple frequency BeiDou signals: estimation, assessment and impact analysis. J Geod 90:593–610
9. Wang J, Stewart M, Sakiri M (1998) Stochastic modeling for static GPS baseline data processing. J Surv Eng 124:171–181
10. Howind J, Kutterer H, Heck B (1999) Impact of temporal correlations on GPS-derived relative point positions. J Geod 73:246–258
11. Liu X (2002) A comparison of stochastic models for GPS single differential kinematic positioning. Proceedings of the ION GPS 2002, pp 1830–1841
12. Teunissen PJG (2007) Influence of ambiguity precision on the success rate of GNSS integer ambiguity bootstrapping. J Geod 81:351–358
13. Euler H, Goad C (1991) On optimal filtering of GPS dual frequency observations without using orbit information. Bull Géod 65:130–143
14. Bona P (2000) Precision, cross correlation, and time correlation of GPS phase and code observations. GPS Solut 4:3–13
15. Han S, Rizos C (1995) Standardization of the variance-covariance matrix for GPS rapid static positioning. Geomat Res Aust 62:37–54
16. Baarda W (1968) A testing procedure for use in geodetic networks, Netherlands Geodetic Commission
17. Teunissen PJG (1998) Minimal detectable biases of GPS data. J Geod 72:236–244
18. Li B, Lou L, Shen Y (2016) GNSS elevation-dependent stochastic modeling and its impacts on the statistic testing. J Surv Eng 142:04015012
19. Teunissen PJG (2000) Testing theory: an introduction. Delft University Press, Delft
20. Neyman J, Pearson ES (1933) On the problem of the most efficient tests of statistical hypotheses. Philos Trans R Soc Lond Ser A Contain Pap Math Phys Charact 231:289–337
21. Förstner W (1983) Reliability and discernability of extended Gauss-Markov models. Proceedings of the SEE N 84-26069 16-43, pp 79–104
22. Yang L, Wang J, Knight NL, Shen Y (2013) Outlier separability analysis with a multiple alternative hypotheses test. J Geod 87:591–604

# Chapter 9
# LRTK: Long-Range RTK

## 9.1 Introduction

The real-time kinematic positioning (RTK) is a high-precision positioning method with Global Navigation Satellite System (GNSS) signals. It is classified into two types, i.e., the single-baseline RTK (SRTK) and the network-based RTK (NRTK). NRTK was originally proposed in middle of 1990s [1, 2]; and has gradually become one of the most popular high-precision real-time GNSS positioning methods since the 2000s. It makes full use of a continuously operating reference stations (CORS) network to generate the observation corrections for the high precision positioning of the rover stations within the area covered by a CORS network [3]. The positioning performance of the rover station is almost location-independent within the coverage of the CORS network since the location-dependent errors contained in the rover observations are well compensated by using the corrections [4]. Hence, NRTK is able to provide high precision positioning service in a relatively large-scale area without any problems about service area division and uneven positioning quality [5]. The inter-station distance of CORS network in the case of Global Positioning System (GPS)-only NRTK is usually from 20 to 90 km and the positioning precision is at the centimeter level [6]. Longer inter-station distance and higher positioning precision is possible by using multi-frequency and multi-system (MFMS) GNSS signals [7]. MFMS GNSS signals double the number of the observations and therefore can improve the positioning performance dramatically [8, 9]. The inter-station distance can be extended to more than 100 km and the precision be enhanced to the few-centimeter level [1]. However, NRTK seems of resource-wasting for a medium-sized city, like Shanghai whose administrative area radius is about 50 km. There are ten CORS distributed in Shanghai currently and most of them are not necessary required for a high-precision RTK service. As an extreme case, only one properly located reference station is required if SRTK can provide comparable positioning service as NRTK based on MFMS GNSS signals.

© The Author(s) 2025

B. Li et al., *GNSS Real-Time Kinematic Positioning*, Navigation: Science and Technology 17, https://doi.org/10.1007/978-981-96-9116-6_9

Traditional GPS-only SRTK can provide high precision RTK service only for the baseline length of typically shorter than 20 km [1]. Otherwise, for long-range RTK (LRTK) situations where the baseline length usually exceeds 20 km, the residual distance-dependent errors contained in the differential observations are too large to be ignored. The modeling of the distance-dependent errors is difficult without any external information or additional observations [10, 11]. MFMS SRTK overcomes this problem to a certain degree since the number of observations is doubled or even tripled. The redundancy of observations increases dramatically and therefore allows introducing more parameters to model the distance-dependent errors in long-baseline cases. Odolinski et al. [12] proposed several models for short- to long-baseline MFMS SRTK (i.e., LRTK) positioning and got sound experimental results in case of baseline length up to 100 km. Li et al. [13] also realized the long-range SRTK with triple-frequency BeiDou Navigation Satellite System (BDS) signals. Hence, SRTK is able to provide high-precision RTK service in a larger area based on MFMS GNSS signals and is more economic and convenient than MFMS NRTK. The only question is whether LRTK can reach the comparable positioning performance including the few-centimeter level positioning precision and quick convergence as NRTK in a large-scale area. This chapter intends to give a positive answer to this question by comprehensively comparing the LRTK to NRTK. Besides the numerical comparison with real data, the theoretical comparison is also carried out for float solution and ambiguity resolution (AR) since they govern the quality of fixed solutions to a certain extent [14, 15].

## 9.2  Mathematical Model

### 9.2.1  Functional and Stochastic Models

The functional model of single-epoch $f$-frequency double-differenced (DD) observations is formulated as [16],

$$\begin{bmatrix} P \\ \Phi \end{bmatrix} = \begin{bmatrix} e_f \otimes A \; e_f \otimes g & \mu \otimes I_s & 0 \\ e_f \otimes A \; e_f \otimes g & -\mu \otimes I_s & \Lambda \otimes I_s \end{bmatrix} \begin{bmatrix} x \\ \tau \\ \iota \\ a \end{bmatrix} + \begin{bmatrix} \varepsilon_P \\ \epsilon_\Phi \end{bmatrix} \tag{9.1}$$

where the subscripts $f$ and $s$ denote the number of frequencies and the number of DD satellite pairs, respectively. $\otimes$ denotes Kronecker product and $e_f$ is a column vector with all $f$ elements of 1. $I_s$ denotes the identical matrix of $s$ dimension. $P = \begin{bmatrix} P_1^T, \ldots, P_f^T \end{bmatrix}^T$ is the vector of $f$-frequency code observations while $P_j$ is the vector of $s$ DD observations on frequency $f_j$. $\Phi$ is for phase observations and has the same structure as $P$. $A$ is the design matrix for baseline parameter $x$. $g$ is the mapping function vector for relative zenith troposphere delay (RZTD) $\tau$ after correcting with the New Brunswick 3 (i.e., UNB3) model [17–19]. $\boldsymbol{\mu} = \left[\mu_1, \ldots, \mu_f\right]^{\mathrm{T}}$ with $\mu_j = f_1^2/f_j^2$ is the scalar vector for DD ionosphere parameters $\iota$. $\boldsymbol{\Lambda} = \mathrm{diag}\left(\left[\lambda_1, \ldots, \lambda_f\right]\right)$ is the diagonal matrix of wavelengths for DD ambiguities $\boldsymbol{a} = \left[\boldsymbol{a}_1^{\mathrm{T}}, \ldots, \boldsymbol{a}_f^{\mathrm{T}}\right]^{\mathrm{T}}$. Other residual errors are lumped into observation noise vectors $\boldsymbol{\varepsilon}_P$ and $\boldsymbol{\epsilon}_{\Phi}$. The stochastic model is generalized as [16],

$$\begin{bmatrix} \boldsymbol{Q}_P & \boldsymbol{0} \\ \boldsymbol{0} & \boldsymbol{Q}_{\Phi} \end{bmatrix} \otimes \boldsymbol{Q} \tag{9.2}$$

where $\boldsymbol{Q}_P = \mathrm{diag}\left(\left[\sigma_{P_1}^2, \ldots, \sigma_{P_f}^2\right]\right)$ and $\boldsymbol{Q}_{\Phi} = \mathrm{diag}\left(\left[\sigma_{\Phi_1}^2, \ldots, \sigma_{\Phi_f}^2\right]\right)$ with $\sigma_{P_j}^2$ and $\sigma_{\Phi_j}^2$ as the variance scalars of undifferenced code and phase on the $j$-th frequency. $\boldsymbol{Q}$ is an $(s \times s)$ cofactor matrix of DD observations with elevation-dependent weighting. The observations on different frequencies are assumed to be independent and have the equal variances for code and phase respectively, i.e., $\boldsymbol{Q}_P = \sigma_P^2 \boldsymbol{I}_f$ and $\boldsymbol{Q}_{\Phi} = \sigma_{\Phi}^2 \boldsymbol{I}_f$.

### 9.2.2  Influences of Tropospheric and Ionospheric Delays

Tropospheric delay and ionospheric delay are two major errors contained in GNSS observations. Inappropriate treatment of the delays will decrease the AR efficiency and degrade the positioning performance [11]. NRTK employ multiple reference stations to model and eliminate the delays accurately based on their distance-dependent property. The residual delays are therefore always ignored in the rover positioning model of NRTK [20]. SRTK employ single reference station to reduce the delays simply through DD operation between observations. The residual delays can also be ignored in short-baseline case due to the strong correlation of the delays between rover station and reference station but become considerable in medium-to long-baseline cases. Ignoring residual delays in such cases will cause system biases on the resolved ambiguity and baseline parameters and dramatically degrade positioning performance. A proper treatment is appending appropriate ionosphere and troposphere parameters in the positioning model of medium- and long-baseline SRTK. The residual delays are absorbed by the appended parameters as long as they describe the properties of the delays well and the system biases are then removed. In such a positioning model the influences of the distance-dependent errors are independent with the baseline length theoretically and consequently out of consideration in the respect of extending service radius of SRTK. However, the appended parameters severely weaken the model strength and therefore slow down the convergence of positioning results and degrade positioning precision. Fortunately, MFMS signals multiply the number of the observations manifold and therefore significantly improve the model strength. Hence, SRTK is possible to provide ideal positioning service in medium- and long-baseline cases based on a proper positioning model and MFMS observations [12]. Before introducing the proper mathematical model employed by the rover positioning of NRTK and SRTK, there are some notes to be mentioned.

The residual ionospheric delay becomes difficult to be modeled or described during periods of active ionospheric conditions of the rover positioning. The positioning performances of both NRTK and SRTK degrade dramatically [20, 21]. At low latitudes where active ionospheric activities frequently occur, NRTK increases the density of reference stations to model the ionospheric delay more accurate. However, in most cities located in the mid-latitude region like Shanghai, we simply avoid employing high-precision positioning service during such periods since active ionospheric activities almost only occur at noon. We consequently ignore the active ionospheric conditions here and concentrate our research attention on whether LRTK can provide comparative positioning performance as NRTK in cities of Shanghai-like medium size and mid-latitude.

The residual tropospheric delay is usually modeled by introducing an RZTD parameter combined with an empirical dynamic model. However, the RZTD parameter is strongly correlated with height and therefore extends the initialization time of precise positioning [7]. We also simply ignore it here to avoid long convergence time in SRTK and the height parameter may bias in medium- and long-baseline cases.

The appended parameters are incapable of describing the properties of ionospheric delay and tropospheric delay perfect especially when we ignore the RZTD parameter and active ionospheric conditions. Therefore, the baseline length limitation of SRTK considering appended parameters still exist but is really loose based on the MFMS observation.

### 9.2.3 Ionosphere-Ignored and -Weighted Models

The distance-dependent errors can be ignored for short-baseline SRTK and NRTK. The functional model (9.1) reduces to ionosphere-ignored (I-I) model as

$$
\begin{bmatrix} P \\ \Phi \end{bmatrix} = \begin{bmatrix} e_f \otimes A & 0 \\ e_f \otimes A & \Lambda \otimes I_s \end{bmatrix} \begin{bmatrix} x \\ a \end{bmatrix} + \begin{bmatrix} \varepsilon_P \\ \epsilon_\Phi \end{bmatrix} \tag{9.3}
$$

where the terms have the same meanings as in (9.1) and the stochastic model as in (9.2). As a common strategy for both SRTK and NRTK, the wide-lane (WL) ambiguities are fixed first and then they are substituted into (9.3) for the narrow-lane (NL) AR [22, 23]. After WL ambiguities are fixed, the I-I model becomes

$$
\begin{bmatrix} P \\ \Phi_w \\ \Phi_1 \end{bmatrix} = \begin{bmatrix} e_f \otimes A & 0 \\ e_w \otimes A & 0 \\ e_1 \otimes A & I_s \end{bmatrix} \begin{bmatrix} x \\ a_1 \end{bmatrix} + \begin{bmatrix} \varepsilon_P \\ \epsilon_{\Phi_w} \\ \epsilon_{\Phi_1} \end{bmatrix} \tag{9.4}
$$

where $\boldsymbol{\Phi}_{\mathrm{w}} = \left[\boldsymbol{\Phi}^{\mathrm{T}}_{(i_1,j_1,k_1)}, \ldots, \boldsymbol{\Phi}^{\mathrm{T}}_{(i_w,j_w,k_w)}\right]^{\mathrm{T}}$ denote $(w \times s)$ ambiguity-fixed WL phase observations. $\boldsymbol{\Phi}_{(i,j,k)} = \frac{if_1\boldsymbol{\Phi}_1 + jf_2\boldsymbol{\Phi}_2 + kf_3\boldsymbol{\Phi}_3}{f_{(i,j,k)}}$ with $f_{(i,j,k)} = if_1 + jf_2 + kf_3$ and $f_1, f_2$ and $f_3$ the triple frequencies, respectively. $\boldsymbol{\Phi}_1$, $\boldsymbol{\Phi}_2$ and $\boldsymbol{\Phi}_3$ denote the undifferenced triple-frequency phase observations. For triple-frequency case, we have two types of WL observations i.e., $w = 2$; while for the dual-frequency case, only one type of WL observation, i.e., $w = 1$. $\boldsymbol{a}_1$ is DD ambiguity vector in meters on frequency L1. $\boldsymbol{\epsilon}_{\boldsymbol{\Phi}_{\mathrm{w}}}$ and $\boldsymbol{\epsilon}_{\boldsymbol{\Phi}_1}$ are the residual errors of $\boldsymbol{\Phi}_{\mathrm{w}}$ and $\boldsymbol{\Phi}_1$ respectively. The stochastic model is then derived via error of propagation law as

$$
\begin{bmatrix}
\sigma_P^2 \boldsymbol{I}_f & \boldsymbol{0} & \boldsymbol{0} \\
\boldsymbol{0} & \boldsymbol{Q}_{\boldsymbol{\Phi}_{\mathrm{w}}} & \boldsymbol{Q}_{\boldsymbol{\Phi}_{\mathrm{w}}\boldsymbol{\Phi}_1} \\
\boldsymbol{0} & \boldsymbol{Q}_{\boldsymbol{\Phi}_1\boldsymbol{\Phi}_{\mathrm{w}}} & \sigma_{\boldsymbol{\Phi}}^2
\end{bmatrix} \otimes \boldsymbol{Q}
\tag{9.5}
$$

where $\boldsymbol{Q}_{\boldsymbol{\Phi}_{\mathrm{w}}} = \sigma_{\boldsymbol{\Phi}}^2 \begin{bmatrix} \alpha^2_{(i_1,j_1,k_1)} & \cdots & \alpha_{1w} \\ \vdots & \ddots & \vdots \\ \alpha_{w1} & \cdots & \alpha^2_{(i_w,j_w,k_w)} \end{bmatrix}_{w \times w}$ with $\alpha^2_{(i,j,k)} = \frac{(if_1)^2 + (jf_2)^2 + (kf_3)^2}{f_{(i,j,k)}^2}$ and

$\alpha_{mn} = \frac{i_m i_n f_1^2 + j_m j_n f_2^2 + k_m k_n f_3^2}{f_{(i_m,j_m,k_m)} f_{(i_n,j_n,k_n)}}$. $\boldsymbol{Q}_{\boldsymbol{\Phi}_1\boldsymbol{\Phi}_{\mathrm{w}}} = \sigma_{\boldsymbol{\Phi}}^2 \left[\frac{i_1 f_1}{f_{(i_1,j_1,k_1)}}, \ldots, \frac{i_w f_1}{f_{(i_w,j_w,k_w)}}\right]$ is the cofactor matrix of $\boldsymbol{\Phi}_1$ and $\boldsymbol{\Phi}_{\mathrm{w}}$. Following [24] and [13], the sum of three coefficients $i$, $j$, and $k$ of WL observations are equal to 0.

We ignore tropospheric delay here to avoid too long convergence. The orbit error is also ignored since it is at most 1 cm for baseline as long as 200 km [25]. The ionospheric delay is estimated by introducing an ionosphere parameter for each DD satellite pair. Besides, to enhance the model strength, we introduce the ionospheric constraints as pseudo-observations

$$
\boldsymbol{\iota}_0 = \boldsymbol{\iota} + \boldsymbol{\varepsilon}_{\iota_0}, \quad \sigma_\iota^2 \boldsymbol{Q}
\tag{9.6}
$$

where $\boldsymbol{\iota}_0$ is the vector of nominated ionosphere delays at first epoch and the variance $\sigma_\iota^2$ is used to specify its uncertainty of the errors $\boldsymbol{\varepsilon}_{\iota_0}$. After the first epoch, the filter results of $\boldsymbol{\iota}$ output from the epoch $k - 1$ can further be used as a time-dependent ionospheric constraint for the current epoch $k$, i.e.,

$$
\hat{\boldsymbol{\iota}}_{k-1} = \boldsymbol{\iota}_k + \boldsymbol{\varepsilon}_{\hat{\iota}_{k-1}} + \boldsymbol{d}_k, \quad \boldsymbol{Q}_{\hat{\iota}_{k-1}} + \boldsymbol{W}_k
\tag{9.7}
$$

where $\hat{\boldsymbol{\iota}}_{k-1}$ is the estimate of $\boldsymbol{\iota}_{k-1}$ output from the epoch $k - 1$ and $\boldsymbol{\varepsilon}_{\hat{\iota}_{k-1}}$ is its error with respect to $\boldsymbol{\iota}_{k-1}$. $\boldsymbol{Q}_{\hat{\iota}_{k-1}}$ is its covariance matrix. $\boldsymbol{d}_k$ is a zero mean process noise with a covariance matrix $\boldsymbol{W}_k = \sigma_d^2 \boldsymbol{Q}$. $\sigma_d^2$ specifies the uncertainty of between-epoch ionospheric constraints. Ignoring the RZTD parameter and considering the ionospheric constraints, the functional model (9.1) becomes the ionosphere-weighted (I-W) model as

$$
\begin{bmatrix} P \\ \Phi \\ \bar{\iota} \end{bmatrix} = \begin{bmatrix} e_f \otimes A & \mu \otimes I_s & 0 \\ e_f \otimes A & -\mu \otimes I_s & \Lambda \otimes I_s \\ 0 & I_s & 0 \end{bmatrix} \begin{bmatrix} x \\ \iota \\ a \end{bmatrix} + \begin{bmatrix} \varepsilon_P \\ \epsilon_\Phi \\ \varepsilon_{\bar{\iota}} \end{bmatrix}
\tag{9.8}
$$

where $\bar{\iota}$ denotes the pseudo-observables derived from ionospheric constraints such as (9.6) and (9.7). $\varepsilon_{\bar{\iota}}$ denotes the residual errors of $\bar{\iota}$. The stochastic model of the I-W model for the first epoch follows as below

$$
\begin{bmatrix} Q_P & 0 & 0 \\ 0 & Q_\Phi & 0 \\ 0 & 0 & \sigma_\iota^2 \end{bmatrix} \otimes Q
\tag{9.9}
$$

With WL ambiguities fixed and substituted into (9.8), the I-W model is transformed to a different form with less parameters but the same number of observations as

$$
\begin{bmatrix} P \\ \Phi_w \\ \Phi_1 \\ \bar{\iota} \end{bmatrix} = \begin{bmatrix} e_f \otimes A & \mu \otimes I_s & 0 \\ e_w \otimes A & -\mu_w \otimes I_s & 0 \\ e_1 \otimes A & -\mu_1 \otimes I_s & I_s \\ 0 & I_s & 0 \end{bmatrix} \begin{bmatrix} x \\ \iota \\ a_1 \end{bmatrix} + \begin{bmatrix} \varepsilon_p \\ \epsilon_{\Phi_w} \\ \epsilon_{\Phi_1} \\ \varepsilon_{\bar{\iota}} \end{bmatrix}
\tag{9.10}
$$

where $\mu_w = \left[ \mu_{(i_1,j_1,k_1)}, \ldots, \mu_{(i_w,j_w,k_w)} \right]^{\mathrm{T}}$ with $\mu_{(i,j,k)} = f_1^2 (1/f_1 + 1/f_2 + 1/f_3)/f_{(i,j,k)}$ is the scalar vector of WL observations to DD ionosphere parameters $\iota$. The positioning precision of I-W model is influenced by the ionosphere constraints. A strong ionosphere constraint with small $\sigma_\iota^2$ and $\sigma_d^2$ can significantly strengthen the model and enhance the positioning performance once the values of $\iota_0$ and $d_k$ are accurate enough. However, the inaccurate values of $\iota_0$ and $d_k$ will bias the resolution dramatically especially when the ionosphere constraints are deemed strongly and therefore degrade the positioning performance [16]. A loose ionosphere constraint can hardly degrade the positioning performance but also hardly benefits the AR and positioning. In the next section we compare the strength of I-I model and I-W model through redundancy analysis to demonstrate that LRTK has comparable model strength with NRTK [12].


## 9.3   Model Strength Analysis

The model strength comparison is divided into three parts including redundancy analysis for float solution, success rate (SR) analysis for WL AR and ambiguity dilution of precision (ADOP) analysis SR for NL AR. Such a division of comparison corresponds to the three main procedures of RTK technique. Each procedure is able to output the coordinate results with different precision and the former two procedures are the pre-requisites for the last. The final precise positioning results are only available after the NL AR was successfully completed. In other words, the analyses of the three main procedures reflect the expected positioning performance of RTK and decide whether MFMS SRTK is comparable with NRTK theoretically. The dilution of precision (DOP) values and redundancy are usually used to evaluate the potential resolving quality of the float solution. However, DOP values only depend on the design matrix while redundancy only depends on the numbers of observations and parameters of the positioning model. They are all unilateral to be employed to compare the model strengths. Hence, we propose to use another indication named as redundant observation components (ROCs) to evaluate the model strength of resolving float solution and derive the related formulae based on a Kalman filter. ROCs reflect the expected quality of resolving float solution but is unable to reflect the possibility of a successful ambiguity fixing. The SR indicates the a priori probability of a successful ambiguity fixing [14] and can be precisely calculated based on the given a priori information in a geometry-free (GF) model. Thanks to the relative long wavelength, the WL ambiguities are usually fixed to integers based on GF model [13]. Hence, we use SR to compare the model strengths for WL AR of MFMS SRTK and NRTK. The fixed WL ambiguities are substituted into Eqs. (9.4) or (9.10) to improve the NL AR and the NL ambiguities are normally resolved based on geometry-based (GB) model due to relatively short wavelength. Since the SR is difficult to be calculated in GB model [26], we calculate the ADOP values that also reflect the model strengths of NL AR for comparison. We also give the approximate transformation from ADOP values to SR for a more intuitive reflection on the possibility of a successful ambiguity fixing.

### 9.3.1 Analysis of Float Solution

The float solution gives an approximate positioning result and is an input for AR. Through evaluating the float solution, one can roughly understand how strong the model strength for further AR and precise positioning is. The redundancy defined as the number of observations minus the number of estimated parameters can, to a certain extent, reflect the model strength and then the expected float solution. With the same set of observations, the positioning model with fewer independent parameters has larger redundancy and consequently stronger model strength. For a given positioning model, more observations result in larger redundancy and also stronger model strength. However, for the different positioning models with different types of observations, it is difficult to evaluate the positioning precision by comparing their model strength with redundancy analysis. For instance, the larger redundancy in the single point positioning model with only code observations does not mean that it can obtain more precise solutions than the RTK model even with fewer redundancies. In this case, one should analyze the ROCs instead of redundancy itself. We therefore calculate the ROCs to compare the model strength of SRTK with NRTK to demonstrate SRTK also has a promising float solution. Since the float solutions are often solved sequentially based on a Kalman filter [17, 27], we derive the formula of ROCs based on the Kalman filter. The Kalman filter-based float solution reads

$$\hat{X}_k = \overline{X}_k + K_k\left(L_k - A_k\overline{X}_k\right) \tag{9.11}$$

where $\overline{X}_k = \boldsymbol{\Phi}_{k,k-1}\hat{X}_{k-1}$ is the predicted solution of parameters at epoch $k$, including baseline, ambiguity and ionosphere parameters. $\boldsymbol{\Phi}_{k,k-1}$ is the transition matrix and it is the identity matrix here. $K_k = \left(A_k^{\mathrm{T}}P_kA_k + P_{\overline{X}_k}\right)^{-1}A_k^{\mathrm{T}}P_k$ is the Kalman gain matrix. $P_{\overline{X}_k}$ is the inverse of covariance matrix of $\overline{X}_k$. $L_k, A_k$ and $P_k$ are the observation vector, design matrix and weight matrix of $L_k$ at epoch $k$, respectively. $\hat{X}_k$ is the final Kalman filter solution. The residuals are then calculated as

$$V_k = \left[\, I - A_kK_k \;\; A_kN_k^{-1}P_{\overline{X}_k} \,\right]\begin{bmatrix} L_k \\ \overline{X}_k \end{bmatrix} \tag{9.12}$$

where $N_k = A_k^{\mathrm{T}}P_kA_k + P_{\overline{X}_k}$. The redundancy matrix is the projection from the observations to residuals and represents the redundancies of observations. However, the residuals in a Kalman filter are calculated from not only the real observations $L_k$ but also the pseudo-observations (predicted solution of parameters) $\overline{X}_k$ as shown in (9.12). We derive the redundancy matrix $R_k$ based on the relationship that redundancy matrix equals to the variance-covariance matrix of the residuals multiplied by the weight matrix of corresponding observations as

$$\begin{aligned} R_k = Q_{V_k}P_k &= \left[\, I - A_kK_k \;\; A_kN_k^{-1}P_{\overline{X}_k} \,\right] \\ &\quad \begin{bmatrix} P_k^{-1} & 0 \\ 0 & P_{\overline{X}_k}^{-1} \end{bmatrix}\left[\, I - A_kK_k \;\; A_kN_k^{-1}P_{\overline{X}_k} \,\right]^{\mathrm{T}}P_k \\ &= I - A_kK_k \end{aligned} \tag{9.13}$$

The diagonal elements of $R_k$ are defined as the ROCs. The redundancy matrix is independent of the real observations, but its computation needs the geometry information of the design matrix and weight matrix of the observations. Hence, to identify the real situation, we use a 100 km baseline to compute the design matrix and weight matrix with the I-W model. To make a comparison, a short baseline of 10 km is computed with the I-I model. Here the long-baseline and short-baseline share the same rover station. During the computation, the dual-frequency GPS and triple-frequency BDS signals are used with a sampling rate of 1 s and the cut-off elevation of 10°. The elevation-dependent stochastic model

$$\sigma = \frac{1.02}{\sin\theta + 0.02}\sigma_{90°} \tag{9.14}$$

is applied for the undifferenced measurements with a zenith precision of $\sigma_{90°} = 2$ mm for phase and 0.2 m for code [28]. For the I-W model of long-baseline, the ionospheric

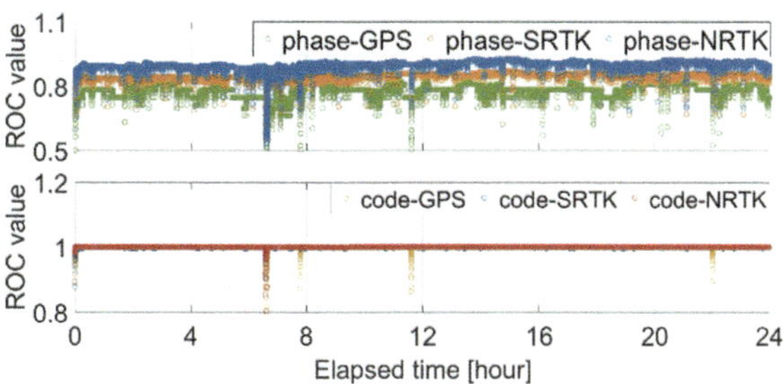

**Fig. 9.1** The average ROCs of all observations



constraints are empirically applied with $\iota_0 = 0$, $\sigma_\iota = 10^{-6} l_{en}$, $\boldsymbol{d}_k = 0$, $\sigma_d = 10^{-8} l_{en}$ where $l_{en}$ is the baseline length in meter.

Figure 9.1 displays the average ROCs of all phase observations and all code observations respectively. The "code-NRTK" and "code-SRTK" represent the average ROCs of code observations for short- and long-baseline, respectively. They are all close to 1 after the first epoch, which means that the contribution of code observations on the final solution is very limited due to their poor precisions except for the first epoch. The "phase-NRTK" and "phase-SRTK" represent the average ROCs of phase observations of short- and long-baseline, respectively.

Since the GPS-only NRTK with the I-I model is traditionally used to provide NRTK services, we calculate the ROCs of NRTK only with dual-frequency GPS signals for further comparison. The results are denoted by "phase-GPS" and "code-GPS". Obviously, due to lack of high precision signals, the ROCs of phase-GPS are relatively smaller than those of phase-NRTK with MFMS signals and even smaller than those of phase-SRTK with the I-W model based MFMS signals. We consequently conclude that the expected performance of the float solution of MFMS SRTK is worse than MFMS NRTK but better than GPS-only NRTK. Such a comparison results of the expected float solution is sufficient to support that LRTK has potential to provide comparable positioning service as NRTK after the ambiguities being correctly fixed since the performance of ambiguity-fixed solutions of GPS-only NRTK is comparable to MFMS NRTK with a denser CORS network. We further need to analyze the AR efficiency to transfer the "potential" to "ability".

### 9.3.2 Analysis of WL AR

The SR is a useful indicator reflecting whether the ambiguity can be fixed to its integer [14]. We analyze the AR efficiency of different WL combinations used in SRTK and NRTK by analyzing their SRs. With the GF model, one can derive the float ambiguity estimate as $\hat{a}_{(i,j,k)} = \frac{P_{(l,m,n)} - \Phi_{(i,j,k)}}{\lambda_{(i,j,k)}}$ with wavelength $\lambda_{(i,j,k)}$, and its standard deviation (STD) as $\sigma_{\hat{a}_{(i,j,k)}} = \frac{\sqrt{\sigma_\Phi^2 \alpha_{(i,j,k)}^2 + \sigma_P^2 \alpha_{(l,m,n)}^2}}{\lambda_{(i,j,k)}}$. For NRTK, the float WL ambiguity is adequately assumed to be unbiased. Its SR is

$$P_{s,u} = 2\Phi\left(\frac{1}{2\sigma_{\hat{a}_{(i,j,k)}}}\right) - 1 \tag{9.15}$$

where $\Phi(x) = \int_{-\infty}^{x} \frac{1}{\sqrt{2\pi}}\exp\left(-\frac{v^2}{2}\right)dv$. For long-range SRTK, the float WL ambiguity could be biased by the unmodelled ionospheric delay. The SR of fixing this biased WL ambiguity reads

$$P_{s,b} = \Phi\left(\frac{1 + 2\delta_{\hat{a}_{(i,j,k)}}}{2\sigma_{\hat{a}_{(i,j,k)}}}\right) + \Phi\left(\frac{1 - 2\delta_{\hat{a}_{(i,j,k)}}}{2\sigma_{\hat{a}_{(i,j,k)}}}\right) - 1 \tag{9.16}$$

where $\delta_{\hat{a}_{(i,j,k)}} = \frac{\mu_{(i,j,k)} + \mu_{(l,m,n)}}{\lambda_{(i,j,k)}}\iota$ represents the bias of $\hat{a}_{(i,j,k)}$ caused by unmodelled ionospheric delay $\iota$. In theory, $P_{s,b}$ is smaller than $P_{s,u}$ for the same WL combination with the same STD $\sigma_{\hat{a}_{(i,j,k)}}$. However, in practice, the SRs would be comparable for WL AR in NRTK with (9.15) and in long-range SRTK with (9.16). The reason is that the STD $\sigma_{\hat{a}_{(i,j,k)}}$ is typically very small with several epochs and in such a case, a small bias $\delta_{\hat{a}_{(i,j,k)}}$ in cycles (due to the large wavelength) can hardly affect the SR [13]. Moreover, there are some ionosphere-free (IF) WL combinations with which the unbiased float solution is also obtainable in long-range SRTK. Table 9.1 presents the SRs of several sets of commonly used WL combinations for NRTK and SRTK, respectively. The SRs are computed for the number of epochs from 1 to 5. In computations, we set $\sigma_{\Phi} = 5$ mm and $\sigma_P = 0.5$ m, and for long-range SRTK, $\iota = 0.3$ m as an error budget representing the baseline length typically from 100 to 200 km [7]. The SRs of WL AR in SRTK are very close to those in NRTK as shown in Table 9.1. Therefore, LRTK has a comparable efficiency for WL AR as NRTK.

### 9.3.3  Analysis of NL AR

The ADOP is proposed by [15] and defined as,

$$\text{ADOP} = \sqrt{\det Q_{\hat{a}}}^{\frac{1}{n}} \tag{9.17}$$

where $Q_{\hat{a}}$ is the covariance matrix of float ambiguity solution with dimension $n$. ADOP is a well-known scalar measure used to infer the strength of the GNSS model for AR [29]. Based on ADOP, the upper bound of SR can be computed [26]

$$P_s \cong \left[2\Phi\left(\frac{1}{2\text{ADOP}}\right) - 1\right]^n \tag{9.18}$$

With ADOP of 0.15, the SR of 99% can be obtained. Although this upper bound could be loose [28], it can give insight into AR capability for the purpose of comparison. Hence, we compare the ADOP values of NL AR for a short baseline with the I-I

**Table 9.1** SR of WL combinations employed by NRTK and MFMS SRTK

| | System | Combination schemes | $\lambda_{(i,j,k)}$ (m) | $\mu_{(i,j,k)} + \mu_{(l,m,n)}$ | $\sigma^{SE}_{\hat{a}_{(i,j,k)}}$ (cycle) | SR (%) | | | | | |
|---|---|---|---|---|---|---|---|---|---|---|---|
| | | | | | | 1 | 2 | 3 | 4 | 5 | 6 |
| SRTK $\iota = 0.3$ m | GPS | $\Phi_{(1,-1,0)}$ | $P_{(1,1,0)}$ | 0.862 | 0 | 0.415 | 77.21 | 91.18 | 96.32 | 98.41 | 99.30 |
| | BDS | $\Phi_{(0,1,-1)}$ | $P_{(0,1,1)}$ | 4.884 | 0 | 0.078 | 100 | 100 | 100 | 100 | 100 |
| | | $\Phi_{(1,4,-5)}$ | $P_{(1,1,1)}$ | 6.371 | 2.015 | 0.143 | 99.77 | 100 | 100 | 100 | 100 |
| NRTK $\iota = 0$ | GPS | $\Phi_{(1,-1,0)}$ | $P_{(1,1,0)}$ | 0.862 | 0 | 0.415 | 77.21 | 91.18 | 96.32 | 98.41 | 99.30 |
| | BDS | $\Phi_{(0,1,-1)}$ | $P_{(0,1,1)}$ | 4.884 | | 0.078 | 100 | 100 | 100 | 100 | 100 |
| | | $\Phi_{(1,4,-5)}$ | $P_{(1,1,1)}$ | 6.371 | | 0.143 | 99.95 | 100 | 100 | 100 | 100 |

*Note* $\sigma^{SE}_{\hat{a}_{(i,j,k)}}$ denotes the formal STD of the single-epoch float ambiguity estimate, based on which SRs of these combination schemes are computed for the number of epochs from 1 to 5

**Fig. 9.2** The ADOP values of NL AR

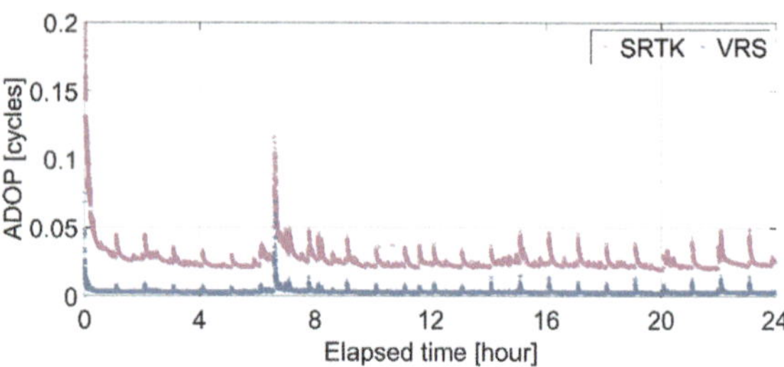



model and a long baseline with the I-W model, which specify the NRTK and SRTK, respectively. The parameter settings in the computations are completely the same as for computing ROCs in Sect. 9.3. The results are shown in Fig. 9.2.

The legends of SRTK and NRTK denote the ADOP values for the long baseline with the I-W model and for the short baseline with the I-I model, respectively. The regular fluctuations are caused by the change of satellite subset for partial NL AR [22, 30]. Although the I-I model of NRTK indeed have much smaller ADOP values than the I-W model of SRTK, the ADOP values of SRTK, whose corresponding upper bound of SRs exceed 99% in a few epochs, are already small enough for achieving comparable NL AR efficiency as NRTK.

The AR efficiency of SRTK with I-W model is comparable to NRTK with I-I model based on MFMS observations. Combined with the conclusion based on the analysis with ROC in Sect. 9.3.1, we roughly conclude that long-range MFMS SRTK has ability to provide comparable positioning service as NRTK with the I-I model when the baseline is no more than 100 km theoretically. In following, we will make a comprehensive comparison with real data processing to verify our rough conclusion.

## 9.4   Results and Discussion

Three experiments with real observations were carried out to compare the performance of long-range SRTK and NRTK in Shanghai. The long-range MFMS SRTK software was developed by the GNSS group at Tongji University, which is named as Tongji real-time kinematic positioning (TJRTK) in the following. The first experiment compares the performance of TJRTK with Trimble-VRS (TVRS) maintained by the Shanghai Institute of Surveying and Mapping. This experiment was conducted on land within the Shanghai area. The second experiment compares TJRTK with Land-Star that is another NRTK system maintained by the Shanghai Center of Maritime Surveying and Mapping (SCMSM), Ministry of Transport. This experiment was conducted in the offshore area of the Yangtze river estuary near Shanghai. Besides, the third experiment comprehensively evaluates the performance of TJRTK in a variety of long baselines, including the positioning accuracy, ratio of fixed solution as well as the initialization time. The initialization time in the following refers to the time to first fix the NL ambiguities. In all experiments, the dual-frequency GPS and

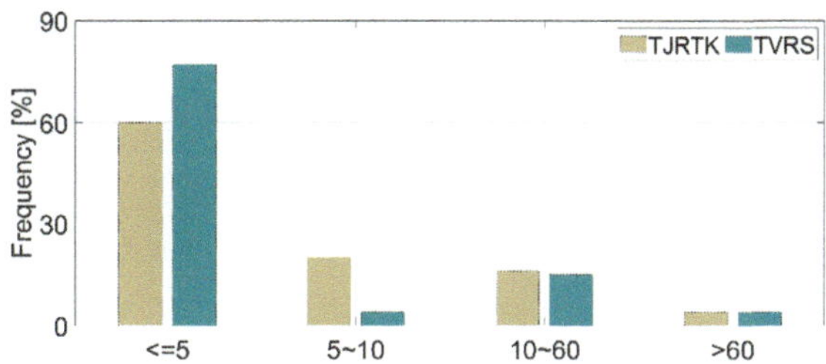

**Fig. 9.3** Initialization time (m) of TJRTK and TVRS

triple-frequency BDS observations were used with the elevation-dependent model (9.14).

### 9.4.1 Comparison Between LRTK and TVRS

For SRTK, we set up a reference station in the center of Shanghai. There are a total 43 points with distances to reference station from 10 to 60 km. We conducted RTK at each of 43 points by using TJRTK and TVRS, respectively. At each point, we restarted both TJRTK and TVRS four times and each time we occupied the point for 15 s after the initialization and recorded the average ambiguity-fixed solution as well as the initialization time. Hence, we have four coordinate solutions and four initialization time results for each software at every point.

Figure 9.3 shows the probability distribution of initialization time of TJRTK and TVRS at each point. The initialization time is less than 60 s by larger than 90% for both TJRTK and TVRS. The average initialization time is about 10 s for TVRS and 12.5 s for TJRTK. The static experiment displays the positioning performance of TJRTK and TVRS resolving 43 points. The statistics indicate that TJRTK can reach equivalent positioning performance including centimetre-level positioning precision and quick convergence compared to TVRS.

### 9.4.2 Comparison Between LRTK and LandStar

Another experiment was carried out in the offshore area of the Yangtze river estuary near Shanghai. In this area, the LandStar system maintained by SCMSM provides the NRTK service with 12 CORS stations. We conducted the kinematic positioning in the area of blue shadow, and changed 5 reference stations (green circles) to test the performance of our TJRTK in varying baseline length. As shown in Fig. 9.4, all of two LandStar receivers and one TJRTK receiver were mounted on a plane, and the TJRTK receiver was exactly at the middle of two LandStar receivers. Such layout guaranteed that the position of TJRTK receiver was the mean position of two LandStar receivers. As a result, one can compare the TJRTK solution with the

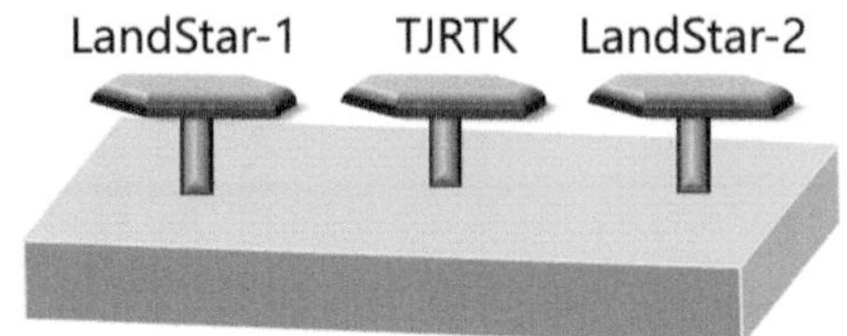



**Fig. 9.4** Illustration of rover stations of TJRTK and LandStar

mean position of two LandStar solutions. All RTK solutions were recorded second by second except that the LandStar results might discontinue sometimes due to the loss of NRTK correction communication.

In the following comparison, the mean value of RTK solutions from two LandStar receivers is used as reference. Then the difference between the TJRTK solution and its reference provided by the LandStar NRTK system is defined as the coordinate differences of TJRTK. To form different baseline lengths, during the RTK experiment, we changed the reference stations of R1, R2, R3, R4 and R5, respectively, and for each reference station, the RTK experiment lasted by about 3 h. Figure 9.5 presents the coordinate differences of TJRTK with different reference stations where the baseline lengths are about 30, 45, 78, 140 and 160 km, respectively. The solution gaps are due to the gaps in the LandStar NRTK corrections and then the lack of LandStar reference solutions. Table 9.2 provides the root-mean-square (RMS) statistics of TJRTK coordinate differences and the initialization time to get the first ambiguity-fixed solutions.

From Fig. 9.5, the coordinate differences are mostly smaller than 5 cm for N and E components on both short and long baselines. The coordinate differences are relatively larger for the U component but all are much smaller than 20 cm and even smaller than 10 cm for most of epochs. With the LandStar NRTK as references, the RMS of TJRTK coordinate differences are all smaller than 2 cm for the two horizontal components. Such precisions are better than those of the first experiment. The reason is that the observation environment on the ocean is very open and much better than that in the urban area. The horizontal RMS are comparable for the different baselines, whereas the vertical RMS are larger in the long-baseline tests than those in the short-baseline tests. The reason is as follows. In our processing the RZTD parameter was not set up and the increased residual tropospheric delays in the long baselines were to a certain degree absorbed by the height parameter [31]. Again, due to the increased residual errors in long baselines, the initialization time of TJRTK increased as well. In terms of positioning precision and convergence time, it can be concluded that the performance of TJRTK is comparable to NRTK for baselines shorter than 50 km. For baselines longer than 100 km, TJRTK can still obtain comparable horizontal precision while the vertical precision and convergence time deteriorate.

**Fig. 9.5** TJRTK coordinate differences with R1, R2, R3, R4 and R5 (from top to bottom) as reference station

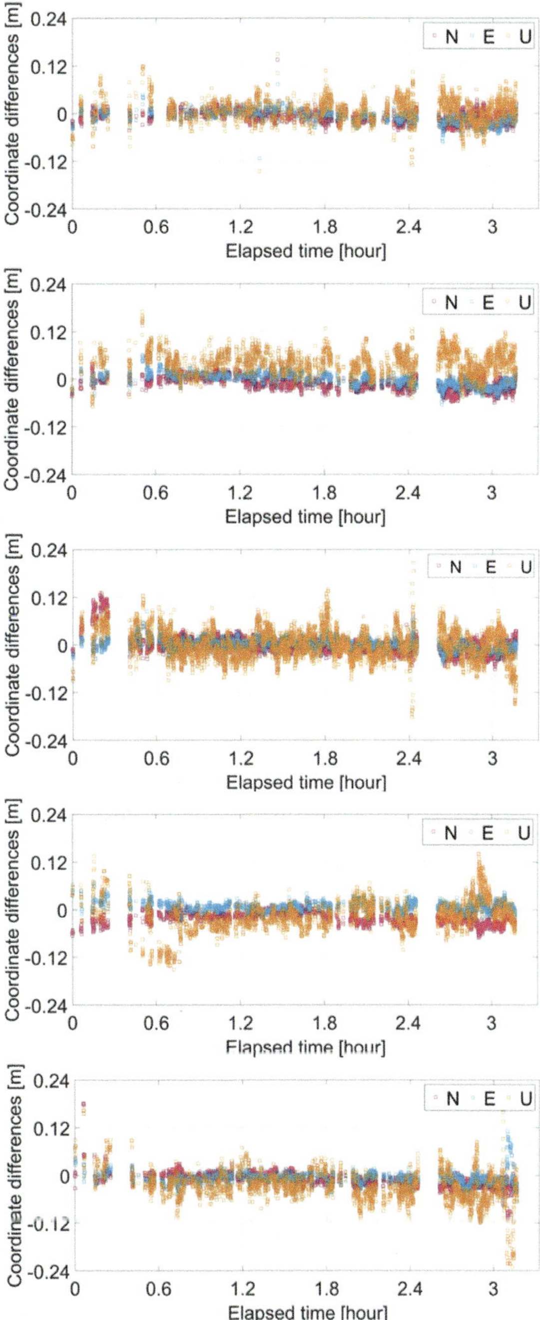

**Table 9.2** The statistics of TJRTK coordinate differences and the initialization time

| Reference station | Baseline length (km) | Initialization time (s) | RMS (cm) | | |
|---|---|---|---|---|---|
| | | | N | E | U |
| R1 | 30 | 10 | 2.8 | 1.9 | 3.4 |
| R2 | 45 | 0 | 2.6 | 2.2 | 4.8 |
| R3 | 78 | 8 | 1.6 | 2.3 | 7.1 |
| R4 | 140 | 41 | 1.8 | 2.5 | 10.3 |
| R5 | 160 | 66 | 2.5 | 1.7 | 5.9 |

### 9.4.3 Performance Assessment of LRTK

In the former two experiments, we evaluated the positioning performance of TJRTK by comparing it with NRTK services in urban and ocean environments. In this experiment, we further evaluate the TJRTK performance in detail to show its capability to provide a high-precision RTK positioning service in Shanghai. Three static stations are located to form two long baselines of 122 km (TJ01-TJ03) and 147 km (TJ02-TJ03) and one short baseline of 27 km (TJ01-TJ02). The observation duration is 24 h and sampling rate is 1 s. The precise coordinates of the three stations were computed with the Bernese GNSS Software Version 5.2 in precise point positioning (PPP) mode with all 24-h observations and precise orbit products. The solution serves as the benchmark in the following analyses. All three baselines were resolved with TJRTK. The coordinate differences shown in Fig. 9.6 are defined as the differences between the TJRTK results and their benchmark. The horizontal axis represents the elapsed time in the unit of Universal Time Coordinated (UTC) hour. The fluctuations of the coordinate differences during the UTC time 4–6 (local time 12–14) of the clock are slightly larger than other periods due to the active ionospheric condition at noon. The RMS statistics after the initialization time are presented in Table 9.3. The horizontal accuracies are about 2 cm. The vertical accuracies of long baselines are inferior relative to those of short baseline since the RZTD parameter was not set up in TJRTK [31]. The ratio of fixed solutions is defined as the proportion of the number of fixed solutions relative to the number of total solutions. The ratios of fixed solution exceeded 99% for all three baselines. It means that the ratio of fixed solutions is independent of the baseline length.

Besides, the three baselines form a synchronous closed loop. The misclosure of integer ambiguities of the same DD satellite pair from three baselines should be equal to 0. With this theoretical condition, we can check two indicators to evaluate the AR efficiency. One is the ratio of correctly fixed ambiguities (RCFA), which is defined as the ratio of the number of correctly fixed ambiguities to the total number of fixed ambiguities. The other indicator is the ratio of correctly fixed solutions (RCFS), which is defined as the ratio of the number of correctly fixed solutions to the total number of fixed solutions and a correctly fixed solution is confirmed only when all ambiguities of one epoch are correctly fixed. Both RCFA and RCFS exceed 99.9%

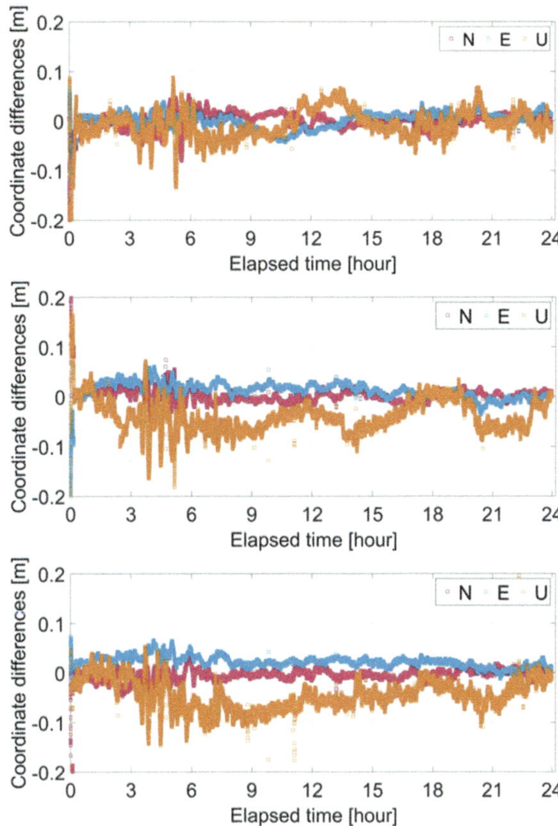

**Fig. 9.6** The coordinate differences of TJ02-TJ01 (top), TJ01-TJ03 (middle) and TJ02-TJ03 (bottom)

**Table 9.3** The RMS accuracies of TJRTK coordinate differences and their corresponding ratios of fixed solutions

| Baseline (length) | Ratio of fixed solutions (%) | RMS (cm) | | |
| --- | --- | --- | --- | --- |
| | | N | E | U |
| TJ02-03 (147 km) | 99.09 | 1.16 | 2.38 | 5.52 |
| TJ01-03 (122 km) | 99.86 | 1.12 | 1.86 | 5.17 |
| TJ02-01 (27 km) | 99.63 | 1.41 | 1.54 | 2.80 |

in this experiment. It means that the correctness of ambiguity fixing in TJRTK does not degrade with increasing baseline length.

However, the condition of zero misclosure of ambiguities from three baselines is not sufficient to confirm the correct ambiguity fixing since the ambiguities of two baselines could be biased with the same integer value although it rarely occurs. Hence, we check the correctness of TJRTK solutions with the misclosure of three baseline solutions since any wrong ambiguity fixing will definitely bias the baseline

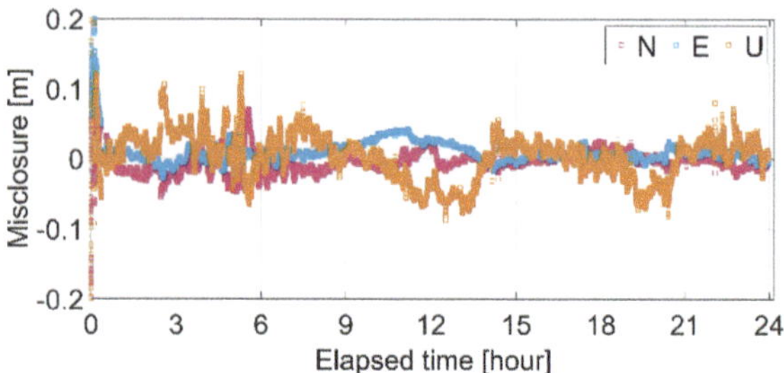

**Fig. 9.7** Closure discrepancy sequences of baseline parameters in geocentric coordinate system

solutions. As shown in Fig. 9.7, except for the beginning epochs with convergence, all misclosures are much smaller than 10 cm and most of them are smaller than 5 cm. The results further confirm the AR correctness of TJRTK.

We finally evaluate the initialization time of TJRTK. We included more stations to form a total of 11 baselines with lengths from 40 to 130 km. For each baseline, we restarted the RTK engine hourly to obtain 24 results. In total 264 results of initialization time were obtained. The probability mass function (PMF) with respect to the initialization time is shown in Fig. 9.8. The results indicate that the initialization time is smaller than 30 s with 100% for short baselines, while it is smaller than 30 s with 85% and 40 s with 90% for long baselines, respectively. Considering that in long-baseline RTK, the MFMS observations are applied, one can not always fix all ambiguities and even does not necessarily need to fix all ambiguities for achieving high-precision positioning. An AR fixing ratio of 80% of is already enough to compute high-precision solutions.

Based on the experimental analysis on the positioning accuracy, AR efficiency and initialization time, it is summarized that the long-range SRTK exhibits the rather comparable performance as NRTK when the baseline is shorter than 50 km. Even for the baseline as long as 100 km, the positioning accuracies are still comparable with NRTK except for the vertical component and relatively longer convergence. Therefore, we can reassure that the long-range SRTK is able to provide comparable positioning service with the NRTK in Shanghai with much reduced expense.

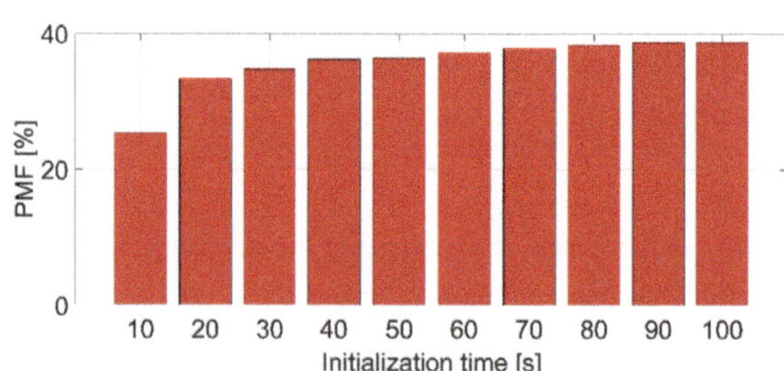

**Fig. 9.8** Initialization time of TJRTK

## 9.5 Conclusion

We explore the capability of LRTK (implemented by TJRTK software) with MFMS observations in high-precision positioning from both theoretical and practical aspects. Regarding the medium city with Shanghai-like area, we try to answer the question whether we can use the long-range SRTK to provide comparable positioning service as NRTK system. The research findings and conclusions are summarized as follows: TJRTK is able to provide centimeter-level positioning service in Shanghai based on RTK instead of NRTK. The costs of the CORS infrastructure maintenance needed by NRTK will be dramatically reduced by TJRTK; the positioning performance of TJRTK is comparable to NRTK when the baseline is generally shorter than 50 km. Even when the baseline is extended to about 100 km, TJRTK can still provide the desirable horizontal solutions with relatively enlarged convergence time; TJRTK has a very promising prospect for offshore applications, where it is rather difficult or very expensive to establish and maintain a CORS network on ocean. With TJRTK, one can employ the reference station on the shore to realize the high-precision RTK on a large area of offshore.

## References


1. Janssen V, Haasdyk J (2011) Assessment of network RTK performance using CORSnet-NSW. Proceedings of international global navigation satellite system society, pp 1–18
2. Wubbena G, Bagge A, Seeber G, Boder V, Hankemeier P (1996) Reducing distance dependent errors for real-time precise DGPS applications by establishing reference station networks. Proceedings of ION GPS, pp 1845–1852
3. Takac F, Zelzer O (2008) The relationship between network RTK solutions MAC, VRS, PRS, FKP and i-MAX. Proceedings of the ION GNSS 2008, pp 348–355
4. Fotopoulos G, Cannon ME (2001) An overview of multi-reference station methods for cm-level positioning. GPS Solut 4:1–10
5. Grejner-Brzezinska DA, Kashani I, Wielgosz P, Smith DA, Spencer PSJ, Robertson DS, Mader GL (2007) Efficiency and reliability of ambiguity resolution in network-based real-time kinematic GPS. J Surv Eng 133:56–65
6. Wanninger L (2002) Virtual reference stations for centimeter-level kinematic positioning. Proceedings of the ION GPS 2002, pp 1400–1407
7. Wang C, Feng Y, Higgins M, Cowie B (2010) Assessment of commercial network RTK user positioning performance over long inter-station distances. J Glob Position Syst 9:78–89
8. Li X, Ge M, Douša J, Wickert J (2014) Real-time precise point positioning regional augmentation for large GPS reference networks. GPS Solut 18:61–71
9. Gao W, Gao C, Pan S, Yu G, Hu H (2017) Method and assessment of BDS triple-frequency ambiguity resolution for long-baseline network RTK. Adv Space Res 60:2520–2532
10. Vollath U, Buecherl A, Landau H, Pagels C, Wagner B (2000) Long-range RTK positioning using virtual reference stations. Proceedings of the ION GPS 2000, pp 1143–1147
11. Grejner-Brzezinska DA, Arslan N, Wielgosz P, Hong C-K (2009) Network calibration for unfavorable reference-rover geometry in network-based RTK: Ohio CORS case study. J Surv Eng 135:90–100
12. Odolinski R, Teunissen PJG, Odijk D (2015) Combined GPS + BDS for short to long baseline RTK positioning. Meas Sci Technol 26:045801

13. Li B, Feng Y, Gao W, Li Z (2015) Real-time kinematic positioning over long baselines using triple-frequency BeiDou signals. IEEE Trans Aerosp Electron Syst 51:3254–3269
14. Teunissen PJG (1999) An optimality property of the integer least-squares estimator. J Geod 73:587–593
15. Teunissen PJG, Odijk D (1997) Ambiguity dilution of precision: definition, properties and application. Proceedings of the ION GPS 1997, pp 891–899
16. Li B, Li Z, Zhang Z, Tan Y (2017) ERTK: extra-wide-lane RTK of triple-frequency GNSS signals. J Geod 91:1031–1047
17. Wang M, Li B (2016) Evaluation of empirical tropospheric models using satellite-tracking tropospheric wet delays with water vapor radiometer at Tongji, China. Sensors (Basel) 16:186
18. Niell AE (1996) Global mapping functions for the atmosphere delay at radio wavelengths. J Geophys Res Solid Earth 101:3227–3246
19. Collins JP, Langley RB (1997) A tropospheric delay model for the user of the wide area augmentation system. University of New Brunswick, New Brunswick
20. Wielgosz P, Kashani I, Grejner-Brzezinska D (2005) Analysis of long-range network RTK during a severe ionospheric storm. J Geod 79:524–531
21. Mohino E, Gende M, Brunini C (2007) Improving long baseline (100–300 km) differential GPS positioning applying ionospheric corrections derived from multiple reference stations. J Surv Eng 133:1–5
22. Li B, Shen Y, Feng Y, Gao W, Yang L (2014) GNSS ambiguity resolution with controllable failure rate for long baseline network RTK. J Geod 88:99–112
23. Tang W, Deng C, Shi C, Liu J (2014) Triple-frequency carrier ambiguity resolution for Beidou navigation satellite system. GPS Solut 18:335–344
24. Cocard M, Bourgon S, Kamali O, Collins P (2008) A systematic investigation of optimal carrier-phase combinations for modernized triple-frequency GPS. J Geod 82:555–564
25. Teunissen PJG, Kleusberg A (1998) GPS for geodesy. Springer, Berlin
26. Teunissen PJG (2003) An invariant upperbound for the GNSS bootstrapped ambiguity success rate. J GPS 2:13–17
27. Li X, Ge M, Dai X, Ren X, Fritsche M, Wickert J, Schuh H (2015) Accuracy and reliability of multi-GNSS real-time precise positioning: GPS, GLONASS, BeiDou, and Galileo. J Geod 89:607–635
28. Verhagen S, Li B, Teunissen PJG (2013) Ps-LAMBDA: ambiguity success rate evaluation software for interferometric applications. Comput Geosci 54:361–376
29. Odijk D, Teunissen PJG (2008) ADOP in closed form for a hierarchy of multi-frequency single-baseline GNSS models. J Geod 82:473–492
30. Brack A (2017) Reliable GPS + BDS RTK positioning with partial ambiguity resolution. GPS Solut 21:1083–1092
31. Li B, Feng Y, Shen Y (2010) Three carrier ambiguity resolution: distance-independent performance demonstrated using semi-generated triple frequency GPS signals. GPS Solut 14:177–184

# Chapter 10
# ERTK: Extra-Wide-Lane RTK

## 10.1  Introduction

So far, all satellites of both BeiDou Navigation Satellite System (BDS) and Galileo systems and partial satellites of Global Positioning System (GPS) have been providing at least triple-frequency signals. In near future, the triple-frequency signals will be fully available. And future Global Navigation Satellite System (GNSS) will transmit three or more frequency signals. It is anticipated that the efficiency and the reliability of carrier ambiguity resolution (AR) for long distance can be significantly enhanced with additional frequency signals, which is rather crucial to realize real-time precise positioning at regional or even global scales. It is an irreversible trend to develop the multi-frequency (with three or even more frequencies) GNSS systems. Compared to dual-frequency GPS signals, the additional frequency signals will speed up the carrier-phase AR [1–3], improve the precision and reliability of positioning [4], mitigate or inverse the various categories of error sources [5, 6], reduce the communication bandwidth of transmission and so on [7], therefore promising for long baseline real-time kinematic positioning (RTK) [8] and large-scale network RTK applications [2, 9].

The previous studies on triple-frequency signals are as following logical sequence. Since the more combinations can be formed by triple-frequency signals with respect to the dual-frequency signals, the very beginning studies were mainly on seeking for optimal combinations of various applications, for instance, fast AR [2, 10, 11], cycle slip detection [12, 13], high-precision positioning and ionosphere inversion etc. [6]. It is concluded that no matter what method was applied, the obtained combinations are nearly the same. Afterwards, many studies focused on the triple-frequency AR based on these combinations. Many methods, typically with three carrier ambiguity resolution (TCAR) and cascading integer resolution (CIR) as representations, were proposed for successively fixing ambiguities in order of wavelengths [1, 10, 11, 14, 15]. Actually, all of these methods are essentially equivalent to a bootstrapping procedure [16]. Regarding triple-frequency AR, a common conclusion advises

© The Author(s) 2025

B. Li et al., *GNSS Real-Time Kinematic Positioning*, Navigation: Science and Technology 17, https://doi.org/10.1007/978-981-96-9116-6_10

that the extra-wide-lane (EWL)/wide-lane (WL) ambiguities can be reliably solved instantaneously or at most with very few epochs nearly without distance restriction, but the fast narrow-lane (NL) AR is still challenging depending on the atmospheric behaviors over long baselines [17]. In addition, as another important benefit, the improvement of observation redundancy and then the reliability gained by triple-frequency signals was numerically investigated by [4], which is rather important in quality control. However, all these works stayed on the theoretical study or numerical analysis based on purely simulated triple-frequency data. Until the end of 2013 when the BeiDou Interface Control Document (ICD) was released, the real triple-frequency BDS data was used to intensively demonstrate the triple-frequency capabilities for AR [18, 19], precise point positioning [20, 21], short-baseline RTK [22–24] as well as the stochastic modelling [25].

As aforementioned, the superiority of triple-frequency GNSS signals with respect to dual-frequency ones is to form more useful combinations, of which the EWL combinations are most useful for instantaneous AR with very high success rate over several tens to hundred kilometer baselines [2, 11, 26]. However, in a long term, we are used to start with the centimetre RTK solutions after all carrier ambiguities are fixed although this process may take many minutes particularly when the baselines are over tens to hundreds of kilometers. In this processing, the ambiguity-fixed EWL observations serve as pseudoranges but with higher precision than actual pseudoranges. It is thus expectable to obtain a better RTK solution directly with EWL than pseudorange, which was foreseen in Feng and Li [26].

This chapter dedicates to fully exploit the RTK capability of virtual EWL signals over long baselines, which is referred to as Extra-wide-lane RTK (ERTK). First of all, the canonical formulae of single-epoch float ambiguity solution are derived for a variety of models, i.e., ionosphere-weighted, -fixed and -float in both geometry-based and geometry-free models. Based on these canonical formulae the easiness of EWL AR and the difficulty of NL AR are shown. Then three ERTK models, ionosphere-ignored, -float and -smoothed models, are formulated and their relationships are discussed. Finally, by using the real triple-frequency BDS data from a 4-station network with baseline lengths from 33 to 75 km, the ERTK performance is numerically demonstrated. The results show that the ionosphere-ignored model is overall better than the ionosphere-float model. Such result can be further improved to centimeter level by ionosphere-smoothed model, which is equivalent to the ionosphere-float model smoothed by adding NL observations but without complicated NL AR.

The chapter is structured as follows. The canonical formulae of single-epoch float ambiguity solution are derived, based on which we show the easiness of EWL AR and the difficulty of NL AR. Then we first formulate two ERTK models, i.e., ionosphere-ignored and -float models. The ERTK equivalence of using any two EWL/WL observations is proved. Besides, the condition for choosing ionosphere-ignored or -float model is presented. Finally, the ionosphere-smoothed ERTK model is introduced. The experiment and analysis are carried out, and some concluding remarks are summarized.

## 10.2  Multi-frequency Observation Model

In this section, the functional and stochastic models of single-epoch $f$-frequency double-differenced (DD) observations together with ionospheric constraints are outlined at first. Based on which, the canonical formulae of covariance matrix of float ambiguity solution are derived. Then the difficulty of NL AR and the easiness of EWL AR are numerically demonstrated.

Considering the residual ionosphere and troposphere effects, the DD observations equations of code and phase on $f$ frequencies read

$$
\mathrm{E}\left(\begin{bmatrix} \boldsymbol{P} \\ \boldsymbol{\phi} \end{bmatrix}\right) = \begin{bmatrix} \boldsymbol{e}_f \otimes \boldsymbol{A} \ \boldsymbol{e}_f \otimes \boldsymbol{g} & \boldsymbol{\mu} \otimes \boldsymbol{I}_s & \boldsymbol{0} \\ \boldsymbol{e}_f \otimes \boldsymbol{A} \ \boldsymbol{e}_f \otimes \boldsymbol{g} & -\boldsymbol{\mu} \otimes \boldsymbol{I}_s & \boldsymbol{\Lambda} \otimes \boldsymbol{I}_s \end{bmatrix} \begin{bmatrix} \boldsymbol{x} \\ \tau \\ \iota \\ \boldsymbol{a} \end{bmatrix}
\tag{10.1}
$$

where $\boldsymbol{P} = \left[ \boldsymbol{P}_1^{\mathrm{T}}, \ldots, \boldsymbol{P}_f^{\mathrm{T}} \right]^{\mathrm{T}}$ is the $f$-frequency code observations with $\boldsymbol{P}_j$ the observations of frequency $f_j$. $\boldsymbol{\phi}$ is the $f$-frequency phase observations with the same structure as $\boldsymbol{P}$. $\boldsymbol{A}$ is the design matrix to baseline parameter $\boldsymbol{x}$; while $\boldsymbol{g}$ is the mapping function vector to relative zenith troposphere delay (RZTD) $\tau$ after correcting with University of New Brunswick 3 (UNB3) model [27]. $\boldsymbol{\mu} = \left[ \mu_1, \ldots, \mu_f \right]^{\mathrm{T}}$ with $\mu_j = f_1^2/f_j^2$ the scalar vector to DD ionosphere parameters $\iota$. $\boldsymbol{\Lambda} = \mathrm{diag}\left(\left[\lambda_1, \ldots, \lambda_f\right]\right)$ is diagonal matrix of wavelengths to DD ambiguities $\boldsymbol{a} = \left[ \boldsymbol{a}_1^{\mathrm{T}}, \ldots, \boldsymbol{a}_f^{\mathrm{T}} \right]^{\mathrm{T}}$. The subscript $s$ denotes the number of DD satellite pairs.

We specify the stochastic model of (10.1) as

$$
\mathrm{D}\left(\begin{bmatrix} \boldsymbol{P} \\ \boldsymbol{\phi} \end{bmatrix}\right) = \begin{bmatrix} \boldsymbol{Q}_P & \boldsymbol{0} \\ \boldsymbol{0} & \boldsymbol{Q}_\phi \end{bmatrix} \otimes \boldsymbol{Q} = \boldsymbol{Q}_f \otimes \boldsymbol{Q}
\tag{10.2}
$$

where $\boldsymbol{Q}_P = \mathrm{diag}\left(\left[\sigma_{P_1}^2, \ldots, \sigma_{P_f}^2\right]\right)$ and $\boldsymbol{Q}_\phi = \mathrm{diag}\left(\left[\sigma_{\phi_1}^2, \ldots, \sigma_{\phi_f}^2\right]\right)$ with $\sigma_{P_j}^2$ and $\sigma_{\phi_j}^2$ the variance scalars of undifferenced code and phase on the $j$th frequency. $\boldsymbol{Q}$ is an $(s \times s)$ cofactor matrix of DD observations with elevation-dependent weighting. In following, the unique variances are assumed respectively for code and phase, namely, $\boldsymbol{Q}_P = \sigma_P^2 \boldsymbol{I}_f$ and $\boldsymbol{Q}_\phi = \sigma_\phi^2 \boldsymbol{I}_f$.

The ionospheric constraints are applied as pseudo-observations to generalize the model (10.1) as

$$
\mathrm{E}(\iota) = \iota_0, \quad \mathrm{D}(\iota) = \sigma_\iota^2 \boldsymbol{Q}
\tag{10.3}
$$

where the variance $\sigma_\iota^2$ is used to model the spatial uncertainty of baseline ionosphere. Incorporating the ionosphere constraints into model (10.1) obtains the ionosphere-weighted model.

### 10.2.1   Canonical Formulae of Float Ambiguity Solution

The tropospheric delays have been corrected by at least 90% with UNB3 model and the residual tropospheric delays are rather small. If they are further compensated by setting up a RZTD parameter, it will take long time for convergence due to high correlation between RZTD and height component [28]. Therefore, from now on, the residual tropospheric delays are ignored after all we emphasize the quick convergence with EWL observations.

Firstly, the ionosphere-weighted model is introduced and deduced. Applying the least-squares (LS) criterion to solve observation models (10.1) with ionospheric constraints (10.3) and stochastic model (10.2), the normal equations of parameters $x$, $\iota$ and $a$ are derived. By reducing the parameters $x$ and $\iota$, we obtain the covariance matrix of float ambiguity solution $\hat{a}$ as

$$Q_{\hat{a}\hat{a}}^{(\text{based})} = \left[\Lambda^{-1}\left(\sigma_\phi^2 I_f + \alpha\mu\mu^{\mathrm{T}}\right)\Lambda^{-1}\right] \otimes Q + \Theta \otimes P_A Q \qquad (10.4)$$

where $P_A = A\left(A^{\mathrm{T}}Q^{-1}A\right)^{-1}A^{\mathrm{T}}Q^{-1}$ and

$$\alpha = \left[\sigma_\iota^{-2} + \sigma_P^{-2}\mu^{\mathrm{T}}\mu\right]^{-1} \qquad (10.5a)$$

$$\Theta = \frac{qq^{\mathrm{T}}}{e_f^{\mathrm{T}}\left(\sigma_P^2 I_f + \sigma_\iota^2\mu\mu^{\mathrm{T}}\right)^{-1}e_f} \qquad (10.5b)$$

$$q = \Lambda^{-1}\left[I_f + \alpha\sigma_P^{-2}\mu\mu^{\mathrm{T}}\right]e_f \qquad (10.5c)$$

In case of geometry-free model, i.e., $A = I_s$, then

$$Q_{\hat{a}\hat{a}}^{(\text{free})} = \left[\Lambda^{-1}\left(\sigma_\phi^2 I_f + \alpha\mu\mu^{\mathrm{T}}\right)\Lambda^{-1} + \Theta\right] \otimes Q \qquad (10.6)$$

Since $Q_{\hat{a}\hat{a}}^{(\text{free})}$ is a rank $-(s-3)$ update of $Q_{\hat{a}\hat{a}}$, in terms of the ambiguity precision, it is easy to prove $Q_{\hat{a}\hat{a}}^{(\text{free})} \geq Q_{\hat{a}\hat{a}}^{(\text{based})}$ for number of DD satellite pairs $s \geq 3$ [29]. The superscripts 'based' and 'free' denote the geometry-based and -free model, respectively.

Next, we present two extreme cases identified by variance of ionospheric constraints, i.e., $\sigma_\iota^2 = \infty$ and 0, which are referred to as ionosphere-float and ionosphere-fixed model, respectively.

Secondly, the ionosphere-float model with $\sigma_\iota^2 = \infty$ is analyzed and presented. The ionosphere-weighted model reduces to the ionosphere-float model when the variance of ionospheric constraints is extremely large (i.e., $\sigma_\iota^2 = \infty$), namely the ionospheric constraints are unavailable. In such case, the variables in (10.4) and (10.6) become

$$\alpha_\infty = \sigma_P^2 \left(\mu^{\mathrm{T}} \mu\right)^{-1} \tag{10.7a}$$

$$\Theta_\infty = \sigma_P^2 \left[ f - \frac{e_f^{\mathrm{T}} \mu \mu^{\mathrm{T}} e_f}{\mu^{\mathrm{T}} \mu} \right]^{-1} q_\infty q_\infty^{\mathrm{T}} \tag{10.7b}$$

$$q_\infty = \Lambda^{-1} \left[ I_f + \frac{\mu \mu^{\mathrm{T}}}{\mu^{\mathrm{T}} \mu} \right] e_f \tag{10.7c}$$

Here the matrix inverse was applied in derivation for the denominator of (10.5b). Substituting these new variables (10.7a)–(10.7c) into (10.4) and (10.6) yields the covariance matrix of float ambiguity solution in geometry-based and geometry-free ionosphere-float model, respectively.

Then the ionosphere-fixed model with $\sigma_\iota^2 = 0$ is derived and analyzed. The ionosphere-weighted model reduces to the ionosphere-fixed model when $\sigma_\iota^2 = 0$. It implies two situations: (i) the ionospheric biases are indeed precisely known, which is the best case to gain the strongest model strength; (ii) the ionospheric biases are completely ignored. This makes sense for short baselines that the ionospheric biases are small enough to be ignored. However, for long baselines discussed in this chapter, it leads to actually the ionosphere-ignored (biased) model. In this case, the variables in (10.4) and (10.6) reduce to

$$\alpha_0 = 0, \quad \Theta_0 = \sigma_P^2 \frac{q_0 q_0^{\mathrm{T}}}{f}, \quad q_0 = \Lambda^{-1} e_f \tag{10.8}$$

Substituting these new variables into (10.4) yields the covariance matrix of float ambiguity solution in geometry-based ionosphere-fixed model

$$Q_{\hat{a}\hat{a}}^{(\text{based})} = \sigma_\phi^2 \Lambda^{-2} \otimes Q + \sigma_P^2 \frac{\Lambda^{-1} e_f e_f^{\mathrm{T}} \Lambda^{-1}}{f} \otimes P_A Q \tag{10.9}$$

while substituting them into (10.6) yields the covariance matrix of float ambiguity solution in geometry-free ionosphere-fixed model

$$Q_{\hat{a}\hat{a}}^{(\text{free})} = \left[ \sigma_\phi^2 \Lambda^{-2} + \sigma_P^2 \frac{\Lambda^{-1} e_f e_f^{\mathrm{T}} \Lambda^{-1}}{f} \right] \otimes Q \tag{10.10}$$

As mentioned, the ionospheric biases would exist on long baselines, denoted by $\iota_b$, the float ambiguity solution will be biased and the bias is derived as

$$b_{\hat{a}}^{(\text{based})} = \left[ -\Lambda^{-1} \mu \otimes I_s - \frac{\Lambda^{-1} e_f e_f^{\mathrm{T}} \mu}{f} \otimes P_A \right] \iota_b \tag{10.11}$$

for geometry-based model and

$$b_{\hat{a}}^{(\text{free})} = -\Lambda^{-1}\left(I_f + e_f e_f^{\mathrm{T}}/f\right)\mu \otimes \iota_b \qquad (10.12)$$

for geometry-free model.

Based on the canonical formulae of covariance matrix of float ambiguity solution, we numerically study the single-epoch AR capabilities of different models by analyzing their success rates. Both geometry-based and geometry-free models are examined, for which the different types of ionosphere models are identified by assigning the corresponding variances of ionospheric constraints. Regarding the computation of success rate, the bootstrapped success rate is employed due to its tightest bound of actual success rate and very efficient computation [30]. The geometry-free model is free of effect of satellite geometry, one can directly compute success rate based on the covariance matrix $Q_{\hat{a}\hat{a}}^{(\text{free})}$. Different from geometry-free model, the geometry-based model is affected by the variant satellite geometry. We therefore compute the single-epoch success rate every minute and total 1440 success rates are obtained over 24 h. Then the mean of success rates is computed. Since the success rate would change for the different number of ambiguities, the different number of satellite pairs from 1 to 4 are analyzed with $\sigma_P = 0.2$ m and $\sigma_\phi = 3$ mm and the varying standard deviation (STD) of ionospheric constraints as $\sigma_\iota = 0, 5, 10$ cm and $\infty$.

Corresponding to the number of satellite pairs from 1 to 4, the number of triple-frequency ambiguities is from 3 to 12. Related study by the authors reveals that (i) for both geometry-based and -free models, the success rates become smaller when the ionospheric constraints become weaker with larger $\sigma_\iota$. (ii) For the geometry-free model, the success rates get definitely smaller when the more satellite pairs are involved except for $\sigma_\iota = 0$. This makes sense because the geometry gain from more satellites has no contribution to geometry-free model. (iii) For the geometry-based model, the variation of success rate depends on the balance of the satellite geometry and the number of ambiguities. When the success rate increased by geometry gain from more satellites is larger than that reduced by more ambiguities, the success rate will increase and vice versa. For instance, the success rate of 3 satellite pairs is larger than that of 2 satellite pairs but smaller than that of 4 satellite pairs.

In summary, the single-epoch full ambiguity resolution (FAR) is possible with 99.99% success rate only in case of very strong ionospheric constraints (ionosphere-fixed model in our case). In other words, the single-epoch FAR is feasible only for short baselines where the ionospheric constraints can indeed be strong. For long baselines with weak ionosphere constraints, single-epoch FAR is impossible for both geometry-based and -free models.

### *10.2.2   Transformed EWL AR*

The result above has indicated the difficulty of single-epoch triple-frequency FAR. One may naturally tend to the partial ambiguity resolution (PAR) that was introduced in [29]. PAR is a flexible method that allows one to fix a subset of ambiguities, instead of aiming to fix the complete ambiguity vector, in terms of user-defined success rate. Many literatures studied the strategies of ambiguity subset selection for PAR for various applications, and promising results were achieved, see e.g. [31–35]. In this study, we select the ambiguity subset for PAR by transforming the triple-frequency ambiguities with a pre-set between-frequency transformation matrix. As a result, the so-called EWL combinations are obtained. The rationale behind using EWL combinations is that their ambiguities, due to longer wavelengths, can be resolved better than the original ambiguities.

With ionosphere-weighted geometry-free model as a case study, we apply the between-frequency transformation matrix $\left(z_E^T \otimes I_s\right)$ with $z_E^T = \begin{bmatrix} z_1 & z_2 & z_3 \end{bmatrix}$ to transform the covariance matrix (10.6). It follows

$$
\left(z_E^T \otimes I_s\right) Q_{\hat{a}\hat{a}}^{(\text{free})} (z_E \otimes I_s) = z_E^T \left[\Lambda^{-1}\left(\sigma_\phi^2 I_f + \alpha \mu \mu^T\right)\Lambda^{-1} + \Theta\right] z_E \otimes Q
$$
$$
= \sigma_{\hat{z}}^2 \otimes Q \tag{10.13}
$$

with

$$
\sigma_{\hat{z}}^2 - z_E^T\left[\Lambda^{-1}\left(\sigma_\phi^2 I_f + \alpha \mu \mu^T\right)\Lambda^{-1} + \Theta\right] z_E \tag{10.14}
$$

Since $Q$ is a constant matrix, one can minimize the scalar $\sigma_{\hat{z}}^2$, i.e., $\sigma_{\hat{z}}^2 = \min$, to obtain the optimal transformation matrix $z_E^T$. The least-squares ambiguity decorrelation adjustment (LAMBDA) method [36] can be applied to solve this minimization problem where the zero-vector plays the role of float solution.

For ionosphere-fixed model of $\sigma_\iota^2 = 0$, the variance scalar becomes

$$
\sigma_{\hat{z}}^2 = z_E^T\left[\sigma_\phi^2 \Lambda^{-2} + \sigma_p^2 \frac{\Lambda^{-1} e_f e_f^T \Lambda^{-1}}{f}\right] z_E \tag{10.15}
$$

In this case, the bias (10.12) is accordingly transformed as:

$$
\begin{aligned}
b_{\hat{z}}^{(\text{free})} &= \left(z_E^T \otimes I_s\right) b_{\hat{a}}^{(\text{free})} \\
&= -\frac{1}{f} z_E^T \Lambda^{-1}\left(f I_f + e_f e_f^T\right) \mu \iota_b = -\frac{1}{f} z_E^T \chi \iota_b
\end{aligned} \tag{10.16}
$$

with $\chi = f\Lambda^{-1}\mu + \Lambda^{-1} e_f e_f^T \mu$. In such ionosphere-fixed model with bias, both variance and bias should be taken into account to obtain the best transformation matrix $z_E^T$. It further assumes that the ionospheric biases are unique for all DD satellites, then $\iota_b = e_s \iota_b$. The mean square error (MSE) is applied as a measure by capturing both the variance and bias:

$$
\sigma_{\hat{z}}^2 + b_{\hat{z}}^{(\text{free})} b_{\hat{z}}^{(\text{free})} = z_E^T\left[\sigma_\phi^2 \Lambda^{-2} + \frac{\sigma_P^2}{f} \Lambda^{-1} e_f e_f^T \Lambda^{-1} + \frac{\iota_b^2}{f^2} \chi \chi^T\right] z_E = \min \tag{10.17}
$$

The LAMBDA method is again used to solve this minimization problem.

By solving minimization problem (10.15) with varying ionospheric constraints, $\sigma_\iota = 5$, 10, 15 and 20 cm, and the minimization problem (10.17) with $\sigma_\iota = 0$ and varying ionospheric biases $\iota_b = 0.1$, 0.2, 0.3 and 0.4 m, three EWL combinations, $[0, -1, 1]$, $[1, 3, -4]$ and $[1, 4, -5]$, display the superiority in obtaining high success rates. Their wavelengths are 4.884, 2.765 and 6.371 m, respectively. To intuitively show the performance of these three combinations, their success rates of one DD satellite pair are computed with varying settings shown in related study by the authors where $\sigma_P = 0.2$ m and $\sigma_\phi = 3$ mm. For ionosphere-weighted (unbiased) model, the success rate of rounding method is applied while for the ionosphere-fixed (biased) model, the bias-affected rounding success rate is used [37].

In general, the success rate can be improved, although the improvement is very slightly, by strengthening the ionospheric constraint in ionosphere-weighted model and by reducing the ionospheric bias in ionosphere-fixed model. The results overall reveal that the single-epoch EWL AR can be done in very high success rate with either ionosphere-weighted or -fixed model even for long baselines with ionospheric constraint of $\sigma_\iota = 20$ cm or ionospheric bias of $\iota_b = 0.4$ m. Therefore, although the single-epoch FAR is impossible, one can still instantaneously fix the EWL ambiguities with very high success rate.

## 10.3   Mathematical Model

Once two EWL ambiguities are fixed, the ambiguity-fixed EWL observation plays the role of pseudorange but with higher precision. One can immediately starts the RTK with ambiguity-corrected EWL observations, which is referred to as ERTK. In this section, two ERTK models, i.e., ionosphere-ignored and -float, are presented first. Then the ERTK equivalence of using any two EWL observations is proven. Finally, the condition of selecting either ionosphere-ignored or ionosphere-float model is discussed.

### 10.3.1   Ionosphere-Ignored and -Float Models

If the DD ionospheric biases are so small to be ignored or the biased solution is acceptable, the ionosphere-ignored ERTK model is formulated

$$\begin{bmatrix} P \\ \phi_E \end{bmatrix} = (e_5 \otimes A)x \tag{10.18}$$

where $\phi_E = \begin{bmatrix} \phi_{(0,-1,1)}^T & \phi_{(1,3,-4)}^T \end{bmatrix}^T$. The stochastic model reads

$$\begin{bmatrix} \sigma_P^2 I_3 & 0 \\ 0 & Q_{\phi_E} \end{bmatrix} \otimes Q \quad \text{with} \quad Q_{\phi_E} = \sigma_\phi^2 \begin{bmatrix} \alpha_{(0,-1,1)}^2 & \alpha_{\text{cross}} \\ \alpha_{\text{cross}} & \alpha_{(1,3,-4)}^2 \end{bmatrix} \tag{10.19}$$

where $\alpha_{(i,j,k)}^2 = \frac{(if_1)^2 + (jf_2)^2 + (kf_3)^2}{f_{(i,j,k)}^2}$ and $\alpha_{\text{cross}} = -\frac{3f_2^2 + 4f_3^2}{f_{(0,-1,1)}f_{(1,3,-4)}}$ with $f_{(i,j,k)} = if_1 + jf_2 + kf_3$.
Applying the LS criterion to solve (10.18) with stochastic model (10.19) yields the covariance matrix of estimate $\hat{x}$ as

$$Q_{\hat{x}\hat{x}}^{(\text{ign})} = \gamma^{-1} \left( A^{\mathsf{T}} Q^{-1} A \right)^{-1} = \gamma^{-1} Q_{\hat{x}\hat{x}} \tag{10.20}$$

and

$$\gamma = 3\sigma_P^{-2} + \sigma_\phi^{-2} \frac{a+b}{\omega} = 3\sigma_P^{-2} + 0.0342\sigma_\phi^{-2} \tag{10.21}$$

where the superscript 'ign' denotes the solution for ionosphere-ignored model. $\omega = \alpha_{(0,-1,1)}^2 \alpha_{(1,3,-4)}^2 - \alpha_{\text{cross}}^2$. $a = \alpha_{(1,3,-4)}^2 - \alpha_{\text{cross}}$ and $b = \alpha_{(0,-1,1)}^2 - \alpha_{\text{cross}}$. For $\sigma_P = 0.2$ m and $\sigma_\phi = 3$ mm, the factor $\sqrt{\gamma^{-1}} = 0.016$ m. In the ionosphere-ignored model, the ionospheric biases are ignored, which will lead to the estimate biased. The bias is derived as

$$b_{\hat{x}} = -\gamma^{-1} \beta Q_{\hat{x}\hat{x}} A^{\mathsf{T}} Q^{-1} \iota_b \tag{10.22}$$

where

$$\beta = \sigma_P^{-2} \mu_\Sigma - \sigma_\phi^{-2} \frac{a\mu_{(0,-1,1)} + b\mu_{(1,3,-4)}}{\omega} = 2.2582\sigma_P^{-2} + 0.0437\sigma_\phi^{-2} \tag{10.23}$$

with $\mu_\Sigma = \mu_1 + \mu_2 + \mu_3$ and $\mu_{(i,j,k)} = \frac{f_1^2(i/f_1 + j/f_2 + k/f_3)}{f_{(i,j,k)}}$. Again for $\sigma_P = 0.2$ m and $\sigma_\phi = 3$ mm, the factor $\beta = 4909.87\ \text{m}^{-2}$ and $b_{\hat{x}} \approx -1.268 Q_{\hat{x}\hat{x}} A^{\mathsf{T}} Q^{-1} \iota_b$. Obviously the bias is governed by not only the satellite geometry and ionospheric bias but also the precision of code and phase.

If the ionospheric biases are large particularly over long baselines, they should be properly compensated to reduce their effects on positioning. A normal way is to parameterize such ionospheric biases, resulting in the ionosphere-float model

$$\begin{bmatrix} P \\ \phi_E \end{bmatrix} = \begin{bmatrix} e_3 \otimes A & \mu \otimes I \\ e_2 \otimes A & -\mu_E \otimes I \end{bmatrix} \begin{bmatrix} x \\ \iota \end{bmatrix} \tag{10.24}$$

With the same stochastic model of (10.19), the covariance matrix of LS estimate reads

$$Q_{\hat{x}\hat{x}}^{(\text{float})} = \left( \gamma - \frac{\beta^2}{\varsigma} \right)^{-1} Q_{\hat{x}\hat{x}} \tag{10.25}$$

$$\varsigma = \sigma_P^{-2}\boldsymbol{\mu}^{\mathrm{T}}\boldsymbol{\mu} + \sigma_\phi^{-2}\frac{\alpha_{(1,3,-4)}^2\mu_{(0,-1,1)}^2 + \alpha_{(0,-1,1)}^2\mu_{(1,3,-4)}^2 - 2\alpha_{\mathrm{cross}}\mu_{(1,3,-4)}\mu_{(0,-1,1)}}{\omega}$$

$$= 1.7935\sigma_P^{-2} + 0.0560\sigma_\phi^{-2} \tag{10.26}$$

It is obvious that $\left(\gamma - \frac{\beta^2}{\varsigma}\right)^{-1} > \gamma^{-1}$. For $\sigma_P = 0.2$ m and $\sigma_\phi = 3$ mm, the factor $\sqrt{\left(\gamma - \frac{\beta^2}{\varsigma}\right)^{-1}} = 0.210$ m is much larger than $\sqrt{\gamma^{-1}} = 0.016$ m. It means that although the bias is eliminated in the ionosphere-float model, its uncertainty is 13-times larger than that in ionosphere-ignored model.

### 10.3.2　ERTK Equivalence for Using Any Two EWLs

The above formulation of two ERTK models are based on the EWL combinations $[0, -1, 1]$ and $[1, 3, -4]$. In principle, only two EWL/WL combinations are independent and any other EWL/WL ambiguity $a_{(i,j,k)}$ can be recovered from these two integer EWL/WL ambiguities [2, 11, 26]. The question is whether the equivalent solution is achievable by using any two kinds of EWL observations.

In terms of [11], the sum of three coefficients of an EWL combination is equal to 0. Let two arbitrary EWL/WL ambiguities $a_{(i,j,-i-j)}$ and $a_{(k,l,-k-l)}$, we have the transformation from $a_{(0,-1,1)}$ and $a_{(1,3,-4)}$ as

$$\begin{bmatrix} a_{(i,j,-i-j)} \\ a_{(k,l,-k-l)} \end{bmatrix} = \begin{bmatrix} 3i - j & i \\ 3k - l & k \end{bmatrix}\begin{bmatrix} a_{(0,-1,1)} \\ a_{(1,3,-4)} \end{bmatrix} \tag{10.27}$$

where the integer coefficients in transformation matrix satisfy with that their associated transformation matrix is unimodular, i.e., $(3i - j)k - (3k - l)i = \pm 1$, in order to retain the integer invertible property. For instance, the coefficients of $i = k = 1, j = -1$ and $l = 0$ derives the combinations $a_{(1,-1,0)}$ and $a_{(1,0,-1)}$.

It is understandable that if a full-rank square matrix is applied to transform an equation system, the derived solutions will be definitely equivalent. In our case, it is easy to find a full-rank square transformation matrix:

$$\boldsymbol{R} = \begin{bmatrix} \boldsymbol{I}_3 & \boldsymbol{0} \\ \boldsymbol{0} & \boldsymbol{R}_E \end{bmatrix} \otimes \boldsymbol{I}_s \quad \text{with} \quad \boldsymbol{R}_E = \begin{bmatrix} \frac{\lambda_{(j,j,-i-j)}}{\lambda_{(0,-1,1)}}(3i - j) & \frac{\lambda_{(j,j,-i-j)}}{\lambda_{(1,3,-4)}}i \\ \frac{\lambda_{(k,l,-k-l)}}{\lambda_{(0,-1,1)}}(3k - l) & \frac{\lambda_{(k,l,-k-l)}}{\lambda_{(1,3,-4)}}k \end{bmatrix} \tag{10.28}$$

By applying the transformation matrix to (10.18) and (10.24), $\boldsymbol{\phi}_{\mathrm{E}}$ is converted to new EWL observations $\boldsymbol{\phi}_{\mathrm{E}} = \begin{bmatrix} \boldsymbol{\phi}_{(i,j,-i-j)}^{\mathrm{T}} & \boldsymbol{\phi}_{(k,l,-k-l)}^{\mathrm{T}} \end{bmatrix}^{\mathrm{T}}$ with transformed EWL ambiguities (10.27). Again, due to the full-rank of transformation matrix $\boldsymbol{R}$, the equivalent ERTK solutions will be definitely obtained by using the transformed equation system. In other words, the ERTK estimate $\hat{\boldsymbol{x}}$ and its corresponding covariance matrix $\boldsymbol{Q}_{\hat{\boldsymbol{x}}\hat{\boldsymbol{x}}}^{(\mathrm{ign})}$

and bias $\boldsymbol{b}_{\hat{x}}$ obtained with ionosphere-ignored model (10.18) are equivalent to the solutions from its transformed system with $\boldsymbol{R}$. The ERTK estimate $\hat{x}$ and its covariance matrix $\boldsymbol{Q}_{\hat{x}\hat{x}}^{(\text{float})}$ obtained with ionosphere-float model (10.24) is equivalent to the solutions from its transformed system with $\boldsymbol{R}$ as well. Therefore, it is equivalent to use any two ambiguity-corrected EWL/WL observations for both ionosphere-ignored and -float ERTK.

### 10.3.3 Analysis of Ionosphere-Ignored and -Float Models

We have formulated two ERTK models and indicated that the ionosphere-ignored solution is biased although with smaller covariance matrix; while the ionosphere-float solution is unbiased but with larger covariance matrix. Although we qualitatively know that ionosphere-float model should be applied if the ionospheric biases are sufficiently large, the question is how large it is? In this section, we will quantitatively answer this question.

To be simple, we again assume that all DD observations are affected by the same magnitude of ionospheric biases, i.e., $\iota_b = \boldsymbol{e}_s \iota_b$. Then the bias (10.22) reduces

$$\boldsymbol{b}_{\hat{x}} = -\gamma^{-1}\beta\iota_b \boldsymbol{Q}_{\hat{x}\hat{x}}\boldsymbol{A}^{\mathrm{T}}\boldsymbol{Q}^{-1}\boldsymbol{e}_s \tag{10.29}$$

The MSE indicator is employed to measure the accuracy of a biased estimate as

$$\boldsymbol{Q}_{\hat{x}\hat{x},b_{\hat{x}}}^{(\text{ign})} = \boldsymbol{Q}_{\hat{x}\hat{x}}^{(\text{ign})} + \boldsymbol{b}_{\hat{x}}\boldsymbol{b}_{\hat{x}}^{\mathrm{T}} = \gamma^{-1}\boldsymbol{Q}_{\hat{x}\hat{x}} + \gamma^{-2}\beta^2\iota_b^2 \boldsymbol{Q}_{\hat{x}\hat{x}}\boldsymbol{A}^{\mathrm{T}}\boldsymbol{Q}^{-1}\boldsymbol{e}_s\boldsymbol{e}_s^{\mathrm{T}}\boldsymbol{Q}^{-1}\boldsymbol{A}\boldsymbol{Q}_{\hat{x}\hat{x}} \tag{10.30}$$

To make comparison, we rewrite the covariance matrix (10.25) of ionosphere-float ERTK solution as

$$\boldsymbol{Q}_{\hat{x}\hat{x}}^{(\text{float})} = \left(\gamma - \frac{\beta^2}{\varsigma}\right)^{-1}\boldsymbol{Q}_{\hat{x}\hat{x}} = \left[\gamma^{-1} + \frac{\beta^2}{\gamma(\varsigma\gamma - \beta^2)}\right]\boldsymbol{Q}_{\hat{x}\hat{x}} \tag{10.31}$$

The difference between the covariance matrix of ionosphere-float unbiased estimate and the MSE of ionosphere-ignored biased estimate is

$$\delta\boldsymbol{Q} = \boldsymbol{Q}_{\hat{x}\hat{x}}^{(\text{float})} - \boldsymbol{Q}_{\hat{x}\hat{x},b_{\hat{x}}}^{(\text{ign})} = \frac{\beta^2}{\gamma(\varsigma\gamma - \beta^2)}\boldsymbol{Q}_{\hat{x}\hat{x}} - \frac{\beta^2\iota_b^2}{\gamma^2}\boldsymbol{Q}_{\hat{x}\hat{x}}\boldsymbol{A}^{\mathrm{T}}\boldsymbol{Q}^{-1}\boldsymbol{e}_s\boldsymbol{e}_s^{\mathrm{T}}\boldsymbol{Q}^{-1}\boldsymbol{A}\boldsymbol{Q}_{\hat{x}\hat{x}} \tag{10.32}$$

Taking the trace to both sides of (10.32) to measure the overall quantity of $\delta\boldsymbol{Q}$ yields

$$\text{trace}(\delta\boldsymbol{Q}) = \frac{\beta^2}{\gamma(\varsigma\gamma - \beta^2)}\text{trace}(\boldsymbol{Q}_{\hat{x}\hat{x}}) - \frac{\beta^2\iota_b^2}{\gamma^2}\text{trace}(\boldsymbol{e}_s^{\mathrm{T}}\boldsymbol{Q}^{-1}\boldsymbol{A}\boldsymbol{Q}_{\hat{x}\hat{x}}^2\boldsymbol{A}^{\mathrm{T}}\boldsymbol{Q}^{-1}\boldsymbol{e}_s) \tag{10.33}$$

It follows then that the ionosphere-ignored model should be applied when the condition of trace$(\delta \boldsymbol{Q}) > 0$ is hold true, i.e.,

$$\iota_b < \sqrt{\frac{\gamma}{5\gamma - \beta^2} \frac{\text{trace}(\boldsymbol{Q}_{\hat{x}\hat{x}})}{\text{trace}(\boldsymbol{e}_s^{\mathrm{T}} \boldsymbol{Q}^{-1} \boldsymbol{A} \boldsymbol{Q}_{\hat{x}\hat{x}}^2 \boldsymbol{A}^{\mathrm{T}} \boldsymbol{Q}^{-1} \boldsymbol{e}_s)}} \tag{10.34}$$

It is clear that the threshold $\iota_b$ of ionospheric bias is a function of satellite geometry, number of satellites and observation precisions. For an extreme case of $\boldsymbol{A} = \boldsymbol{I}_s$ with elevation-independent weighting, then

$$\iota_b < 2\sqrt{\frac{\gamma}{5\gamma - \beta^2}} \tag{10.35}$$

With $\sigma_P = 0.2$ m and $\sigma_\phi = 3$ mm, it follows that $\iota_b < 0.33$ m for which the ionosphere-ignored model will obtain the better ERTK solution than the ionosphere-float model. In this case, the bias effect on ERTK solution in ionosphere-ignored model is smaller than the effect of model weakness induced by ionosphere-float model. The numerical experience indicates that for the baseline length as long as tens to hundred kilometer, the condition of $\iota_b < 0.33$ m is generally satisfied and the ionosphere ignored model outperforms the ionosphere-float model.

### 10.3.4   ERTK Improved by Adding NL Observations

We have shown in Sect. 10.3.1 that the ionosphere-ignored model achieves the small uncertainty of biased ERTK solution, while the ionosphere-float model the much larger uncertainty of unbiased solution. Moreover, in most cases of several tens kilometer baselines, the ionosphere-ignored model outperforms the ionosphere-float model. The question is that what should be done if one wants to further improve the ERTK solution.

In theory, in order to improve the ERTK solution, the effect of ionospheric biases should be significantly reduced and meantime the model strength should not be decreased so much. For reducing the effect of ionospheric biases, we have to use the ionosphere-float model unless the ionospheric biases are precisely known (impossible in real RTK applications). Now, the key is to improve the model strength of ionosphere-float model. Additional to the observations used in ionosphere-float ERTK model, we extend the ionosphere-float ERTK model by incorporating the L1 phase observations:

$$\begin{bmatrix} \boldsymbol{P} \\ \boldsymbol{\phi}_{\mathrm{E}} \\ \boldsymbol{\phi}_1 \end{bmatrix} = \begin{bmatrix} \boldsymbol{e}_3 \otimes \boldsymbol{A} & \boldsymbol{\mu} \otimes \boldsymbol{I}_s & \boldsymbol{0} \\ \boldsymbol{e}_2 \otimes \boldsymbol{A} & -\boldsymbol{\mu}_{\mathrm{E}} \otimes \boldsymbol{I}_s & \boldsymbol{0} \\ \boldsymbol{A} & -\mu_1 \otimes \boldsymbol{I}_s & \boldsymbol{I}_s \end{bmatrix} \begin{bmatrix} \boldsymbol{x} \\ \iota \\ a \end{bmatrix}, \quad \begin{bmatrix} \sigma_P^2 \boldsymbol{I}_3 & \boldsymbol{0} & \boldsymbol{0} \\ \boldsymbol{0} & \boldsymbol{Q}_{\phi_{\mathrm{E}}} & \boldsymbol{Q}_{\phi_{\mathrm{E}}\phi_1} \\ \boldsymbol{0} & \boldsymbol{Q}_{\phi_1\phi_{\mathrm{E}}} & \sigma_\phi^2 \end{bmatrix} \otimes \boldsymbol{Q} = \boldsymbol{Q}_{yy} \otimes \boldsymbol{Q} \tag{10.36}$$

where $a$ is ambiguity vector in meter. $Q_{\phi_1 \phi_E} = \begin{bmatrix} 0, & \sigma_\phi^2 \end{bmatrix}$. For a single epoch, L1 observations will not contribute to the ERTK solution since the number of L1 observations is equal to the number of newly introduced ambiguity parameters. But when the multiple epoch data is continuously applied, the ambiguities keep constant and thus the L1 observations can improve the ERTK solutions. Essentially, the very precise between-epoch single-differenced L1 observation information is used to smooth the unbiased ionosphere-float solution so as to reduce its uncertainty. Therefore, the model (10.36) is referred also to as ionosphere-smoothed ERTK model.

Let us now mathematically analyze how the ionosphere-smoothed model improves the ERTK solutions. In terms of the equivalence theory, the ionospheric parameters can be equivalently eliminated by transforming the ionosphere-smoothed model (10.36) with the following transformation matrix

$$\overline{T} = \begin{bmatrix} \frac{\mu_2}{\mu_2-\mu_1} & \frac{-\mu_1}{\mu_2-\mu_1} & 0 & 0 & 0 & 0 \\ \frac{\mu_3}{\mu_3-\mu_1} & 0 & \frac{-\mu_1}{\mu_3-\mu_1} & 0 & 0 & 0 \\ 0 & 0 & 0 & \frac{\mu_{E2}}{\mu_{E2}-\mu_{E1}} & \frac{-\mu_{E1}}{\mu_{E2}-\mu_{E1}} & 0 \\ 0 & 0 & 0 & \frac{\mu_1}{\mu_1-\mu_{E1}} & 0 & \frac{-\mu_{E1}}{\mu_1-\mu_{E1}} \\ \frac{1}{2} & 0 & 0 & 0 & 0 & \frac{1}{2} \end{bmatrix} \otimes I_s = T \otimes I_s \qquad (10.37)$$

which yields

$$\overline{T}y = \begin{bmatrix} e_5 \otimes A \ \Upsilon \otimes I_s \end{bmatrix} \begin{bmatrix} x \\ a \end{bmatrix}, \quad Q_{\overline{yy}} \otimes Q \qquad (10.38)$$

where $y = \begin{bmatrix} P^T & \phi_E^T & \phi_1^T \end{bmatrix}^T$, $\Upsilon = \begin{bmatrix} 0_{1\times3} & \frac{-\mu_{E1}}{\mu_1-\mu_{E1}} & 0.5 \end{bmatrix}^T$ and $Q_{\overline{yy}} = TQ_{yy}T^T$. Let the subscript $k$ denote the $k$th epoch, its normal matrix of $x$ and $a$ reads

$$\begin{bmatrix} pQ_{\hat{x}_k}^{-1} & tA_k^T Q^{-1} \\ tQ^{-1}A_k & \vartheta Q^{-1} \end{bmatrix} \qquad (10.39)$$

where $p = e_5^T Q_{\overline{yy}}^{-1} e_5$, $t = e_5^T Q_{\overline{yy}}^{-1} \Upsilon$, $\vartheta = \Upsilon^T Q_{\overline{yy}}^{-1} \Upsilon$ and $Q_{\hat{x}_k}^{-1} = A_k^T Q^{-1} A_k$. The covariance matrix of single-epoch ionosphere-smoothed ERTK estimate is

$$N_{\hat{x}_k}^{(\text{smooth})} = \left( p - \frac{t^2}{\vartheta} \right) Q_{\hat{x}_k}^{-1} \qquad (10.40)$$

which is equivalent to the ionosphere-float ERTK solution (10.25), i.e., $p - t^2/\vartheta = \gamma - \beta^2/\varsigma$. However, it is emphasized that three variables are not individually equal. For instance, with $\sigma_P = 0.2$ m and $\sigma_\phi = 3$ mm, $p = 18404.453$, $t = 10303.738$ and $\vartheta = 5775.643$, while $\gamma = 3872.315$, $\beta = 4909.875$ and $\varsigma = 6261.992$.

By reducing the parameter $x$, one obtains the single-epoch normal matrix of $a$ as $N_{\hat{a},k} = \vartheta Q^{-1} - \frac{t^2}{p} Q^{-1} A_k Q_{\hat{x}_k} A_k^T Q^{-1}$. The associated sum normal matrix over $K$

epochs reads

$$N_{\Sigma a} = K\vartheta Q^{-1} - \frac{t^2}{p}Q^{-1}\left(\sum_{k=1}^{K} A_k Q_{\hat{x}_k} A_k^{\mathrm{T}}\right)Q^{-1} \tag{10.41}$$

For the $(K+1)$th epoch, the normal matrix of smoothed ERTK solution is derived

$$N_{\hat{x}_{K+1}}^{(\mathrm{smooth})} = pQ_{\hat{x}_{K+1}}^{-1} - t^2 A_{K+1}^{\mathrm{T}} Q^{-1}\left[(K+1)\vartheta Q^{-1} - \frac{t^2}{p}Q^{-1}\left(\sum_{k=1}^{K} A_k Q_{\hat{x}_k} A_k^{\mathrm{T}}\right)Q^{-1}\right]^{-1} Q^{-1} A_{K+1} \tag{10.42}$$

It is obvious that $N_{\hat{x}_{K+1}}^{(\mathrm{smooth})} > N_{\hat{x}_{K+1}}^{(\mathrm{float})} = \left(p - \frac{t^2}{\vartheta}\right)Q_{\hat{x}_{K+1}}^{-1}$. Hence the covariance matrix of smoothed solution is smaller than that of ionosphere-float solution. The improvement depends on the gain obtained from $N_{\Sigma a}$ as a function of the number of epochs $K$ for smoothing, the satellite geometry as well as the observation precisions.

To intuitively show how the ionosphere-smoothed model improves the ERTK solution, we again take a special case of geometry-free model, i.e., $A_k = I_s$, for $k = 1, \ldots, K$. In this case, the covariance matrix of (10.42) reduces to

$$Q_{\hat{x}_{K+1}}^{(\mathrm{smooth})} = \left[p - \frac{t^2}{\vartheta + K\left(\vartheta - t^2/p\right)}\right]^{-1} Q_{\hat{x}_{K+1}} \tag{10.43}$$

It is clear that when $K$ gets larger the solution becomes better and gets comparable to ionosphere-fixed solution but promisingly without bias. When $K$ is sufficiently large, $Q_{\hat{x}_{K+1}}^{(\mathrm{smooth})} \approx p^{-1} Q_{\hat{x}_{K+1}}$, which is even better than the ionosphere-fixed solution because of $p > \gamma$. Compared to ionosphere-fixed model, it is highlighted that the betterment of ionosphere-smoothed solution comes from both its unbiased property and smaller (or at least comparable) uncertainty.

Some comments are given on two alternative implementations of ionosphere-smoothed model (10.36). The first implementation starts with computing the ionospheric biases with two ambiguity-fixed EWL observations. These ionospheric biases are very noisy with uncertainty of $89\sigma_\phi$. They can be smoothed following Hatch filter [38] by using very precise epoch-differenced ionospheric biases solved with epoch-differenced geometry-free L1-L2 phase observations. Since the uncertainty of this epoch-differenced ionospheric bias is as precise as $3.5\sigma_\phi$, the smoothed ionospheric biases will be precise as well. Finally, one can conduct the ionosphere-fixed ERTK using observations corrected with smoothed ionospheric biases. Instead of solving and smoothing ionospheric biases, in the second alternative implementation, one can directly form the ionosphere-free combination with two EWL observations. Again this ionosphere-free combination is rather noisy with uncertainty of $114\sigma_\phi$. Similarly, these noisy ionosphere-free combination is smoothed by using epoch-differenced

L1-L2 ionosphere-free phase combination whose uncertainty is $3\sigma_\phi$. As a result the precise ionosphere-free EWL observations are obtained for precise ERTK solution.

## 10.4  Results and Discussion

In this section, the experiment and analysis are carried out. Firstly, the experiment setup is introduced. Then the results of ionosphere-ignored and -float ERTK and the results of ionosphere-smoothed ERTK are presented and analyzed.

### *10.4.1  Experiment Setup*

The daily triple-frequency BDS data was collected by using ComNav M300Pro multi-GNSS receivers with sampling interval of 1 s at a network of four stations in Shanghai area. The date is on day of year (Doy) of 198, 2016. For station A, there is no data for the first 2 h due to some abnormality. From these four stations, six baselines are formed with the baseline lengths from 36 to 75 km. The coordinates of four stations are precisely known, serving as references. The proposed ERTK models were implemented in "TJRTK" software that is a self-developed software in Tongji University for the multi-frequency multi-GNSS RTK processing and relevant engineering and scientific applications.

Total 14 satellites are tracked, of which all five Geostationary Earth Orbit (GEO) satellites at the south side and five Inclined Geosynchronous Orbit (IGSO) satellites at the south side for most of time. Very few satellites were tracked in the north especially in the northeast sky due to the current distribution of BDS constellations, which may lead to the degraded solutions in north as seen in DOP values. Keep in mind that the latter results reflect the ERTK performance only of the current BDS constellations, which can be definitely improved with further development of BDS constellations.

In data processing, the cut-off elevation is set to $10°$ and the elevation-dependent stochastic model

$$\sigma = \frac{1.02}{\sin\theta + 0.02}\sigma_{90°} \tag{10.44}$$

is applied for the undifferenced measurements with the zenith precision $\sigma_{90°} = 2$ mm for phase and 0.2 m for code. Although the data is post-processed, the processing is completely analogous to the real-time processing, namely, the data loading and all computations are implemented epoch by epoch.

### 10.4.2	Results of Ionosphere-Ignored and -Float ERTK

The capability of instantaneous EWL AR has been numerically demonstrated in many literatures with real triple-frequency BDS data [17, 18, 24]. We do not repeat the similar result of EWL AR anymore in this monograph. With ambiguity-fixed EWL observations, we immediately study the performance of ionosphere-ignored and -float ERTK, following the theory in Sect. 10.3.2.

The errors of ionosphere-ignored ERTK results are computed for all 6 baselines. Here N, E and U denote the North, East and Height components in topocentric coordinate system. Note the three baselines connected with station A have no data for the first 2 h. The variations of horizontal errors are all within 0.5 m and mostly within 20 cm for N and E component, respectively; while the variations of height errors are all within 1 m and mostly within 40 cm. The accumulated probabilities as function of absolute positioning errors are computed for all three components of 6 baselines. The results show that the errors are smaller than 10 cm by 80% and 90% and smaller than 20 cm by 90% and 95% for N and E component, respectively; while the height errors are smaller than 40 cm by 90%. Overall, the decimeter-level results are obtained although some systematic errors exhibit, especially at the duration of UTC time 16–18 h. The statistics of positioning errors, STD and root mean square error (RMS), are estimated. Both STD and RMS of E component are better than 10 cm; while they are slightly larger for N component with averagely around 10 cm. The accuracies of height component is 2-times worse than horizontal component.

The significant systematic biases exhibit during the period of Coordinated Universal time (UTC) time 16–18 h for all baselines. This can be mainly attributed to the residual ionospheric biases ignored in the ionosphere-ignored ERTK model. To confirm this issue, we fix all DD ambiguities of L1 and L2 phase observations by post-processing, and then compute the DD ionospheric biases by using ambiguity-fixed L1–L2 geometry-free phase combinations. The results show that the duration of large ionospheric biases coincide to the duration of large ERTK errors. Besides the multipath could be another attribution to these large ERTK errors due to the special constellation of GEO satellites.

Let us now analyze the ionosphere-float ERTK solutions computed according to the formulae in Sect. 10.3.2. Comparing with the ionosphere-ignored model, it is apparent that the systematic errors basically vanish and the random noises are significantly enlarged, which is consistent to the theoretical analysis that the noise is enlarged by approximate 13 times, see the context after (10.26). Therefore, there is a trade-off for choosing the ionosphere-ignored and ionosphere-float model. When the effect of ionospheric biases on ionosphere-ignored solutions is less than that of enlarged noises on ionosphere-float solutions, one should adopt the ionosphere-ignored model and vice versa. In terms of theoretical analysis in Sect. 10.3.4 that one should use ionosphere-ignored model when the ionospheric bias is smaller than the threshold of 0.33 m. Moreover, considering the ionospheric biases where almost all DD ionospheric biases are smaller than this threshold, the ionosphere-ignored ERTK solutions should be better than the ionosphere-float ones.

The accumulated probabilities of absolute positioning errors and the STD and RMS statistics are estimated for ionosphere-float ERTK model. In this case, only 30% errors are smaller than 10 cm for horizontal components, and the error reaches to 50 cm with the accumulated probability of 90%. The height component of ionosphere-float model is much worse than that of ionosphere-ignored model. The mean RMS accuracies are about 25, 30 and 80 cm for N, E and U, respectively. Such accuracies are approximately 3-times worse than those of ionosphere-ignored ERTK solutions.

It is noticed that the systematic errors still remain in ionosphere-float ERTK solutions. This can be attributed to the multipath and also probably the second-order ionospheric biases. The existence of severe multipath has been demonstrated by Wu et al. [39] and Odolinski [40] in BDS RTK solutions, and also by Li et al. [17] through examining the DD geometry-free and ionosphere-free (GIF) combinations. Especially, in ionosphere-float model, the multipath is significantly enlarged together with the enlarged noises. Here, let us simply analyze the effect of the second-order ionospheric biases. It is assumed the first- and second-order ionospheric biases as $\iota_1 = K_1/f_1^2$ and $\iota_2 = K_2/2f_1^3$ at L1 frequency with $K_1$ and $K_2$ the functions of total electronic contents. In ionosphere-float model, only the first-order ionospheric biases are compensated by parameterization which is similar to using the ionosphere-tree model to eliminate the first-order ionospheric biases. In this case, the remained ionospheric biases are derived as follows. Different from the first-order ionospheric bias, the combination coefficient of the second-order ionospheric bias is

$$\tilde{\mu}_{(i,j,k)} = \frac{f_1^3 \left( i/f_1^2 + j/f_2^2 + k/f_3^2 \right)}{f_{(i,j,k)}} \tag{10.45}$$

Then in the ionosphere-free EWL observation, the remained second-order ionospheric bias is

$$\left( \frac{\mu_{(0,-1,1)}\tilde{\mu}_{(1,3,-4)}}{\mu_{(1,3,-4)} - \mu_{(0,-1,1)}} - \frac{\mu_{(1,3,-4)}\tilde{\mu}_{(0,-1,1)}}{\mu_{(1,3,-4)} - \mu_{(0,-1,1)}} \right) \iota_2 \approx -1.6\iota_2 \tag{10.46}$$

Since the second-order ionospheric bias is about one-percent of the first-order one [41], the remained ionospheric bias is about $0.016\iota_1$. That can introduce about 1 cm error at ionosphere-float ERTK model.

### 10.4.3 Results of Ionosphere-Smoothed ERTK

As stated in Sect. 10.3.4, although the ionosphere-float ERTK is overall worse than the ionosphere-ignored ERTK, it provides a possibility for users to further improve its solution by additional smoothing processing with incorporating the L1 phase observations. The positioning errors of ionosphere-smoothed ERTK solutions are computed for all 6 baselines after 2 min smoothing. The accumulated probabilities and STD and RMS statistics of positioning errors are estimated.

The very exciting results are obtained. In general, 90% positioning errors are within 4 cm and nearly all within 10 cm for horizontal components. The improvement is even more significant for height. In case of accumulated probability of 90%, the error magnitude is reduced to 10 cm from larger than 1 m of ionosphere-float model and from 35 cm of ionosphere-ignored model, respectively. For horizontal components, the STD and RMS statistics are comparable, which means that the ionospheric biases are indeed well eliminated. The accuracies are improved to 3 cm from 10 cm of ionosphere-ignored model and from 25 cm of ionosphere-float model, respectively. The height accuracies are averagely improved to better than 10 cm. Such accuracy positioning is comparable to the traditional short-baseline RTK.

As more GNSS systems, such as BDS and Galileo, are capable of transmitting five or even six frequency signals, integer ambiguity resolution over distances ranging from tens to hundreds of kilometers can be further enhanced. More combination observables with long wavelength and small noise are provided to speed up NL ambiguity convergence. This is rather crucial to fix NL integer ambiguities within a few epochs, thus enabling the real-time high-precision positioning. In this section, the ionosphere-smoothed ERTK model is extended to be compatible with multiple frequencies. The multi-frequency and multi-GNSS data of station LEIJ, BAUT, FFMJ, WARN and WTZZ were assessed by IGS with sampling interval of 30 s. The date is on Doy of 300, 2024. Three baselines are formed with baseline length from 150.94 to 510.99 km. The positioning errors of 3 baselines are presented in Figs. 10.1, 10.2 and 10.3. The STD statistics of positioning errors and Time To First Fix (TTFF) are shown in Fig. 10.4. It can be found that once the NL ambiguities are fixed, centimetre-level positioning can be achieved even over distances of hundreds of kilometres. For all three baselines, the magnitude of horizontal positioning errors is less than 1 cm, and it is no more than 2 cm in the vertical. Moreover, the NL ambiguities are correctly fixed with a few epochs, benefitting from the multiple frequency signals and lots of precise combination observables. Compared to traditional RTK model, the TTFF is significantly shortened while maintaining the equivalent positioning performance.

## 10.5   Conclusion

The most benefit of triple-frequency GNSS is to form the EWL combinations whose ambiguities can be instantaneously fixed for tens to hundred kilometer baselines. We focused in this contribution on exploiting this benefit for real-time positioning and the ERTK technique was proposed. Based on the comprehensively theoretical and numerical analysis, the conclusions are summarized as follows: The instantaneous EWL AR is rather easy even for several tens kilometer baselines although the corresponding NL AR is still difficult. This is the most benefit of EWL combination of triple-frequency GNSS signals; Two ERTK models, i.e., ionosphere-ignored and

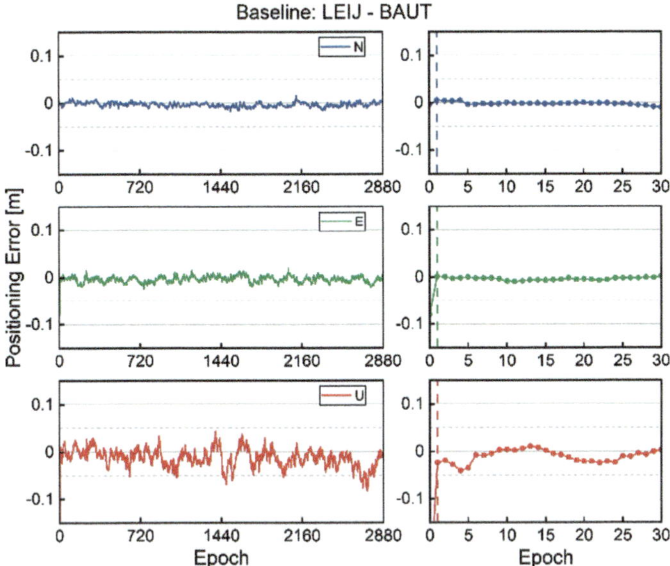

**Fig. 10.1**  Positioning errors of baseline LEIJ BAUT of which the baseline length is 150.94 km. Positioning errors of the first 30 epochs are depicted on the right

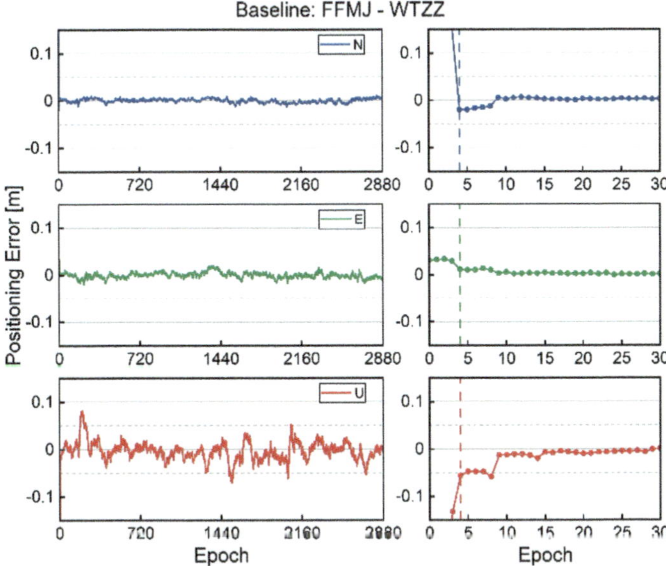

**Fig. 10.2**  Positioning errors of baseline FFMJ-WTZZ of which the baseline length is 322.12 km. Positioning errors of the first 30 epochs are depicted on the right

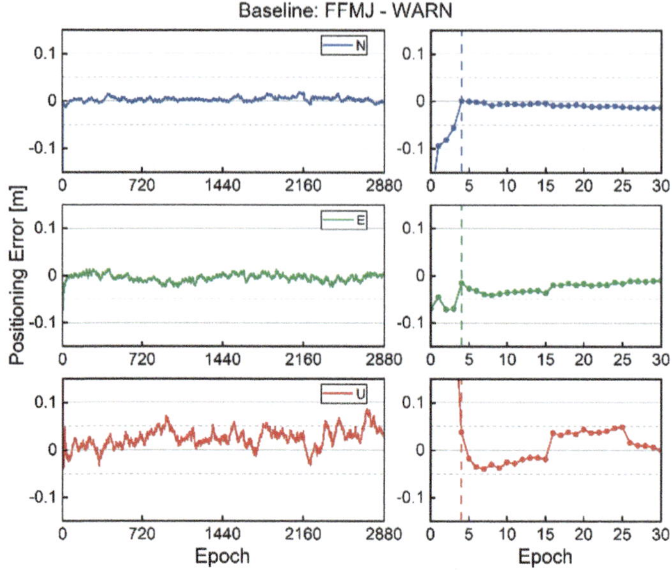


**Fig. 10.3** Positioning errors of baseline FFMJ-WARN of which the baseline length is 510.99 km. Positioning errors of the first 30 epochs are depicted on the right

**Fig. 10.4** STD statistics of positioning errors and TTFF of the three baselines

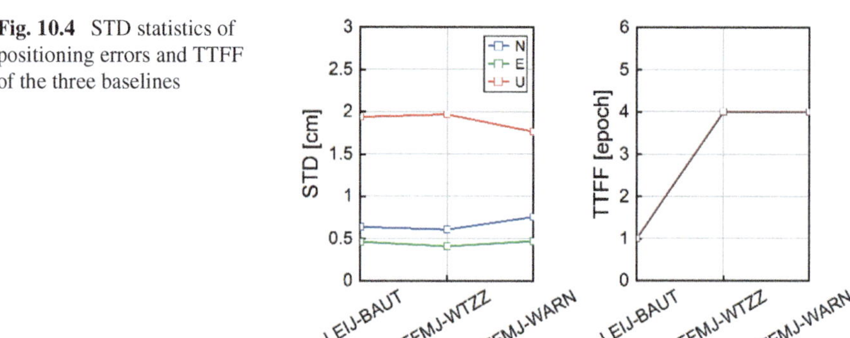


ionosphere-float model, were presented and mathematically compared. It was theoretically clarified that no matter which two EWL/WL observations were applied, the ERTK solutions will be equivalent individually for ionosphere-ignored and ionosphere-float model, respectively. Besides, a rule-of-thumb threshold of ionospheric bias was derived for selecting either ERTK model. When the ionospheric bias is smaller than 0.33 m, one should use ionosphere-ignored model and otherwise the ionosphere-float model; The results based on our experiments of 30–70 km baselines indicated that with ionosphere-ignored model one can achieve ERTK solutions of 10 cm horizontal accuracy. Although the ionosphere-float ERTK is 2-times worse than ionosphere-ignored case, it can be further improved by incorporating the

L1 phase observations where the precise epoch-differenced observation information was essentially employed.

It is emphasized that the achievement of such ERTK results is purely instantaneous without complicated NL AR, thus the ERTK is promising and can already satisfy for many applications. To the best of our knowledge, this monograph is the comprehensive study from both theoretical and practical aspects on making full use of EWL observations of triple-frequency GNSS signals for decimeter to centimeter RTK.

# References


1. Hatch R, Jung J, Enge P, Pervan B (2000) Civilian GPS: the benefits of three frequencies. GPS Solut 3:1–9
2. Feng Y (2008) GNSS three carrier ambiguity resolution using ionosphere-reduced virtual signals. J Geod 82:847–862
3. Geng J, Bock Y (2013) Triple-frequency GPS precise point positioning with rapid ambiguity resolution. J Geod 87:449–460
4. Li B, Shen Y, Zhang X (2013) Three frequency GNSS navigation prospect demonstrated with semi-simulated data. Adv Space Res 51:1175–1185
5. Weber R, Karabatic A (2009) Potential improvements in GNSS-based troposphere monitoring by use of upcoming GALILEO-signals, Eigenverlag
6. Spits J (2012) Total electron content reconstruction using triple frequency GNSS signals. Universite de Liege, de Liege
7. Richert T, El-Sheimy N (2007) Optimal linear combinations of triple frequency carrier phase data from future global navigation satellite systems. GPS Solut 11:11–19
8. Feng Y, Li B (2010) Wide area real time kinematic decimetre positioning with multiple carrier GNSS signals. Sci China Earth Sci 53:731–740
9. Chen X, Vollath U, Landau H, Sauer K (2004) GALILEO/modernized GPS obsolete network RTK. Proceedings of the ENC-GNSS 2004, pp 2523–2530
10. Vollath U, Birnbach S, Landau H, Fraile-Ordoñez JM, Martin-Neira M (1998) Analysis of three-carrier ambiguity resolution (TCAR) technique for precise relative positioning in GNSS-2. Navig 46:13–23
11. Cocard M, Bourgon S, Kamali O, Collins P (2008) A systematic investigation of optimal carrier-phase combinations for modernized triple-frequency GPS. J Geod 82:555–564
12. de Lacy MC, Reguzzoni M, Sansò F (2012) Real-time cycle slip detection in triple-frequency GNSS. GPS Solut 16:353–362
13. Zhang X, Li P (2016) Benefits of the third frequency signal on cycle slip correction. GPS Solut 20:451–460
14. Fernández-Plazaola U, Martín-Guerrero TM, Entrambasaguas JT (2008) A new method for three-carrier GNSS ambiguity resolution. J Geod 82:269–278
15. Henkel P, Günther C (2010) Reliable integer ambiguity resolution with multi-frequency code carrier linear combinations. Proceedings of the ION GNSS 2010, pp 185–195
16. Teunissen P, Joosten P, Tiberius C (2003) A comparison of TCAR, CIR and LAMBDA GNSS ambiguity resolution. Proceedings of the ION GPS 2002, pp 2799–2808
17. Li B, Feng Y, Gao W, Li Z (2015) Real-time kinematic positioning over long baselines using triple-frequency BeiDou signals. IEEE Trans Aerosp Electron Syst 51:3254–3269
18. Wang K, Rothacher M (2013) Ambiguity resolution for triple-frequency geometry-free and ionosphere-free combination tested with real data. J Geod 87:539–553
19. Zhang X, He X (2016) Performance analysis of triple-frequency ambiguity resolution with BeiDou observations. GPS Solut 20:269–281

20. Gu S, Lou Y, Shi C, Liu J (2015) BeiDou phase bias estimation and its application in precise point positioning with triple-frequency observable. J Geod 89:979–992
21. Guo F, Zhang X, Wang J, Ren X (2016) Modeling and assessment of triple-frequency BDS precise point positioning. J Geod 90:1223–1235
22. Shi C, Zhao Q, Hu Z, Liu J (2013) Precise relative positioning using real tracking data from COMPASS GEO and IGSO satellites. GPS Solut 17:103–119
23. Odolinski R, Teunissen P, Odijk D (2013) An analysis of combined COMPASS/BeiDou-2 and GPS single- and multiple-frequency RTK positioning. Proceedings of the ION 2013 Pacific PNT Meeting, pp 69–90
24. He H, Li J, Yang Y, Xu J, Guo H, Wang A (2014) Performance assessment of single- and dual-frequency BeiDou/GPS single-epoch kinematic positioning. GPS Solut 18:393–403
25. Li B, Zhang L, Verhagen S (2017) Impacts of BeiDou stochastic model on reliability: overall test, w-test and minimal detectable bias. GPS Solut 21:1095–1112
26. Li B, Feng Y, Shen Y (2010) Three carrier ambiguity resolution: distance-independent performance demonstrated using semi-generated triple frequency GPS signals. GPS Solut 14:177–184
27. Collins JP, Langley RB (1997) A tropospheric delay model for the user of the wide area augmentation system. University of New Brunswick, New Brunswick
28. Li B, Feng Y, Shen Y, Wang C (2010) Geometry-specified troposphere decorrelation for subcentimeter real-time kinematic solutions over long baselines. J Geophys Res Solid Earth 115:B11404
29. Teunissen PJG, Joosten P, Tiberius CCJM (1999) Geometry-free ambiguity success rates in case of partial fixing, Proceedings of the 1999 national technical meeting of the institute of navigation, pp 201 207
30. Verhagen S, Li B, Teunissen PJG (2013) Ps-LAMBDA: ambiguity success rate evaluation software for interferometric applications. Comput Geosci 54:361–376
31. Dai L, Eslinger D, Sharpe T (2007) Innovative algorithms to improve long range RTK reliability and availability. Proceedings of the ION 2007, pp 860–872
32. Takasu T, Yasuda A (2010) Kalman-filter-based integer ambiguity resolution strategy for long-baseline RTK with ionosphere and troposphere estimation, Proceedings of the ION GNSS 2010, pp 161–171
33. Parkins A (2011) Increasing GNSS RTK availability with a new single-epoch batch partial ambiguity resolution algorithm. GPS Solut 15:391–402
34. Li B, Teunissen PJG (2014) GNSS antenna array-aided CORS ambiguity resolution. J Geod 88:363–376
35. Hou Y, Verhagen S, Wu J (2016) A data driven partial ambiguity resolution: two step success rate criterion, and its simulation demonstration. Adv Space Res 58:2435–2452
36. Teunissen PJG (1995) The least-squares ambiguity decorrelation adjustment: a method for fast GPS integer ambiguity estimation. J Geod 70:65–82
37. Teunissen PJG (2001) Integer estimation in the presence of biases. J Geod 75:399–407
38. Hatch R (1982) The synergism of GPS code and carrier measurements. Proceedings of the International geodetic symposium on satellite doppler positioning, pp 1213–1231
39. Wu X, Zhou J, Wang G, Hu X, Cao Y (2012) Multipath error detection and correction for GEO/IGSO satellites. Sci China Phys Mech Astron 55:1297–1306
40. Odolinski R, Teunissen PJG, Odijk D (2015) Combined BDS, Galileo, QZSS and GPS single-frequency RTK. GPS Solut 19:151–163
41. Leick A, Rapoport L, Tatarnikov D (2015) GPS satellite surveying. Wiley, New Jersey

# Chapter 11
# SMC-RTK: RTK with BDS Short-Message Communication

## 11.1 Introduction

High-precision positioning based on Global Navigation Satellite Systems (GNSSs) is generally limited in marine applications due to the lack of communication access and reference stations in ocean environments [1]. Precise point positioning (PPP), as a standalone high-precision positioning technique, is possibly available in ocean environments [2]. However, it typically takes approximately 10–30 min to achieve centimeter-level accuracy even with multi-frequency multi-GNSS signals in static applications [3, 4] and longer in real-time applications [5]. In addition, communication is a bottleneck in marine applications. Some commercial companies provide PPP services globally by transmitting augmented information via satellite communication at high costs. If there is no regional reference station network, the performance of these PPP products is equivalent to real-time PPP. A low-earth-orbit (LEO) satellite-augmented PPP would be promising for marine applications, but the LEO constellation is still in development.

Real-time kinematic (RTK) is another high-precision positioning technique possibly applicable to the ocean applications. Network-based RTK can provide service in large-scale areas but requires a reference station network [6]. Single-based RTK (SRTK) only needs one reference station and is thus more applicable to ocean applications where very few reference stations are available. The coverage radius of a single reference station is traditionally approximately 20 km [7] and can be extended to 100 km using multi-frequency multi-GNSS signals [8, 9] and even to 1000 km using precise ephemerides [10]. For such long baselines, SRTK can still provide centimeter-level positioning service once the ambiguities are correctly fixed. However, the probability of incorrect fixing increases with baseline length extension. In addition, SRTK is rarely implemented at sea since there is generally a lack of nearby reference stations, and it is difficult to access real-time precise ephemeris due to limited communication means. Traditional terrestrial radio only provides





B. Li et al., *GNSS Real-Time Kinematic Positioning*, Navigation: Science and Technology 17, https://doi.org/10.1007/978-981-96-9116-6_11

communication services within a radius of 50 km, and satellite communication is expensive.

As an initial study, [1] proposed a technique called ocean-RTK that employs the short message service (SMS) of the BeiDou navigation satellite system (BDS) to transmit the differential corrections for SRTK on the ocean. The experimental results showed that for a baseline longer than 300 km, the horizontal positioning accuracy was higher than 10 cm. However, there are the following simplifications and limitations in [1]. (1) To overcome the narrow communication bandwidth of the SMS, ionosphere-free (IF) combined corrections were used instead of uncombined corrections. This means that the IF observations must be applied for positioning at the user side, which reduces the observation redundancy and eliminates the possibility of further applying available ionospheric constraints. (2) The broadcast ephemeris was employed, and its effects on long-baseline positioning were ignored. (3) Due to the weakness of the IF model as well as the considerable effects of orbit errors, ambiguity resolution was not conducted and the float solutions with subdecimeter to centimeter accuracy were used directly, depending on the convergence time.

In [11], we substantially upgrade ocean-RTK by technically solving its aforementioned limitations. First, uncombined corrections instead of IF corrections are applied to increase the observation redundancy. Meanwhile, ultrarapid precise ephemerides are assimilated into the uncombined corrections to eliminate the effects of orbit errors on long-baseline positioning. Here, the ultrarapid precise ephemeris is provided by the Tongji BeiDou Analysis Center (TJBAC) with centimeter accuracy. For more details of its quality evaluation, one can refer to http://www.igmas.org/product. Importantly, a more efficient encoding strategy is proposed to compress the uncombined corrections that have much larger data volumes than the IF corrections. Second, an ionosphere-weighted RTK model is formulated with uncombined corrections to improve the model strength and positioning performance. Third, again due to the narrow bandwidth of the SMS, it is not able to transmit continuous corrections, and thus, the corrections of a single epoch must be used for several epochs at the user side. As a result, asynchronously differential observations are used for positioning where between-epoch time-correlations must exist. An asynchronous and time-differenced filter is employed to assimilate the time-delays and time-correlations of these corrections. Finally, a partial ambiguity resolution (PAR) strategy is employed to fix the narrow-lane ambiguities, further improving the positioning performance.

The rest of the chapter starts with the RTK with BDS short-message communication (SMC-RTK) infrastructure, emphasizing the generation of uncombined corrections with assimilated precise ephemerides and the new efficient encoding strategy. Then an ionosphere-weighted RTK model is formulated using the uncombined corrections followed by an asynchronous and time-differenced filter. The numerical experiments are presented. Finally, some concluding remarks are given.

## 11.2  SMC-RTK Infrastructure at a Reference Station

The new version of SMC-RTK contains three components/steps: (1) the uncombined corrections are generated and encoded at the reference station; (2) the encoded corrections are transferred to users through the BDS SMS; and (3) the users decode and apply the corrections to realize high-precision RTK. The key difference between SMC-RTK and SRTK is that the corrections are transferred through the BDS SMS. Therefore, both encoding and broadcast strategies need to be carefully designed to fulfill the narrow bandwidth of the BDS SMS.

### 11.2.1  Generating the Uncombined Corrections

For code division multiple access (CDMA)-type GNSS signals, the pseudorange $P_{r,j}^s$ and carrier phase $\Phi_{r,j}^s$ observation equations at reference station $r$ and epoch time $t_0$ are

$$
P_{r,j}^s(t_0) - \left\| X_{r,\mathrm{b}}^s(t_0) - X_r \right\| = \tau_r^s(t_0) + \mu_j \iota_r^s(t_0) + dt_{r,j}(t_0)
$$
$$
+ \, \delta t_j^s(t_0) + o_{r,\mathrm{b}}^s(t_0) + \varepsilon_{P_{r,j}^s(t_0)}
$$
$$
\Phi_{r,j}^s(t_0) - \left\| X_{r,\mathrm{b}}^s(t_0) - X_r \right\| = \tau_r^s(t_0) - \mu_j \iota_r^s(t_0) + dt_{r,j}(t_0)
$$
$$
+ \, \delta t_j^s(t_0) + o_{r,\mathrm{b}}^s(t_0) - \lambda_j a_{r,j}^s + \varepsilon_{\Phi_{r,j}^s(t_0)} \tag{11.1}
$$

where the subscripts $j$ and $s$ denote the frequency and satellite, respectively. $\| * \|$ is the Euclidean norm of $*$. $X_{r,\mathrm{b}}^s$ is the satellite coordinate calculated with the corresponding broadcast ephemeris. Its calculation relates to the receiver since the receiver position together with the epoch time determine the transmission time of the satellite signal. $X_r$ is the coordinate of the reference station, which is precisely known. $\tau_r^s$ and $\iota_r^s$ represent the tropospheric and ionospheric delays, respectively. The coefficient $\mu_j = f_1^2 / f_j^2$ with $f_1$ and $f_j$ being the first and $j$th frequencies, respectively. $dt_{r,j}$ and $\delta t_j^s$ are the clock errors of receiver $r$ and satellite $s$ on frequency $j$, respectively. $a_{r,j}^s$ is the time-independent unknown ambiguity with wavelength $\lambda_j$. $o_{r,\mathrm{b}}^s$ is the orbit error introduced by the broadcast ephemeris. $\varepsilon$ denotes the residual observation errors.

The effect of the orbit error $o_{r,\mathrm{b}}^s$ can be up to 10 cm for baselines as long as several hundred kilometers [12]. It can be reduced to a few millimeters by applying an (ultrarapid) precise ephemeris. In other words, the orbit error $o_{r,\mathrm{b}}^s$ at epoch time $t_0$ at the reference station can be computed as

$$
o_{r,\mathrm{b}}^s(t_0) = \left\| X_{r,\mathrm{p}}^s(t_0) - X_r \right\| - \left\| X_{r,\mathrm{b}}^s(t_0) - X_r \right\| \tag{11.2}
$$

where $X_{r,\mathrm{p}}^s$ is the satellite coordinate from the precise ephemeris. Since it is difficult to directly transmit the precise ephemeris to a user, the orbit error at epoch time $t_1$ at the user end can be computed at the reference station in advance as

$$o_{u,b}^{s}(t_1) = \left\| X_{u,p}^{s}(t_1) - X_u(t_1) \right\| - \left\| X_{u,b}^{s}(t_1) - X_u(t_1) \right\| \tag{11.3}$$

where the subscript $u$ denotes the user station and $X_u$ is its approximate coordinate. The user orbit error $o_{u,b}^{s}(t_1)$ cannot be replaced by $o_{r,b}^{s}(t_1)$ due to the long baseline length. Hence, $X_u$ must be sent from the user back to the computation center at the reference station, which requires two-way communication between the reference station and user. Fortunately, the user can either receive or send a short message to the computation center through the BDS SMS. In addition, $X_u$ can be continuously used until the coordinate variation reaches tens of kilometers. Hence, we send $X_u$ to the computation center one time per hour, and thus, the communication burden can be ignored. Here, we directly compute the satellite coordinates at the nominal epoch time instead of the actual time when the satellite signal is broadcast. The reason is that the orbit error variation is less than 1 mm over 1 s. The difference between the nominal time and the broadcast time is only approximately 0.075 s. Such processing does not practically affect the orbit error. Moreover, the orbit error is assumed to vary linearly over a short period as

$$d_o^{s} = \frac{o_{u,b}^{s}(t_1) - o_{u,b}^{s}(t_0)}{t_1 - t_0} \tag{11.4}$$

where $d_o^{s}$ denotes the variation rate of the orbit error at the user station. One can then compute the code and phase corrections at epoch time $t_0$ at the reference station as

$$\begin{aligned}
\tilde{P}_{r,j}^{s}(t_0) &= P_{r,j}^{s}(t_0) - \left\| X_{r,p}^{s}(t_0) - X_r \right\| + o_{u,b}^{s}(t_0) \\
&\quad - \hat{\tau}_r^{s} - \mu_j \hat{\imath}_r^{s} - \hat{dt}_{r,j} + \hat{\delta t}^{s} \\
\tilde{\Phi}_{r,j}^{s}(t_0) &= \Phi_{r,j}^{s}(t_0) - \left\| X_{r,p}^{s}(t_0) - X_r \right\| + o_{u,b}^{s}(t_0) - \hat{\tau}_r^{s} \\
&\quad + \mu_j \hat{\imath}_r^{s} - \hat{dt}_{r,j} + \hat{\delta t}^{s} - \lambda_j \check{N}_{r,j}^{s}
\end{aligned} \tag{11.5}$$

The corrections at epoch time $t_1$ at the user station are recovered as

$$\begin{aligned}
\tilde{P}_{r,j}^{s}(t_1) &= \tilde{P}_{r,j}^{s}(t_0) + d_o^{s} \times (t_1 - t_0) \\
\tilde{\Phi}_{r,j}^{s}(t_1) &= \tilde{\Phi}_{r,j}^{s}(t_0) + d_o^{s} \times (t_1 - t_0)
\end{aligned} \tag{11.6}$$

All the terms, without epoch time for simplification, on the right side with a hat or check denote approximations, which are subtracted from the observations to minimize the absolute values of the corrections to save communication resources. The nominal tropospheric delay $\hat{\tau}_r^{s}$ is computed using the New Brunswick 3 (UNB3) model [13, 14] together with the Niell mapping function [15]. The nominal ionospheric delay $\hat{\imath}_r^{s}$ is computed through the Klobuchar model [16]. The satellite clock error $\hat{\delta t}^{s}$ is calculated with the broadcast ephemerides. The approximate receiver clock error is determined by averaging the residuals of all $m$ satellites

$$\hat{dt}_{r,j} = \frac{1}{m}\left(\sum_{s=1}^{s=m} P^s_{r,j}(t_0) - \left\|X^s_{r,\mathrm{p}}(t_0) - X_r\right\| - \hat{\tau}^s_r - \mu_j \hat{\imath}^s_r + \hat{\delta t}^s\right) \tag{11.7}$$

Here, it is important to reduce the phase correction by subtracting an integer from the phase observation. The integer is computed as

$$\check{N}^s_{r,j} = \left[\frac{\Phi^s_{r,j}(t_0) - \left\|X^s_{r,\mathrm{p}}(t_0) - X_r\right\| - \hat{\tau}^s_r + \mu_j \hat{\imath}^s_r - \hat{dt}_{r,j} + \hat{\delta t}^s}{\lambda_j}\right] \tag{11.8}$$

where [*] denotes rounding variable * to its nearest integer.

After deducing these terms, the absolute values of the corrections are normally smaller than 20 m for satellites with elevations higher than $10°$. Therefore, as shown in Table 11.1, if $\tilde{P}^s_{r,j}$ is located outside its given range, it will be abandoned since in this case, there is a high probability it is incorrect. If $\tilde{\Phi}^s_{r,j}$ is located outside its given range, a new integer $\check{N}^s_{r,j}$ will be computed to make $\tilde{\Phi}^s_{r,j}$ inside the given range, and an indicator of cycle slip is marked to indicate the integer change. Otherwise, $\check{N}^s_{r,j}$ remains constant. Hence, the absolute values of the corrections are rigorously restricted and can be encoded to a few characters. In our previous study, the corrections were further reduced by forming the corresponding IF corrections. Although the IF corrections alleviate the communication burden, they degrade the positioning performance as explained above. Hence, in this contribution, we directly utilize these uncombined corrections instead of the IF corrections.

### 11.2.2  Encoding the Uncombined Corrections

A new encoding strategy is proposed to efficiently compress the uncombined corrections. In SMC-RTK, only dual-frequency Global Positioning System (GPS) and triple-frequency BDS observations are used due to the limited communication resources. The encoding is identical for the observations at each frequency of each satellite and each system. Hence, we only present the overall structure of the encoding strategy and the encoding details of observations at the first frequency of the first BDS satellite in Table 11.1. First, the epoch time in seconds is encoded with 6 bits, which will be used by users to compute the time difference between user observations and the corrections. Then, the number of BDS satellites is encoded with 4 bits. For the first satellite, the pseudo random number (PRN), the integer hour of its ephemeris epoch time and its orbit error variation rate are encoded with 18 bits, followed by the phase and pseudorange corrections of all frequencies. For the first frequency, the corrections and the cycle slip indicator are encoded with 29 bits. With the same strategy, the corrections of the other frequencies and other BDS or GPS satellites are encoded. Table 11.2 presents an example of the encoded corrections for a single epoch, including the encoded binary bits and their corresponding correction contents.

**Table 11.1**  Uncombined correction encoding strategy

| | | | Size (bits) | Range | Resolution | Encoding method |
|---|---|---|---|---|---|---|
| Observation time (seconds in one minute) | | | 6 | $0 \sim 59$ | 1 s | d2b (seconds of observation time) |
| BDS | Number of BDS satellites | | 4 | $0 \sim 15$ | 1 | d2b (number of BDS satellites) |
| | 1st sat. | Satellite PRN | 5 | $1 \sim 32$ | 1 | d2b (satellite PRN) |
| | | Ephemeris time | 5 | $0 \sim 23$ | 1 h | d2b (integer hour of ephemeris time) |
| | | Variation rate of orbit error $d_o^s$ | 8 | $-1.28 \sim 1.27$ mm | 0.01 mm | d2b $([100 \times d_o + 128])$ |
| | | 1st frequency $\tilde{\Phi}_{r,1}^1$ | 16 | $-32.768 \sim 32.767$ m | 1 mm | d2b $([1000 \times \tilde{\Phi}_{r,1}^1 + 32,768])$ |
| | | $\tilde{P}_{r,1}^1$ | 12 | $-40.96 \sim 40.94$ m | 2 cm | d2b $\left( \left[ 50 \times \tilde{P}_{r,1}^1 + 2048 \right] \right)$ |
| | | CS indicator | 1 | 0 or 1 | 1 | 0 for CS absent, and 1 for CS existent |
| | | Other frequencies | …… | | | |
| | Other sat. | …… | | | | |
| GPS | …… | | | | | |

*Note* d2b (*) is a function for converting a variable from a decimal system to a binary system. CS represents cycle slip

**Table 11.2**  An example of encoded corrections

| Encoded binary bits | Correction contents |
|---|---|
| 000000 | Observation time $= 0$ s |
| 1100 | No. of BDS satellites $= 12$ |
| 00001 | Satellite PRN $= 1$ |
| 00000 | Ephemeris time $= 0$ h |
| 10001110 | Variation rate $= 0.14$ mm/s |
| 1001010100111110 | $\tilde{\Phi}_{r,1}^1 = 5.438$ m |
| 100011011001 | $\tilde{P}_{r,1}^1 = 4.340$ m |
| 0 | No cycle slip occurs on $\tilde{\Phi}_{r,1}^1$ |
| …… | …… |

If $m_1$ GPS satellites (with dual-frequency observations) and $m_2$ BDS satellites (with triple-frequency observations) are simultaneously tracked, the corrections of a single epoch will be encoded to a character string whose data volume is

$$n_{cs} = 6 + 4 + (18 + 29 \times 2)m_1 + 4 + (18 + 29 \times 3)m_2$$

$$= 14 + 76m_1 + 105m_2 \tag{11.9}$$

The data volume is computed according to the encoding strategy in Table 11.1. Here, 6 bits are for the observation time, and 4 bits are for the number of GPS or BDS satellites. The factors of 2 and 3 in the first and second brackets indicate the dual-frequency GPS signals and triple-frequency BDS signals. The data volume of our encoded corrections in the binary stream is only approximately 40% of the standard RTK corrections. After encoding, the corrections are divided into several short messages and sent to users through the BDS SMS. Moreover, for each short message, a preamble is appended at the message head to denote its serial number.

### *11.2.3  The Broadcast Strategy*

The encoded corrections are broadcast to users through the BDS SMS. The SMS working mechanisms are as follows; see also [1].

1.  One BDS SMS card is allowed to send one message within 1 min. The transmission time is so short that it can be omitted, but the same card will be deactivated for 1 min after sending a message.
2.  One message is limited to 78.5 bytes, i.e., 628 bits.
3.  Several cards can be integrated into a multicard machine to send messages in turn. The minimal sending interval can be 1 s if 60 SMS cards are integrated.

According to these mechanisms, the maximum bandwidth of the BDS SMS is 628 bps. It is obviously not practical to send the uncombined corrections per second. Considering the case of $m_1$ GPS satellites and $m_2$ BDS satellites in uncombined corrections, the corrections of a single epoch require $n_{\mathrm{sm}} = \mathrm{floor}\left(\frac{n_{\mathrm{cs}}}{628} + 1\right)$ short messages for transmitting. If $n_{\mathrm{cd}}$ cards are integrated, $n_{\mathrm{ep}} = \mathrm{floor}\left(\frac{n_{\mathrm{cd}}}{n_{\mathrm{sm}}}\right)$ epochs of corrections can be sent to users at most within 1 min. One can of course send the corrections of these $n_{\mathrm{ep}}$ epochs to users successively until all $n_{\mathrm{cd}}$ cards are deactivated. After that, the corrections of the $(n_{\mathrm{ep}} + 1)$th epoch will be sent in the next minute. In this way, the time delay of the corrections, that is, the time difference between the user observations and the most recently received corrections, will exceed $(60 - n_{\mathrm{cd}} + 2n_{\mathrm{sm}})$ s at most. If $n_{\mathrm{cd}}$ is small, the maximum time delay will be too long and significantly degrade the positioning performance. An improved strategy sends the corrections with a sampling interval as

$$\Delta t = \mathrm{floor}\left(\frac{60}{n_{\mathrm{ep}}}\right) = \mathrm{floor}\left(\frac{60}{\mathrm{floor}\left(\dfrac{n_{\mathrm{cd}}}{\mathrm{floor}\left(\frac{8+68m_1+97m_2}{628}+1\right)}\right)}\right) \tag{11.10}$$

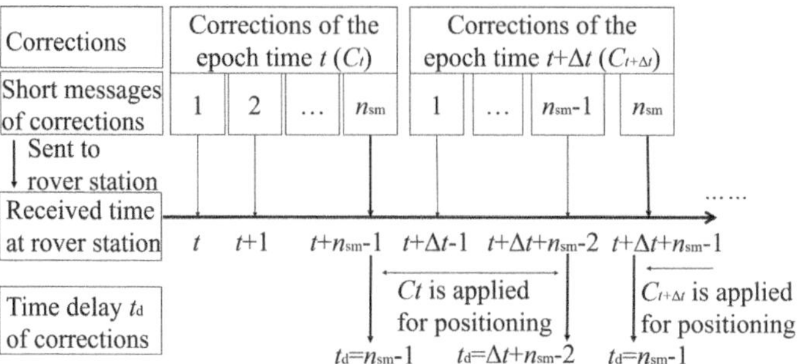


**Fig. 11.1** Time delay of the corrections for the improved broadcast strategy

As shown in Fig. 11.1, the corrections at epoch time $t$ are completely received by users at epoch time $(t + n_{sm} - 1)$ and applied for user positioning continuously to epoch time $(t + \Delta t + n_{sm} - 2)$. Then, at epoch time $(t + \Delta t + n_{sm} - 1)$, the next corrections at epoch time $(t + \Delta t)$ are completely received and applied for user positioning. As a result, the maximum time delay of the corrections is minimized to $(\Delta t + n_{sm} - 2)$ s. In brief, the broadcast strategy is defined by the number of satellites $m_1$ and $m_2$, the number of ID cards $n_{cd}$, and the sampling interval $\Delta t$. If $\Delta t$ is fixed, increasing the number of satellites can improve the positioning performance but requires more ID cards and more communication costs and vice versa. If the number of satellites is fixed, increasing the number of ID cards requires more communication costs but can shorten the time delay and improve the positioning performance and vice versa. If $n_{cd}$ is fixed, increasing the number of satellites increases the redundancy of observations but extends the time delay. In our design, we empirically set $m_1 \leq 9$, $m_2 \leq 9$ and $n_{cd} = 18$; then, $n_{cs} \leq 1643$, $n_{sm} \leq 3$ and $\Delta t = 10$, which leads to a time delay from 2 to 11 s. The communication costs are approximately $18 \times 600 = 10,800$ Chinese yuan for a reference station and 600 Chinese yuan for a single user per year. Here, 600 yuan is the annual fee of a single BDS ID card. The strategy has already satisfied our application. One can also find a better broadcast strategy to improve the positioning performance with less communication costs by changing $m_1$, $m_2$, $n_{cd}$ and $\Delta t$, which remains for further research.

## 11.3   Rover Station Positioning Model

Since the uncombined corrections are received by users, in this contribution, we employ the ionosphere-weighted model instead of the IF positioning model used in [1]. In addition, the time delay and repeat use of the corrections are both considered.

### *11.3.1  Ionosphere-Weighted Model with Asynchronous Time-Correlated Observations*

The asynchronous double-differenced (DD) pseudorange and phase observations at epoch time $t_1$ are formulated as

$$
\begin{aligned}
P_{ur,j}^{sv}(t_1) &= P_{u,j}^{sv} - \left\| X_{u,b}^{s}(t_1) - X_u(t_1) \right\| \\
&\quad + \left\| X_{u,b}^{v}(t_1) - X_u \right\| - \tilde{P}_{r,j}^{sv}(t_1) - \hat{\tau}_u^{sv} + \hat{\delta t}_j^{sv} \\
\Phi_{ur,j}^{sv}(t_1) &= \Phi_{u,j}^{sv} - \left\| X_{u,b}^{s}(t_1) - X_u \right\| \\
&\quad + \left\| X_{u,b}^{v}(t_1) - X_u \right\| - \tilde{\Phi}_{r,j}^{sv}(t_1) - \hat{\tau}_u^{sv} + \hat{\delta t}_j^{sv}
\end{aligned}
\tag{11.11}
$$

where $(*)_{*,j}^{sv} = (*)_{*,j}^{s} - (*)_{*,j}^{v}$. $P_{u,j}^{sv}(t_1)$ and $\Phi_{u,j}^{sv}(t_1)$ are the single-differenced (SD) pseudorange and phase observations of the user at epoch time $t_1$. $\tilde{P}_{r,j}^{sv}(t_1)$ and $\tilde{\Phi}_{r,j}^{sv}(t_1)$ are the SD pseudorange and phase corrections at epoch time $t_1$. $\hat{\tau}_u^{sv}$ is the SD tropospheric delay computed based on the same model and mapping function as used at the reference station. $\hat{\delta t}_j^{sv}$ denotes the SD satellite clock error calculated using the same broadcast ephemeris as used at the reference station.

The observation equations of the asynchronous DD observations are

$$
\begin{aligned}
P_{ur,j}^{sv}(t_1) &= Bx_u + \tau_{ur}^{sv} + \mu_j \iota_{ur}^{sv} + \delta t_j^{sv}(t_1, t_0) \\
&\quad + o_{u,b}^{sv}(t_1) - o_{u,b}^{sv}(t_0) - d_o^{sv}(t_1 - t_0) + \varepsilon_{P_{ur,j}^{sv}(t_i)} \\
&= Bx_u + \tau_{ur}^{sv} + \mu_j \iota_{ur}^{sv} + \delta t_j^{sv}(t_1, t_0) + \varepsilon_{P_{ur,j}^{sv}(t_1)} \\
\Phi_{ur,j}^{sv}(t_1) &= Bx_u + \tau_{ur}^{sv} + \mu_j \iota_{ur}^{sv} + \lambda_j a_{r,j}^{sv} + \delta t_j^{sv}(t_1, t_0) \\
&\quad + o_{u,b}^{sv}(t_1) - o_{u,b}^{sv}(t_0) - d_o^{sv}(t_1 - t_0) + \varepsilon_{P_{ur,j}^{sv}(t_i)} \\
&= Bx_u + \tau_{ur}^{sv} + \mu_j \iota_{ur}^{sv} + \delta t_j^{sv}(t_1, t_0) + \lambda_j a_{r,j}^{sv} + \varepsilon_{P_{ur,j}^{sv}(t_1)}
\end{aligned}
\tag{11.12}
$$

where the orbit error is basically eliminated, i.e., $o_{u,b}^{sv}(t_1) - o_{u,b}^{sv}(t_0) - d_o^{sv}(t_1 - t_0) = 0$. The asynchronous DD residual satellite clock error $\delta t_j^{sv}(t_1, t_0) = \delta t_j^{sv}(t_i) - \delta t_j^{sv}(t_0)$ can be precisely corrected with the coefficients of the satellite clock error from the broadcast ephemerides [1]. Omitting the subscripts of satellites and receivers and the epoch time, the asynchronous DD observation equations are

$$
\begin{aligned}
P_j &= Bx + \mu_j \iota + \varepsilon_{P_j} \\
\Phi_j &= Bx - \mu_j \iota + \lambda_j a_j + \varepsilon_{\Phi_j}
\end{aligned}
\tag{11.13}
$$

where $\iota$ is the asynchronous DD ionospheric delay. $a_j = a_{u,j}^{sv} - a_{r,j}^{sv} + \check{N}_{r,j}^{sv}$ is the DD integer ambiguity, while $a_{u,j}^{sv}$ and $a_{r,j}^{sv}$ are the SD ambiguities of the user and reference station, respectively; $\check{N}_{r,j}^{sv}$ is the SD subtracted integer. If no cycle slip occurs, $a_j$ is constant. $\varepsilon_*$ is the random error including all residual errors. The residual tropospheric delay can be absorbed by the residual zenith tropospheric delay (RZTD) parameter [17, 18]. However, the RZTD parameter is strongly correlated with the vertical coordinate [19]. Hence, we simply ignore it in this chapter to speed up the positioning convergence considering its minor effect on ambiguity integers [20]. The single-epoch positioning model with asynchronous observations can be expressed in matrix form as

$$\begin{bmatrix} \boldsymbol{p}_k \\ \boldsymbol{\phi}_k \end{bmatrix} = \begin{bmatrix} \boldsymbol{e}_f \otimes \boldsymbol{B}_k & \boldsymbol{\mu} \otimes \boldsymbol{I}_s & \boldsymbol{0} \\ \boldsymbol{e}_f \otimes \boldsymbol{B}_k & -\boldsymbol{\mu} \otimes \boldsymbol{I}_s & \boldsymbol{\Lambda} \otimes \boldsymbol{I}_s \end{bmatrix} \begin{bmatrix} \boldsymbol{x}_k \\ \boldsymbol{\iota}_k \\ \boldsymbol{a} \end{bmatrix} + \begin{bmatrix} \boldsymbol{\varepsilon}_{p_k} \\ \boldsymbol{\varepsilon}_{\phi_k} \end{bmatrix} \tag{11.14}$$

where $k$ indicates the epoch number. $\boldsymbol{p}_k = \begin{bmatrix} \boldsymbol{p}_{1,k}, \ldots, \boldsymbol{p}_{f,k} \end{bmatrix}^{\mathrm{T}}$ and $\boldsymbol{\phi}_k = \begin{bmatrix} \boldsymbol{\phi}_{1,k}, \ldots, \boldsymbol{\phi}_{f,k} \end{bmatrix}^{\mathrm{T}}$ are the asynchronous DD observation vectors of the pseudorange and phase, respectively. $f$ is the number of frequencies. $\boldsymbol{e}_f$ is an $f$-dimensional column vector with all elements equal to 1. $\boldsymbol{B}_k$ is the design matrix of the coordinate parameters $\boldsymbol{x}_k$. $\boldsymbol{\mu} = \begin{bmatrix} \mu_1, \ldots, \mu_f \end{bmatrix}^{\mathrm{T}}$ is the vector of the scalar coefficients for the ionosphere parameters $\boldsymbol{\iota}_k$. $\boldsymbol{I}_s$ is the $s$-dimensional identity matrix, with $s$ being the number of DD satellite pairs. $\boldsymbol{a} = \begin{bmatrix} \boldsymbol{a}_1, \ldots, \boldsymbol{a}_f \end{bmatrix}^{\mathrm{T}}$ is the ambiguity vector, and $\boldsymbol{\Lambda} = \mathrm{diag}\left( \begin{bmatrix} \lambda_1, \ldots, \lambda_f \end{bmatrix} \right)$. $\boldsymbol{\varepsilon}_*$ is the vector of the residual observation errors. The stochastic model of (11.14) is

$$\boldsymbol{Q}_{\varepsilon_k} = \mathrm{blkdiag}\left( \begin{bmatrix} \sigma_p^2, \sigma_\phi^2 \end{bmatrix} \right) \otimes \boldsymbol{I}_f \otimes \boldsymbol{Q}_k \tag{11.15}$$

where $\sigma_p^2$ and $\sigma_\phi^2$ are the variance scalars of the undifferenced pseudorange and phase observations at zenith, respectively. $\boldsymbol{Q}_k$ is an $(s \times s)$ cofactor matrix of DD observations with elevation-dependent weighting [21]. To enhance the model strength, the initial and time-variant ionospheric constraints are applied as

$$\begin{aligned} \iota_0 &= \bar{\iota}, & \boldsymbol{Q}_{\bar{\iota}} &= \sigma_{\bar{\iota}}^2 \boldsymbol{Q}_0 \\ \iota_k &= \iota_{k-1} + \boldsymbol{w}_{\iota_k}, & \boldsymbol{Q}_{w_{\iota_k}} &= \sigma_{w_\iota}^2 \boldsymbol{Q}_k \end{aligned} \tag{11.16}$$

where $\bar{\iota}$ is the prior DD ionospheric delay at the first epoch and $\sigma_{\bar{\iota}}^2$ is the variance scalar of the undifferenced ionospheric delay at zenith. $\boldsymbol{w}_{\iota_k}$ is the transition noise vector with zero mean and known covariance matrix $\boldsymbol{Q}_{w_{\iota_k}}$ with $\sigma_{w_\iota}^2$ as the variance scalar of the transition noise of the undifferenced ionospheric delay at zenith. The sequential Kalman filter solutions are derived based on the least squares (LS) criterion. The ambiguity parameters are incorporated in the filter state with extremely small variances to characterize their constant property [22], namely, $\boldsymbol{a}_k = \boldsymbol{a}_{k-1} + \boldsymbol{w}_{a_k}$, where $\boldsymbol{w}_{a_k}$ is zero mean with the variance matrix $\boldsymbol{Q}_{w_{a_k}} = \mathrm{e}^{-16} \boldsymbol{I}_{sf}$.

Due to the time delay of communication, the corrections of a given epoch from the reference station need to be continuously used for several user epochs. Hence, the DD observations of multiple epochs are correlated to each other when the same corrections are repeatedly utilized in (11.14). This time correlation should be properly captured in stochastic modeling; otherwise, the ambiguity resolution efficiency will be degraded [23]. To derive the filter solution with time correlation, we organize observation Eq. (11.14) as

$$l_k = A_k \xi_k + \varepsilon_k + \eta_k$$

$$Q_{\varepsilon_k} = \text{blkdiag}\left(\left[\sigma_p^2, \sigma_\phi^2\right]\right) \otimes I_f \otimes Q_{c_k}$$

$$Q_{\eta_k} = \text{blkdiag}\left(\left[\sigma_p^2, \sigma_\phi^2\right]\right) \otimes I_f \otimes \left(Q_k - Q_{c_k}\right) \tag{11.17}$$

where $l_k = \left[p_k^{\mathrm{T}}, \phi_k^{\mathrm{T}}\right]^{\mathrm{T}}$, $A_k = \begin{bmatrix} e_f \otimes B_k & \mu \otimes I_s & 0 \\ e_f \otimes B_k & -\mu \otimes I_s & \lambda \otimes I_s \end{bmatrix}$ and $\xi_k = \left[x_k^{\mathrm{T}}, \iota_k^{\mathrm{T}}, a_k^{\mathrm{T}}\right]^{\mathrm{T}}$.

Here, the observation errors in (11.14) are decomposed into two parts. One is the time-independent observation noise $\eta_k$ introduced by the user SD observations. The other is the linearly time-dependent noise $\varepsilon_k$ introduced by the repeatedly used SD corrections. $Q_{c_k}$ is an $(s \times s)$ cofactor matrix of SD corrections with elevation-dependent weighting. Since $\varepsilon_k$ is linearly time dependent, its transition equation is

$$\varepsilon_k = S_{k,k-1}\varepsilon_{k-1}, \quad Q_{c_k} = S_{k,k-1}Q_{c_{k-1}}S_{k,k-1}^{\mathrm{T}} \tag{11.18}$$

where $S_{k,k-1}$ is the transition matrix and taken as an identity matrix here. In addition, the random walk process is applied for the state transition as

$$\xi_k = \xi_{k-1} + w_k, \quad Q_{\xi_k} = Q_{\xi_{k-1}} + Q_{w_k} \tag{11.19}$$

where

$$w_k = \left[w_{x_k}^{\mathrm{T}}, w_{\iota_k}^{\mathrm{T}}, w_{a_k}^{\mathrm{T}}\right]^{\mathrm{T}}, \quad Q_{w_k} = \text{blkdiag}\left(\sigma_x^2 I_3, \sigma_{w_\iota}^2 Q_k, \mathrm{e}^{-16} I_{sf}\right) \tag{11.20}$$

and $w_{x_k}$ is the transition noise vector for the position parameters and is assumed to have a zero mean and covariance matrix $\sigma_x^2 I_3$. Here, we take an extremely large value for $\sigma_x^2$ to conservatively characterize the position variation, i.e., the true kinematic situation. A standard Kalman filter can be applied to solve Eqs. (11.17)–(11.20) if the observations $l_k$ and $l_{k-1}$ are time-independent, namely, $Q_{c_k} = 0$ and $S_{k,k-1} = 0$. Otherwise, the filter solution should consider the observation time-correlations [24]. By applying the time-differencing method introduced in [25], we derive the time-correlated filter solution as

$$\begin{aligned}
\bar{\xi}_k &= \hat{\xi}_{k-1} + w_k \\
Q_{\bar{\xi}_k} &= Q_{\hat{\xi}_{k-1}} + Q_{w_k} \\
\hat{\xi}_k &= \bar{\xi}_k + J_k\left(z_k - H_k\bar{\xi}_k\right) \\
Q_{\hat{\xi}_k} &= (I - J_k H_k)Q_{\bar{\xi}_k} - J_k C_k^{\mathrm{T}}
\end{aligned} \tag{11.21}$$

where $\bar{\xi}_k$ is the predicted state vector and

$$z_k = l_k - S_{k,k-1} l_{k-1}$$

$$J_k = \left( Q_{\bar{\xi}_k} H_k^T + C_k \right) \left( H_k Q_{\bar{\xi}_k} H_k^T + Q_{\nu_k} + H_k C_k + C_k^T H_k^T \right)^{-1}$$

$$H_k = A_k - S_{k,k-1} A_{k-1}$$

$$v_k = S_{k,k-1} A_{k-1} w_k + \eta_k - S_{k,k-1} \eta_{k-1} \tag{11.22}$$

$$C_k = Q_{w_k} A_{k-1}^T S_{k,k-1}^T$$

$$Q_{\nu_k} = S_{k,k-1} A_{k-1} Q_{w_k} A_{k-1}^T S_{k,k-1}^T + Q_{\eta_k} + S_{k,k-1} Q_{\eta_{k-1}} S_{k,k-1}^T$$

Note that once the corrections of the new epoch from the reference station are first used at epoch $k$, one needs only to set $S_{k,k-1} = 0$, $\varepsilon_k = 0$, $Q_{c_k} = 0$, and the time-correlated filter reduces to a standard Kalman filter.

### 11.3.2  Strategies for Partial Ambiguity Resolution

The ionosphere-weighted model with asynchronous and time-correlated observations resolves the problems of delayed corrections and long baselines for SMC-RTK. However, the SMC-RTK ambiguity resolution problem remains unresolved. The ambiguity resolution is difficult for baselines as long as several hundred kilometers since the DD residual atmospheric delays will be considerable. Therefore, [1] only fixed the wide-lane ambiguities, and [26] only provided the float solutions. In this study, multi-frequency multi-GNSS uncombined corrections, ionospheric constraints and precise ephemerides are applied to improve the float solutions, which make the narrow-lane ambiguity resolution efficient. The wide-lane and extra wide-lane ambiguities are first fixed based on a geometry-free model [27]. Then, we apply PAR to solve the narrow-lane ambiguities. The ambiguity subset $\hat{a}_1$ is chosen based on the accumulated tracking time of signals and their satellite elevations. The tracking time threshold is set to 5 s (i.e., the ambiguities continuously tracked for at least 5 s will be added to the ambiguity fixing subset). The satellite elevation threshold is empirically set to $25°$ initially and raised to $45°$ in $10°$ increments successively if the previous ambiguity subset is not fixed. Once the ambiguity subset is successfully fixed, the remaining ambiguities $\hat{a}_2$ are updated through the following relationships:

$$\tilde{a}_2 = \hat{a}_2 - Q_{\hat{a}_2 \hat{a}_1} Q_{\hat{a}_1}^{-1} (\hat{a}_1 - \check{a}_1)$$

$$Q_{\tilde{a}_2} = Q_{\hat{a}_2} - Q_{\hat{a}_2 \hat{a}_1} Q_{\hat{a}_1}^{-1} Q_{\hat{a}_1 \hat{a}_2} \tag{11.23}$$

where $\hat{a}_1$ and $\hat{a}_2$ denote the float solutions of the selected ambiguities and the remaining ambiguities with variance matrices $Q_{\hat{a}_1}$ and $Q_{\hat{a}_2}$, respectively. $Q_{\hat{a}_2 \hat{a}_1} = Q_{\hat{a}_1 \hat{a}_2}^T$ is the covariance matrix between $\hat{a}_1$ and $\hat{a}_2$. The integer vector $\check{a}_1$ represents the fixed solution of the selected ambiguities. $\tilde{a}_2$ is the updated float solution of the remaining ambiguities with the variance matrix $Q_{\tilde{a}_2}$. The remaining ambiguities whose satellite elevations are higher than $25°$ are further selected to be fixed. The ambiguities whose satellite elevations are lower than $25°$ remain unfixed to control the ambiguity resolution reliability due to their larger DD residual atmospheric delays. The narrow-lane ambiguities are fixed by employing the least-squares ambiguity decorrelation adjustment method [28]. The decision of accepting the fixed solution is made according to the ratio test with dimension-dependent thresholds advised in [22].

## 11.4 Experimental Analysis of SMC-RTK

Two experiments, static and kinematic situations, are carried out. In SMC-RTK, we process the static data by purely simulating the real-time kinematic situation. Triple-frequency BDS and dual-frequency GPS observations are used. The cutoff elevation is $7°$, and an elevation-dependent stochastic model of undifferenced observations $\sigma = \frac{1.02}{\sin\theta+0.02}\sigma_{90°}$ is applied with a zenith precision of $\sigma_{90°} = 2$ mm for the phase and 0.2 m for the pseudorange. We conservatively take $\bar{\iota} = 0$ and $\sigma_{\bar{\iota}} = 2b_{\text{len}}$ mm/km, where $b_{\text{len}}$ is the baseline length in kilometers and $\sigma_{w_\iota}^2 = 1.5\Delta t\,\text{cm}^2/\text{s}$, where $\Delta t$ is the elapsed time in seconds.

### *11.4.1 Baseline Experiment*

A 72-h dataset was collected on a 320 km baseline with a sampling interval of 1 s. The observation duration was from day of year (DOY) 006 to 008 in 2020. The baseline reference stations, named TJ01 and TJ02, are located in the cities Shanghai and Nanjing, respectively. Two Trimble Alloy receivers are used to decode the satellite signals received by choke ring antennas on the roofs. The coordinates of TJ01 and TJ02 are precisely post solved using the static PPP mode in Bernese GNSS Software (version 5.2) based on the daily dual-frequency GPS observations selected from the baseline dataset on DOY 008 in 2020. In PPP processing, the sampling interval is 30 s, and the cutoff elevation is set to $7°$. The final International GNSS Service (IGS) products are used, and the antenna offsets are corrected with IGS14/igs14.atx. The static PPP results serve as the benchmark for computing the root mean square error (RMSE) of the positioning.

First, we reinitialize the positioning engine hourly and record the positioning solutions and their corresponding convergence time. Then, the statistics of the positioning accuracy and convergence time are computed. The convergence time is defined as the time to correctly fix (TTCF) ambiguities. The ambiguities are considered to be correctly fixed when the positioning error of both horizontal components are smaller than 5 cm. After TTCF, the positioning accuracy is defined as the RMSE of the positioning (i.e., the difference between the positions estimated by SMC-RTK in kinematic mode and Bernese in static PPP mode).

The SMC-RTK errors with hourly reinitialization from DOY 006 to 008 are calculated. The ambiguities are instantaneously and correctly fixed after most reinitializations. Sometimes the ambiguities are wrongly fixed at first and lead to decimeter-level positioning errors, such as the positioning errors at the 2nd hour on DOY 006 (a) and the 12th to 14th hours on DOY 007 (b). The incorrect fixing is caused by the unmodeled errors and will make the ratio values close to 1. Hence, a strict ratio test can be used after ambiguity resolution to determine whether to inherit the ambiguities for subsequent epochs so that the incorrectly fixed ambiguities will not be inherited and degrade all subsequent solutions.

The mean values of the convergence time and RMSE are given in Table 11.3. The statistical sample includes the positioning results of all 72 reinitializations. As a comparison, the results of SMC-RTK estimating the RZTD parameter with a conservative processing noise of 0.1 mm/s are also displayed.

When the RZTD parameter is ignored, the RMSE along the north and east directions of SMC-RTK are smaller than 3 cm and are mostly approximately 1 cm, which is similar to the case of estimating the RZTD parameter. The results show that the convergence time of ignoring the RZTD parameter is smaller than 1 min (which means the ambiguities are instantaneously and correctly fixed) for more than 80% of reinitializations. In addition, the RMSE along the up direction when ignoring the RZTD parameter sometimes exceeds 10 cm. As shown in Table 11.3, estimating the RZTD parameter improves the positioning accuracy in the up direction but raises the convergence time by 28%. The experimental results are in line with our expectations because the RZTD parameter is strongly correlated to the vertical coordinate and requires more time to be precisely estimated. Hence, ignoring the RZTD parameter will degrade the vertical precision of positioning but shorten the convergence time. However, ignoring the RZTD parameter is not suitable for all long-baseline cases, especially when the vertical precision is important for the application. In addition, a strict and precise constraint of estimating the RZTD parameter can improve the positioning performance. Hence, the SMC-RTK software keeps the option of ignoring or estimating the RZTD parameter, while this study ignores the RZTD parameter to shorten the convergence time.

We further investigate the positioning performance of SMC-RTK without reinitialization based on the same dataset. We calculated the SMC-RTK errors without reinitialization on DOY 006 (a), 007 (b), and 008 (c) in 2020. The TTCF and the positioning accuracy are given in Table 11.4.

**Table 11.3** Mean values of convergence time and RMSE of SMC-RTK

| | Mean RMSE of positioning (cm) | | | Mean convergence time (s) |
|---|---|---|---|---|
| | North | East | Up | |
| Ignoring the RZTD parameter | 0.9 | 1.2 | 4.6 | 91 |
| Estimating the RZTD parameter | 0.9 | 1.1 | 2.8 | 117 |

**Table 11.4**  RMSE and TTCF of SMC-RTK

| DOY | TTCF (s) | RMSE of SMC-RTK (cm) | | |
|---|---|---|---|---|
| | | North (N) | East (E) | Up (U) |
| 006 | 6 | 1.1 | 2.0 | 4.3 |
| 007 | 6 | 1.0 | 1.1 | 9.6 |
| 008 | 6 | 1.2 | 1.2 | 6.6 |

The horizontal positioning errors are a few centimeters after the ambiguities are correctly fixed. The vertical errors obviously suffer systematic errors due to the ignored residual tropospheric delays. They are sometimes larger than 10 cm even with ambiguities fixed. We zoom-in on the positioning results of the first minutes on DOY 006 (a), 007 (b), and 008 (c). In these cases, the ambiguities are immediately fixed at the sixth second once the PAR is activated (where the ambiguities that are only tracked for at least 5 s are considered for fixing). Actually, once the PAR is activated at the sixth second, the ambiguities are immediately fixed with larger than 50% probability.

It should be noted that using the ultrarapid ephemeris is important for long-baseline resolution because the orbit error of the broadcast ephemerides cannot be eliminated by double-differencing. We computed the orbit errors contained in the asynchronous double-difference observations if the broadcast ephemerides are used. The orbit error is the difference between the asynchronous DD pseudorange calculated using the ultrarapid ephemeris and the broadcast ephemeris.

The results show that many of them are larger than 2 cm and even 3 cm. Considering the wavelengths of satellite signals, such errors will lead to float ambiguities biased by 0.1–0.2 cycles, which dramatically degrades the performance of successful ambiguity resolution. In addition, the BDS satellite orbit errors are comparable to GPS satellites. However, for the geostationary satellites, the orbit errors exceed 10 cm, which means that the related ambiguity biases exceed 0.5 cycles.

### *11.4.2 Kinematic Experiment*

A kinematic experiment was carried out in the field for an hour on DOY 013 in 2020. The CHCNAV X90F all-in-one receiver receives GNSS observations and transmits them to the SMC-RTK software. The observation environment is open but surrounded by the ocean. The BDS Radio Determination Satellite System (RDSS) terminal receives BDS short messages and transmits them to the SMC-RTK software. The workboat moves along the coastline near reference station TJ01. TJ01 and TJ02 both serve as the reference stations in this experiment. TJ02 is located 320 km away from the user station and is utilized to form an extralong baseline for SMC-RTK. TJ01 is located 2 km away from the user station to form a short baseline. The precise

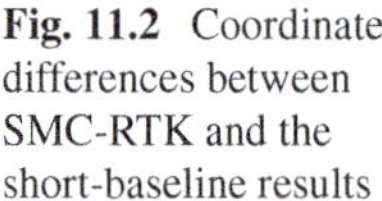

**Fig. 11.2** Coordinate differences between SMC-RTK and the short-baseline results

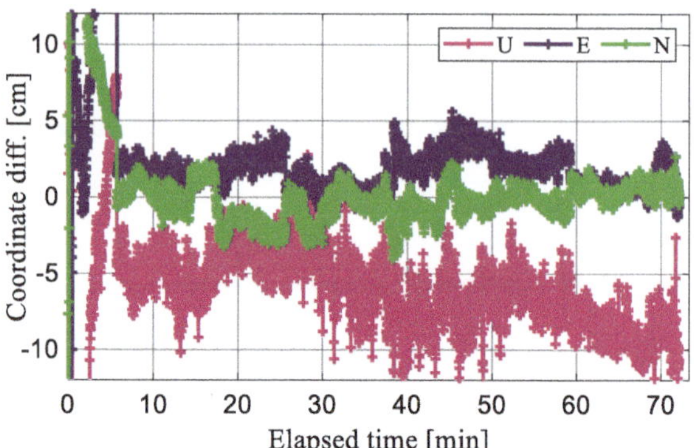



solutions of the short baseline are employed as the references for evaluating the SMC-RTK solutions.

The RTK solutions of the short baseline correctly fixed the ambiguities at all epochs. The coordinate differences along the north, east, and up directions between the SMC-RTK results and the references are shown in Fig. 11.2. The SMC-RTK TTCF is 6 s. After TTCF, the root mean square values of the coordinate differences arc 2.4 cm, 5.6 cm, and 5.5 cm along the north, east and up directions, respectively. The vertical coordinate difference has a systematic bias of approximately 7 cm due to ignoring the residual tropospheric delay. In addition, the coordinate differences are slightly larger than those of the baseline experiment because it involves both the positioning errors of the SMC-RTK and the short-baseline RTK.

## 11.5  Conclusion

This study introduces the SMC-RTK method, which can realize high-precision positioning at sea in real time, and makes significant modifications to the method. In this chapter, the SMC-RTK technique overcomes the problem of communication at sea by sending corrections through the BDS SMS based on an efficient encoding and broadcasting strategy. Moreover, SMC-RTK reduces the dependence on reference stations by using only a single reference station. The service radius of the single reference station is extended to 300 km by applying an asynchronous, time-differenced, precise ephemerides-aided and ionosphere-weighted positioning model. The SMC-RTK TTCF is a few seconds. After TTCF, the horizontal accuracy of SMC-RTK is approximately 1 cm, and the vertical accuracy is approximately 10 cm.

SMC-RTK has several advantages compared to other GNSS positioning methods on the ocean. (1) It can provide positioning results in real time, which fulfills the demands of navigation on the ocean and can enhance the efficiency of engineering works, such as water course surveys. (2) The positioning accuracy is at the centimeter level, and the convergence time is a few seconds even with a baseline length exceeding

300 km. (3) The cost of SMC-RTK is low, while the service fee for each user is minimally only several hundred Chinese yuan per year since one reference station can simultaneously serve 200 users with the help of command ID cards.

There will be additional possible solutions for high-precision positioning on the ocean in the future. BDS satellites have begun to broadcast the PPP-B2b signal and have enabled global PPP service. Additionally, the BDS SMS is available for transmitting corrections of real-time PPP. However, considering its high-precision, quick convergence, and low cost, SMC-RTK will still be a great option.

The new BDS SMS generation (BeiDou-3 SMS) has recently become available. The BeiDou-3 SMS allows users to send a single message containing a maximum of 1750 bytes every 30 s. It can reduce the number of required ID cards and the costs of SMC-RTK but still does not satisfy RTK in virtual-reference-station mode. In addition, the actual service frequency and the maximum length of a single message are bounded to the registration parameters of users, which means that the maximum bandwidth of the BeiDou-3 SMS is not available to most people. Thus, with consideration of the BeiDou-3 SMS, the SMC-RTK method is still superior to RTK on the ocean and is more favorable due to the reduced communication costs.

# References


1. Li B, Zhang Z, Zang N, Wang S (2019) High-precision GNSS ocean positioning with BeiDou short-message communication. J Geod 93:125–139
2. Geng J, Teferle FN, Meng X, Dodson AH (2010) Kinematic precise point positioning at remote marine platforms. GPS Solut 14:343–350
3. Li X, Ge M, Dai X, Ren X, Fritsche M, Wickert J, Schuh H (2015) Accuracy and reliability of multi-GNSS real-time precise positioning: GPS, GLONASS, BeiDou, and Galileo. J Geod 89:607–635
4. Xiang Y, Gao Y, Li Y (2020) Reducing convergence time of precise point positioning with ionospheric constraints and receiver differential code bias modeling. J Geod 94:1–13
5. Nie Z, Gao Y, Wang Z, Ji S, Yang H (2018) An approach to GPS clock prediction for real-time PPP during outages of RTS stream. GPS Solut 22:14
6. Grejner-Brzezinska DA, Kashani I, Wielgosz P, Smith DA, Spencer PSJ, Robertson DS, Mader GL (2007) Efficiency and reliability of ambiguity resolution in network-based real-time kinematic GPS. J Surv Eng 133:56–65
7. Wanninger L (1997) Virtual reference stations (VRS). GPS Solut 7:143–144
8. Odolinski R, Teunissen PJG, Odijk D (2015) Combined GPS + BDS for short to long baseline RTK positioning. Meas Sci Technol 26:045801
9. Zhang Z, Li B, Zou J (2020) Can long-range single-baseline RTK provide service in Shanghai comparable to network RTK? J Surv Eng 146:05020007
10. Takasu T, Yasuda A (2010) Kalman-filter-based integer ambiguity resolution strategy for long-baseline RTK with ionosphere and troposphere estimation. Proceedings of the ION GNSS 2010, pp 161–171
11. Zhang Z, Li B, Gao Y, Zhang Z, Wang S (2023) Asynchronous and time-differenced RTK for ocean applications using the BeiDou short message service. J Geod 97:1–15
12. Teunissen PJG, Montenbruck O (2017) Springer handbook of global navigation satellite systems. Springer, Germany
13. Collins JP, Langley RB (1997) A tropospheric delay model for the user of the wide area augmentation system, Technical report

14. Wang M, Li B (2016) Evaluation of empirical tropospheric models using satellite-tracking tropospheric wet delays with water vapor radiometer at Tongji, China. Sensors 16:186
15. Niell AE (1996) Global mapping functions for the atmosphere delay at radio wavelengths. J Geophys Res Solid Earth 101:3227–3246
16. Klobuchar JA (1987) Ionospheric time-delay algorithm for single-frequency GPS users. IEEE Trans Aerosp Electron Syst 3:325–331
17. Dodson A, Shardlow P, Hubbard L, Elgered G, Jarlemark P (1996) Wet tropospheric effects on precise relative GPS height determination. J Geod 70:188–202
18. Zhang J, Lachapelle G (2001) Precise estimation of residual tropospheric delays using a regional GPS network for real-time kinematic applications. J Geod 75:255–266
19. Alkan RM, Öcalan T (2013) Usability of the GPS precise point positioning technique in marine applications. J Navig 66:579–588
20. Li B, Feng Y, Shen Y, Wang C (2010) Geometry-specified troposphere decorrelation for subcentimeter real-time kinematic solutions over long baselines. J Geophys Res 115:B11404
21. Li B (2016) Stochastic modeling of triple-frequency BeiDou signals: estimation, assessment and impact analysis. J Geod 90:593–610
22. Li B, Shen Y, Feng Y, Gao W, Yang L (2014) GNSS ambiguity resolution with controllable failure rate for long baseline network RTK. J Geod 88:99–112
23. Wang J (1999) Stochastic modeling for RTK GPS/GLONASS positioning. J Inst Navig 46:297–305
24. Brown RG, Hwang PYC (1992) Introduction to random signals and applied Kalman filtering. Wiley, New York
25. Petovello MG, O'Keefe K, Lachapelle G, Cannon ME (2009) Consideration of time-correlated errors in a Kalman filter applicable to GNSS. J Geod 83:51–56
26. Ji S, Sun Z, Weng D, Chen W, Wang Z, He K (2019) High-precision ocean navigation with a single set of BeiDou short-message device. J Geod 93:1589–1602
27. Li B, Li Z, Zhang Z, Tan Y (2017) ERTK: extra-wide-lane RTK of triple-frequency GNSS signals. J Geod 91:1031–1047
28. Teunissen PJG (1995) The least-squares ambiguity decorrelation adjustment: a method for fast GPS integer ambiguity estimation. J Geod 70:65–82

# Chapter 12
# ARTK: Antenna-Array Aided RTK

## 12.1 Introduction

Integer ambiguity resolution is the key to high-precision global navigation satellite system (GNSS) applications. It enables the transformation of the ambiguous carrier phases to ultra precise pseudoranges, thus making high-precision parameter estimation possible. The success of ambiguity resolution depends on the strength of the underlying GNSS model. The weaker the model, the more data needs to be accumulated before ambiguity resolution can be successful and the longer it therefore takes before one can take advantage of the ultra-precise carrier signals. Clearly, the aim is to shorten the time to convergence, preferably zero, thereby enabling truly instantaneous GNSS, integer ambiguity resolved, parameter estimation.

In continuously operating reference system (CORS) network applications, fast and successful resolution of the ambiguities is important as it enables improved availability of the network provided ambiguity-fixed parameter outputs, such as the ionospheric delays. Between-station ambiguity resolution is usually based on data of stations equipped with a single antenna only. In this contribution, we study the potential improvements that can be realized when stations would be equipped with an array of antennas instead of only a single antenna. This array-aided precise point positioning (APPP) concept, proposed in [1], is a measurement concept that uses GNSS data, from multiple antennas in an array of known geometry, to realize improved GNSS parameter estimation. Although we focus on ambiguity resolution in this contribution, integrity improves, since with the known array geometry, redundancy increases, thus allowing improved error detection and multipath mitigation [2, 3].

Consider Fig. 12.1, in which two antenna-array equipped stations, $b_0$ and $u_0$, are shown. The two antenna arrays, with known geometry, $b_1, \ldots, b_r$ and $u_1, \ldots, u_r$, are mounted on the platforms at $b_0$ and $u_0$, respectively. The known platform antenna-array geometry enables one to reduce the platform observations of all its antennas to a single set of platform observations. This set of reduced observations can be interpreted as if it belongs to one single virtual antenna with a better precision than



B. Li et al., *GNSS Real-Time Kinematic Positioning*, Navigation: Science and Technology 17, https://doi.org/10.1007/978-981-96-9116-6_12

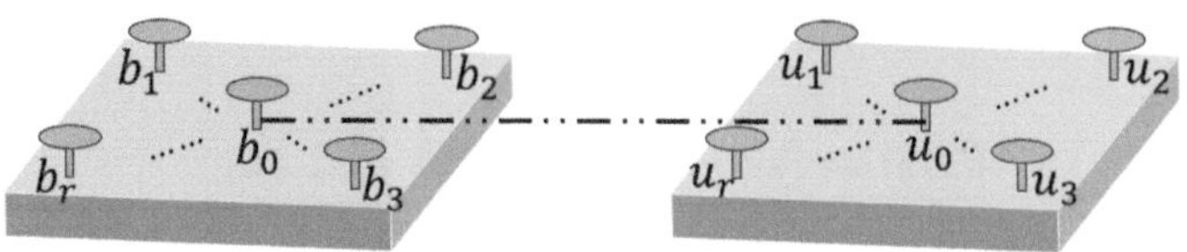

**Fig. 12.1** Two antenna-array equipped platforms $b_0$ and $u_0$

the original observations coming from the individual antennas. Therefore, improved between-platform ($b_0 - u_0$) ambiguity resolution and parameter estimation become possible as compared to the between-platform single-antenna case. This improvement has been initially demonstrated on long-distance real time kinematic (RTK) campaign [4]. Although the reduction of the platform observations also requires ambiguity resolution, namely on the platform, this can be shown possible with high success rates due to the known antenna-array geometry [5].

One of the potential applications of APPP is to speed up the CORS ambiguity resolution. In this contribution, we explore the potential benefits of APPP to the long-range RTK, which is referred to as array-aided RTK (ARTK). An 80 km baseline experiment was conducted for which both stations were equipped with a 4-antenna array platform. The newly formed model for observation reduction with multiple antennas on the platform was solved using the principle of multivariate mixed integer least squares estimation and the reduced data was generated. Then the reduced data was processed to demonstrate the superior performance of ARTK in integer ambiguity resolution (IAR), precise RTK solutions, as well as high robustness, comparing with the conventional RTK (CRTK) with 1-antenna, equipped baseline stations. This contribution is organized as follows. First, we formulate the platform array model and show how its data can be reduced. Second, we describe three different ionosphere-weighted differential CORS array models and present closed form formulae for their ambiguity variance matrices. They determine the success rates with which the integer ambiguities can be estimated. Third, the long-range RTK model between platforms is outlined. Finally. the 80 km baseline experiment is presented.

In following, $I_n$ denotes the identity matrix of order $n$ and $e_n$ the $n$-column vector of ones. $c_1 = [1, 0, \ldots, 0]^T$ is a unit vector with its 1 in the first slot. $D_n^T = \begin{bmatrix} -e_{n-1}, I_{n-1} \end{bmatrix}$ is the differencing matrix. $\otimes$ and **vec** are Kronecker product and vectorization operators. **E** and **D** denote the expectation and dispersion operators. **diag(a)** is the operator to form a square matrix with elements of $a$ as diagonal elements.

## 12.2   Platform Array Model and Its Data Reduction

As an important part of ARTK, the platform array models are studied comprehensively in this section, including the functional model and the stochastic model. Then we show how its data can be reduced.

### *12.2.1  Functional Model*

We start with the linearized single-frequency, between-satellite single-differenced (SD) observation equations of phase and code,

$$E(\boldsymbol{\phi}_{r,j}) = \boldsymbol{G}_r\boldsymbol{x}_r + \boldsymbol{g}_r\tau_r - \mu_j\iota_r - \delta\boldsymbol{t}_{.j} + \lambda_j\boldsymbol{a}_{r,j}$$
$$E(\boldsymbol{p}_{r,j}) = \boldsymbol{G}_r\boldsymbol{x}_r + \boldsymbol{g}_r\tau_r + \mu_j\iota_r - \boldsymbol{dt}_{.j} \tag{12.1}$$

where the subscripts $r$ and $j$ denote the antenna and the frequency $f_j$ (wavelength $\lambda_j$), which are used to emphasize the antenna-specific and frequency-specific terms, respectively. Assuming that $(s+1)$ satellites are simultaneously tracked, $\boldsymbol{\phi}_{r,j} = \left[\phi_{r,j}^1, \ldots, \phi_{r,j}^s\right]^{\mathrm{T}}$ and $\boldsymbol{p}_{r,j} = \left[p_{r,j}^1, \ldots, p_{r,j}^s\right]^{\mathrm{T}}$ are the $(s \times 1)$ SD phase and code observation vectors; $\boldsymbol{G}_r$ is the $(s \times 3)$ design matrix of the unknown baseline increment vector $\boldsymbol{x}_r$; $\tau_r$ is the zenith tropospheric delay (ZTD), with its mapping matrix $\boldsymbol{g}_r$; $\iota_r = \left[\iota_r^1, \ldots, \iota_r^s\right]^{\mathrm{T}}$ is the $(s \times 1)$ vector of SD ionospheric delays on frequency $f_1$ with $\mu_j = f_1^2/f_j^2$; $\delta\boldsymbol{t}_{.j} = \left[\delta t_{.j}^1, \ldots, \delta t_{.j}^s\right]^{\mathrm{T}}$ and $\boldsymbol{dt}_{.j} = \left[dt_{.j}^1, \ldots, dt_{.j}^s\right]^{\mathrm{T}}$ are the SD satellite clock errors for phase and code, respectively; $\boldsymbol{a}_{r,j} = \left[a_{r,j}^1, \ldots, a_{r,j}^s\right]^{\mathrm{T}}$ is the $(s \times 1)$ SD ambiguity vector with the $s$th element $a_{r,j}^s = z_{r,j}^s - \varphi_{.j}^s(t_0)$, where $z_{r,j}^s$ is integer and $\varphi_{.j}^s(t_0)$ is real-valued.

For $f$ frequencies, we define the vectors, $\boldsymbol{y}_r = \left[\boldsymbol{\phi}_r^{\mathrm{T}}, \boldsymbol{p}_r^{\mathrm{T}}\right]^{\mathrm{T}}$ and $\Delta\boldsymbol{t} = \left[\delta\boldsymbol{t}^{\mathrm{T}}, \boldsymbol{dt}^{\mathrm{T}}\right]^{\mathrm{T}}$, with $\boldsymbol{\phi}_r = \left[\boldsymbol{\phi}_{r,1}^{\mathrm{T}}, \ldots, \boldsymbol{\phi}_{r,f}^{\mathrm{T}}\right]^{\mathrm{T}}$, $\delta\boldsymbol{t} = \left[\delta\boldsymbol{t}_{.1}^{\mathrm{T}}, \ldots, \delta\boldsymbol{t}_{.f}^{\mathrm{T}}\right]^{\mathrm{T}}$; where $\boldsymbol{p}_r$ and $\boldsymbol{dt}$ have the same structure as $\boldsymbol{\phi}_r$ and $\delta\boldsymbol{t}$, respectively. Furthermore, $\boldsymbol{\mu} = \left[\mu_1, \ldots, \mu_f\right]^{\mathrm{T}}$, $\boldsymbol{a}_r = \left[\boldsymbol{a}_{r,1}^{\mathrm{T}}, \ldots, \boldsymbol{a}_{r,f}^{\mathrm{T}}\right]^{\mathrm{T}}$ and $\boldsymbol{\Lambda} = \mathrm{diag}(\lambda_1, \ldots, \lambda_f)$.

Since we assume the distances between the antennas on the platform to be very short, we may assume that $\boldsymbol{G}_i = \boldsymbol{G}$, $\boldsymbol{g} = \boldsymbol{g}_i$, $\tau = \tau_i$ and $\iota = \iota_i$ for $i = 1, \ldots, r$. This implies that we may write the $r$ antenna array set of SD observation equations of (12.1) in multivariate form as

$$E(\boldsymbol{Y}) = \boldsymbol{MX} + \boldsymbol{NA} + \boldsymbol{e}_r^{\mathrm{T}} \otimes \left(\boldsymbol{v} \otimes \iota + \boldsymbol{e}_{2f} \otimes (\boldsymbol{g}\tau) - \Delta\boldsymbol{t}\right) \tag{12.2}$$

where $\boldsymbol{Y} = \left[\boldsymbol{y}_1, \ldots, \boldsymbol{y}_r\right]$, $\boldsymbol{M} = \boldsymbol{e}_{2f} \otimes \boldsymbol{G}$, $\boldsymbol{X} = [\boldsymbol{x}_1, \ldots, \boldsymbol{x}_r]$, $\boldsymbol{N} = \boldsymbol{\Gamma} \otimes \boldsymbol{I}_s$, $\boldsymbol{\Gamma} = [\boldsymbol{\Lambda}, 0]^{\mathrm{T}}$, $\boldsymbol{A} = [\boldsymbol{a}_1, \ldots, \boldsymbol{a}_r]$ and $\boldsymbol{v} = \left[-\boldsymbol{\mu}^{\mathrm{T}}, \boldsymbol{\mu}^{\mathrm{T}}\right]^{\mathrm{T}}$.

If we now post-multiply (12.2) with the invertible matrix $\boldsymbol{R}_r = [\boldsymbol{c}_1, \boldsymbol{D}_r]$, we obtain with $\left[\boldsymbol{y}_1, \ddot{\boldsymbol{Y}}\right] = \boldsymbol{Y}\boldsymbol{R}_r$,

$$E\left(\begin{bmatrix} \boldsymbol{y}_1 \\ \tilde{\boldsymbol{Y}} \end{bmatrix}\right) = \begin{bmatrix} \boldsymbol{Mx}_1 + \boldsymbol{Na}_1 + \boldsymbol{v} \otimes \iota + \left(\boldsymbol{e}_{2f} \otimes \boldsymbol{g}\right)\tau - \Delta\boldsymbol{t} \\ \boldsymbol{M}\tilde{\boldsymbol{X}} + \boldsymbol{NZ} \end{bmatrix} \tag{12.3}$$

where $\tilde{Y} = \left[y_{12}, \ldots, y_{1r}\right]$ is the transformed double-differenced (DD) observation matrix, $\tilde{X} = XD_r = [x_{12}, \ldots, x_{1r}]$ is the baseline matrix, and $Z = AD_r = [z_{12}, \ldots, z_{1r}]$ is the integer DD ambiguity matrix.

### 12.2.2   Stochastic Model

We specify the stochastic model of $Y = \left[y_1, \ldots, y_r\right]$ as

$$\mathrm{D}(\mathrm{vec}(Y)) = Q_r \otimes Q \text{ with } Q = Q_f \otimes \left(D_{s+1}^{\mathrm{T}} Q_s D_{s+1}\right) \tag{12.4}$$

where $Q_r$ captures the antenna-specific precision contribution, $Q_s$ is the satellite elevation-dependent cofactor matrix of the $(s + 1)$ undifferenced observations, and $Q_f = \mathrm{blockdiag}\left(Q_\phi, Q_p\right)$ captures the frequency-specific precision contribution, with $Q_\phi = \mathrm{diag}\left(\sigma_{\phi;1}^2, \ldots, \sigma_{\phi;f}^2\right)$ and $Q_p = \mathrm{diag}\left(\sigma_{p;1}^2, \ldots, \sigma_{p;f}^2\right)$, where $\sigma_{\phi;j}^2$ and $\sigma_{p;j}^2$ are the variance scalars of the undifferenced phase and code on frequency $j$, respectively.

Application of the variance propagation law to $\left[y_1, \tilde{Y}\right] = YR_r$ gives the stochastic model of (12.3) as

$$\mathrm{D}\left(\begin{bmatrix} y_1 \\ \mathrm{vec}(\tilde{Y}) \end{bmatrix}\right) = \begin{bmatrix} c_1^{\mathrm{T}} Q_r c_1 & c_1^{\mathrm{T}} Q_r D_r \\ D_r^{\mathrm{T}} Q_r c_1 & D_r^{\mathrm{T}} Q_r D_r \end{bmatrix} \otimes Q \tag{12.5}$$

### 12.2.3   Array Data Reduction

From (12.5) it follows that $y_1$ and $\tilde{Y}$ of (12.3) are correlated. As shown by [1], application of the invertible transformation

$$\begin{bmatrix} 1 - c_1^{\mathrm{T}} Q_r D_r \left(D_r^{\mathrm{T}} Q_r D_r\right)^{-1} \\ 0 \ I_{r-1} \end{bmatrix} \otimes I_{2fs} \tag{12.6}$$

to (12.3) results in the equivalent but decorrelated version

$$\mathrm{E}\left(\begin{bmatrix} \bar{y} \\ \tilde{Y} \end{bmatrix}\right) = \begin{bmatrix} M\bar{x} + N\bar{z} + \left(e_{2f} \otimes g\right)\tau + v \otimes \iota - \Delta t \\ M\tilde{X} + NZ \end{bmatrix} \tag{12.7}$$

where $\left[\bar{y}, \bar{x}\right] = [Y, X]Q_r^{-1} e_r \left(e_r^{\mathrm{T}} Q_r^{-1} e_r\right)^{-1}$ and $\bar{z} = a_1 - ZD_r^+ c_1$, with $D_r^+ = \left(D_r^{\mathrm{T}} Q_r D_r\right)^{-1} D_r^{\mathrm{T}} Q_r$. The dispersion of the reduced observation vector $\bar{y}$ and the DD observation matrix $\tilde{Y}$ is given as

$$D\left(\begin{bmatrix} \bar{y} \\ \text{vec}(\tilde{Y}) \end{bmatrix}\right) = \begin{bmatrix} \left(e_r^{\mathrm{T}} Q_r^{-1} e_r\right)^{-1} 0 \\ 0\ D_r^{\mathrm{T}} Q_r D_r \end{bmatrix} \otimes Q \tag{12.8}$$

showing that $\bar{y}$ is uncorrelated with $\tilde{Y}$.

In the following, we assume that the same type of antennas are used. Thus $Q_r = I_r$, from which it follows that $\bar{y} = \frac{1}{r}\sum_{i=1}^{r} y_i$, $\bar{x} = \frac{1}{r}\sum_{i=1}^{r} x_i$, $\bar{z} = a_1 + \frac{1}{r} Z e_{r-1}$, and $\left(e_r^{\mathrm{T}} Q_r^{-1} e_r\right)^{-1} = 1/r$. The reduced observation vector $\bar{y}$ is thus $r$ times more precise than that of a single antenna.

In case of an APPP-CORS platform, the barycentric position vector $\bar{x}$ is known, since the position vectors $x_i$ of the platform antennas are assumed known. Furthermore, the known geometry $\tilde{X}$ of the antenna configuration on the platform enables one to determine the integer matrix estimator $\check{Z}$ of $Z$ with a very high success rate, see [1]. Hence, for all practical purposes one may also assume the DD integer matrix $Z$ in $\bar{z} = a_1 + \frac{1}{r} Z e_{r-1}$ known. Therefore, with $\bar{x}$ and $Z$ known, the first equation of (12.7) can now be written as

$$E(y') = \left(e_{2f} \otimes g\right)\tau + v \otimes \iota + N a_1 - \Delta \iota \tag{12.9}$$

where $y' = \bar{y} - M\bar{x} - N\tilde{z}$, with $\tilde{z} = \frac{1}{r} Z e_{r-1}$. This is the reduced system of observation equations for a single CORS platform equipped with multiple antennas.

## 12.3  Ambiguity Resolution Between Arrays

In this section, the ionosphere-weighted differential CORS array model is studied. Then the ionosphere-weighted CORS ambiguity resolution is discussed. We determine the multi-epoch ambiguity variance matrix for three different scenarios, including geometry-fixed, geometry-free and geometry-based, sits in between the geometry-fixed one and the geometry-free one.

### 12.3.1  Ionosphere-Weighted Differential Array Model

To determine the differential CORS array model for two CORS platforms equipped with multiple antennas, we can take the difference between their single CORS system of equations. For two CORS platforms, say $b$ and $u$, having the reduced observations $\bar{y}_b$ and $\bar{y}_u$, the between-platform system of observation equations, therefore, reads

$$E(y'_{bu}) = \left(e_{2f} \otimes g_b\right)\tau_{bu} + v \otimes \iota_{bu} + N a_{bu} \tag{12.10}$$

with $\mathbf{y}'_{bu} = \bar{\mathbf{y}}_u - \bar{\mathbf{y}}_b - \mathbf{M}_u \bar{\mathbf{x}}_u + \mathbf{M}_b \bar{\mathbf{x}}_b - \mathbf{N}(\bar{\mathbf{z}}_u - \bar{\mathbf{z}}_b)$, $\tau_{bu} = \tau_u - \tau_b (\mathbf{g}_b \approx \mathbf{g}_u)$, and $\mathbf{a}_{bu} = \mathbf{a}_{1;u} - \mathbf{a}_{1;b}$. Note that $\mathbf{a}_{bu}$ is now again a DD integer ambiguity vector.

With the assumption that the same type of receivers is used on the two platforms (i.e. $\mathbf{Q}_{f;b} = \mathbf{Q}_{f;u} = \mathbf{Q}_f$), the variance matrix of $\mathbf{y}'_{bu}$ is given as

$$\mathrm{D}(\mathbf{y}'_{bu}) = \frac{1}{r} \mathbf{Q}_f \otimes \mathbf{W}^{-1}, \quad \mathbf{W}^{-1} = \mathbf{D}^{\mathrm{T}}_{s+1} \mathbf{Q}_0 \mathbf{D}_{s+1} \tag{12.11}$$

where $\mathbf{Q}_0 = \mathbf{Q}_{s;u} + \mathbf{Q}_{s;b}$ is the cofactor matrix of the between-platform SD observations. This shows how between-platform parameter estimation can benefit from the antenna array and in particular from $r$, the number of antennas in the array.

The ionosphere-weighted version of (12.10) and (12.11) is obtained if we add the ionospheric pseudo-observation equations

$$\mathrm{E}(\iota^0_{bu}) = \iota_{bu}, \quad \mathrm{D}(\iota^0_{bu}) = \sigma_\iota^2 \otimes \mathbf{W}^{-1} \tag{12.12}$$

in which the variance $\sigma_\iota^2$ is used to model the between-platform spatial uncertainty of the ionosphere, i.e. $\sigma_\iota^2$ is small for short baselines and large for long baselines. The two extreme cases, $\sigma_\iota^2 = 0$ and $\sigma_\iota^2 = \infty$ are referred to as the ionosphere-fixed and ionosphere-float model, respectively.

### 12.3.2   Ionosphere-Weighted Ambiguity Resolution

If we use (12.12) to eliminate the unknown ionospheric delays from (12.10), the single-epoch ionosphere-weighted model may also be written as

$$\mathrm{E}(\mathbf{y}) = \left[ \mathbf{e}_{2f} \otimes \mathbf{g} \;\; \mathbf{\Gamma} \otimes \mathbf{I}_s \right] \begin{bmatrix} \tau \\ \mathbf{a} \end{bmatrix}$$

$$\mathrm{D}(\mathbf{y}) = \left( \frac{1}{r} \mathbf{Q}_f + \sigma_\iota^2 \mathbf{v}\mathbf{v}^{\mathrm{T}} \right) \otimes \mathbf{W}^{-1} \tag{12.13}$$

where $\mathbf{y} = \mathbf{y}'_{bu} - \mathbf{v} \otimes \iota^0_{bu}$, $\mathbf{\Gamma} = [\mathbf{\Lambda}, \mathbf{0}]^{\mathrm{T}}$. The short-hands $\mathbf{g}$, $\tau$ and $\mathbf{a}$ have been used instead of $\mathbf{g}_b$, $\tau_{bu}$ and $\mathbf{a}_{bu}$. When we solve the ionosphere-weighted model for the multi-epoch case, we assume no time correlation between the observables and the ambiguity vector $\mathbf{a}$ to be time constant.

We now determine the multi-epoch ambiguity variance matrix for three different scenarios. In the first scenario, referred to as geometry-fixed, all the tropospheric delays are assumed known. The corresponding ambiguity variance matrix is given as

$$\mathbf{Q}^{(\mathrm{fixed})}_{\hat{a}\hat{a}} = \frac{1}{r} \frac{1}{k} \left[ \mathbf{\Lambda}^{-1} \left( \mathbf{Q}_\phi + \alpha \mu \mu^{\mathrm{T}} \right) \mathbf{\Lambda}^{-1} \right] \otimes \overline{\mathbf{W}}^{-1} \tag{12.14}$$

with the time-average weight matrix $\overline{W} = \frac{1}{k}\sum_{t=1}^{k} W_t$ and ionosphere weighting scalar

$$\alpha = \left[ \left(r\sigma_t^2\right)^{-1} + \left(\mu^{\mathrm{T}} Q_p^{-1}\mu\right) \right]^{-1} \tag{12.15}$$

In the second scenario, referred to as geometry-free, all the slant tropospheric delays are assumed unknown. That is, no mapping is applied (i.e. $g$ is replaced by $I_s$) and the delays are assumed to change overtime. The corresponding ambiguity variance matrix is given as

$$Q_{\hat{a}\hat{a}}^{(\text{free})} = Q_{\hat{a}\hat{a}}^{(\text{fixed})} + \frac{1}{k}c_{\hat{\tau}_{\text{free}}}^2 qq^{\mathrm{T}} \otimes \overline{W}^{-1} \tag{12.16}$$

with

$$q = \Lambda^{-1}\left(I_f + \alpha\mu\mu^{\mathrm{T}}Q_p^{-1}\right)e_f$$
$$c_{\hat{\tau}_{\text{free}}}^2 = \left[ e_f^{\mathrm{T}}\left(\frac{1}{r}Q_p + \sigma_t^2\mu\mu^{\mathrm{T}}\right)^{-1} e_f \right]^{-1} \tag{12.17}$$

Finally, the third scenario, referred to as geometry-based, sits in between the geometry-fixed one and the geometry-free one. It is the scenario in which the ZTD is considered unknown, but constant in time. The corresponding ambiguity variance matrix is given as

$$Q_{\hat{a}\hat{a}} = Q_{\hat{a}\hat{a}}^{(\text{fixed})} + \frac{1}{k}c_{\hat{\tau}}^2 qq^{\mathrm{T}} \otimes P_{\overline{g}}\overline{W}^{-1} \tag{12.18}$$

with $\overline{g} = \left(\sum_{t=1}^{k} W_t\right)^{-1}\left(\sum_{t=1}^{k} W_t g_t\right)$, the weighted average ZTD map, the orthogonal projector $P_{\overline{g}} = \overline{g}\left(\overline{g}^{\mathrm{T}}\overline{W}\overline{g}\right)^{-1}\overline{g}^{\mathrm{T}}\overline{W}$, and

$$c_{\hat{\tau}}^2 = c_{\hat{\tau}_{\text{free}}}^2 \left[ 1 + \frac{c_{\hat{\tau}_{\text{free}}}^2}{c_{\hat{\tau}|a}^2} \frac{\frac{1}{k}\sum_{t=1}^{k}\left(g_t - \overline{g}\right)^{\mathrm{T}}W_t\left(g_t - \overline{g}\right)}{\overline{g}^{\mathrm{T}}\overline{W}\overline{g}} \right]^{-1}$$
$$c_{\hat{\tau}|a}^2 = \left[ e_{2f}^{\mathrm{T}}\left(\frac{1}{r}Q_f + \sigma_t^2 vv^{\mathrm{T}}\right)^{-1} e_{2f} \right]^{-1} \tag{12.19}$$

Note that $c_{\hat{\tau}}^2 \leq c_{\hat{\tau}_{\text{free}}}^2$, with equality in the single-epoch case, and that $Q_{\hat{a}\hat{a}}$ and $Q_{\hat{a}\hat{a}}^{(\text{free})}$ are a rank-1 update and a rank-$s$ update of $Q_{\hat{a}\hat{a}}^{(\text{fixed})}$. Hence, in terms of the ambiguity precision, the three cases can be ordered as

$$Q_{\hat{a}\hat{a}}^{\text{(fixed)}} \leq Q_{\hat{a}\hat{a}} \leq Q_{\hat{a}\hat{a}}^{\text{(free)}} \tag{12.20}$$

The geometry-fixed model gives the most precise ambiguities, while the geometry-free model, being the weakest, gives the most imprecise ambiguities.

### 12.3.3  Some Important Derivations

Assuming all matrices and vectors involved have appropriate dimensions, the following properties of the Kronecker product and vectorization operator **vec** [6]:

$$(AB) \otimes (CD) = (A \otimes C)(B \otimes D) \tag{12.21}$$

$$\text{vec}(ABC) = \left(C^{\mathrm{T}} \otimes A\right)\text{vec}(B) \tag{12.22}$$

and the projector identity [7]

$$QD_r\left(D_r^{\mathrm{T}}QD_r\right)^{-1}D_r^{\mathrm{T}} = I_r - e_r\left(e_r^{\mathrm{T}}Q^{-1}e_r\right)^{-1}e_r^{\mathrm{T}}Q^{-1} \tag{12.23}$$

with $D_r^{\mathrm{T}}e_r = 0$, will be frequently applied in the derivations.

For the derivation of some formulae in Sect. 12.2. To derive (12.5), we apply the error propagation law to

$$\begin{bmatrix} y_1 \\ \text{vec}(\tilde{Y}) \end{bmatrix} = \text{vec}(YR_r) = \left(R_r^{\mathrm{T}} \otimes I_{2fs}\right)\text{vec}(Y) \tag{12.24}$$

where $R_r = [c_1, D_r]$. This gives

$$\begin{aligned}
\mathrm{D}\left(\begin{bmatrix} y_1 \\ \text{vec}(\tilde{Y}) \end{bmatrix}\right) &= \left(R_r^{\mathrm{T}} \otimes I_{2fs}\right)\mathrm{D}(\text{vec}(Y))\left(R_r \otimes I_{2fs}\right) \\
&= \left(R_r^{\mathrm{T}} \otimes I_{2fs}\right)\left(Q_r \otimes Q\right)\left(R_r \otimes I_{2fs}\right) \\
&= \begin{bmatrix} c_1^{\mathrm{T}}Q_rc_1 & c_1^{\mathrm{T}}Q_rD_r \\ D_r^{\mathrm{T}}Q_rc_1 & D_r^{\mathrm{T}}Q_rD_r \end{bmatrix} \otimes Q
\end{aligned} \tag{12.25}$$

To derive the first equation of (12.8), we apply the inverible transformation (12.6) to (12.3). It follows:

$$\begin{aligned}
&\left[\left(1, -c_1^{\mathrm{T}}Q_rD_r\left(D_r^{\mathrm{T}}Q_rD_r\right)^{-1}\right) \otimes I_{2fs}\right]\begin{bmatrix} y_1 \\ \text{vec}(\tilde{Y}) \end{bmatrix} \\
&= y_1 - \left(c_1^{\mathrm{T}}Q_rD_r\left(D_r^{\mathrm{T}}Q_rD_r\right)^{-1} \otimes I_{2fs}\right)\text{vec}(\tilde{Y})
\end{aligned}$$

$$
\begin{aligned}
&= y_1 - \mathrm{vec}\left( \tilde{Y}\left(D_r^{\mathrm{T}}Q_rD_r\right)^{-1}D_r^{\mathrm{T}}Q_r c_1 \right) \\
&= y_1 - YD_r\left(D_r^{\mathrm{T}}Q_rD_r\right)^{-1}D_r^{\mathrm{T}}Q_r c_1 \\
&= y_1 - Y\left[ I_r - Q_r^{-1}e_r\left(e_r^{\mathrm{T}}Q_r^{-1}e_r\right)^{-1}e_r^{\mathrm{T}} \right]c_1 \\
&= YQ_r^{-1}e_r\left(e_r^{\mathrm{T}}Q_r^{-1}e_r\right)^{-1}
\end{aligned}
\tag{12.26}
$$

where the identity (12.21) was applied. One can easily work out the variance matrix (12.8) using this identity.

For the derivation of variance matrix (12.16). In the geometry-free model, we replace $g$ with $I_s$ in (12.13) and further use the differencing matrix $D_{2f}^{\mathrm{T}} \otimes I_s$ to eliminate troposphere design matrix:

$$
\left[ D_{2f}^{\mathrm{T}}\Gamma \otimes I_s \right] \quad \text{and} \quad D_{2f}^{\mathrm{T}}\tilde{Q}D_{2f} \otimes W^{-1}
\tag{12.27}
$$

with $\tilde{Q} = Q_f + \sigma_t^2 v v^{\mathrm{T}}$. This gives the normal matrix of $k$ epochs for ambiguities as

$$
\left( \Gamma^{\mathrm{T}}D_{2f}\left(D_{2f}^{\mathrm{T}}\tilde{Q}D_{2f}\right)^{-1}D_{2f}^{\mathrm{T}}\Gamma \right) \otimes \left(k\overline{W}\right) = \tilde{N} \otimes \left(k\overline{W}\right)
\tag{12.28}
$$

We now concentrate on the first part $\tilde{N}$ only and use the identity (12.21) to rewrite it as

$$
\tilde{N} = \Gamma^{\mathrm{T}}\tilde{Q}^{-1}\Gamma - \Gamma^{\mathrm{T}}\tilde{Q}^{-1}e_{2f}\left(e_{2f}^{\mathrm{T}}\tilde{Q}^{-1}e_{2f}\right)^{-1}e_{2f}^{\mathrm{T}}\tilde{Q}^{-1}\Gamma
\tag{12.29}
$$

Using matrix inversion lemma gives:

$$
\tilde{N}^{-1} = \left(\Gamma^{\mathrm{T}}\tilde{Q}^{-1}\Gamma\right)^{-1} + c_{\hat{t}_{\mathrm{free}}}^2 q q^{\mathrm{T}}
\tag{12.30}
$$

with

$$
c_{\hat{t}_{\mathrm{free}}}^{-2} = e_{2f}^{\mathrm{T}}\tilde{Q}^{-1}e_{2f} - e_{2f}^{\mathrm{T}}\tilde{Q}^{-1}\left(\Gamma^{\mathrm{T}}\tilde{Q}^{-1}\Gamma\right)^{-1}\Gamma^{\mathrm{T}}\tilde{Q}^{-1}e_{2f}
\tag{12.31}
$$

$$
q = \left(\Gamma^{\mathrm{T}}\tilde{Q}^{-1}\Gamma\right)^{-1}\Gamma^{\mathrm{T}}\tilde{Q}^{-1}e_{2f}
\tag{12.32}
$$

We first work out $c_{\hat{t}_{\mathrm{free}}}^{-2}$. Using the analogous projector identity

$$
I_{2f} - \Gamma\left(\Gamma^{\mathrm{T}}\tilde{Q}^{-1}\Gamma\right)^{-1}\Gamma^{\mathrm{T}}\tilde{Q}^{-1} = \tilde{Q}\Gamma_{\perp}\left(\Gamma_{\perp}^{\mathrm{T}}\tilde{Q}\Gamma_{\perp}\right)^{-1}\Gamma_{\perp}^{\mathrm{T}}
\tag{12.33}
$$

with $\Gamma_{\perp} = \left[0, I_f\right]^{\mathrm{T}}$, we get

$$c_{\hat{\tau}_{\text{free}}}^{-2} = e_{2f}^{\mathrm{T}} \boldsymbol{\Gamma}_{\perp} (\boldsymbol{\Gamma}_{\perp}^{\mathrm{T}} \tilde{\boldsymbol{Q}} \boldsymbol{\Gamma}_{\perp})^{-1} \boldsymbol{\Gamma}_{\perp}^{\mathrm{T}} e_{2f}$$

$$= e_f^{\mathrm{T}} \left( \frac{1}{r} \boldsymbol{Q}_p + \sigma_\iota^2 \boldsymbol{\mu} \boldsymbol{\mu}^{\mathrm{T}} \right)^{-1} e_f \tag{12.34}$$

with $\tilde{\boldsymbol{Q}} = \boldsymbol{Q}_f + \sigma_\iota^2 \boldsymbol{v} \boldsymbol{v}^{\mathrm{T}}$.

Now we work out the expression for $\boldsymbol{q}$. Premultiplying the matrix identity (12.31) with $(\boldsymbol{\Gamma}^{\mathrm{T}} \boldsymbol{\Gamma})^{-1} \boldsymbol{\Gamma}^{\mathrm{T}}$ gives

$$\left( \boldsymbol{\Gamma}^{\mathrm{T}} \tilde{\boldsymbol{Q}}^{-1} \boldsymbol{\Gamma} \right)^{-1} \boldsymbol{\Gamma}^{\mathrm{T}} \tilde{\boldsymbol{Q}}^{-1} = (\boldsymbol{\Gamma}^{\mathrm{T}} \boldsymbol{\Gamma})^{-1} \boldsymbol{\Gamma}^{\mathrm{T}} \left( \boldsymbol{I}_{2f} - \tilde{\boldsymbol{Q}} \boldsymbol{\Gamma}_{\perp} (\boldsymbol{\Gamma}_{\perp}^{\mathrm{T}} \tilde{\boldsymbol{Q}} \boldsymbol{\Gamma}_{\perp})^{-1} \boldsymbol{\Gamma}_{\perp}^{\mathrm{T}} \right)$$

$$= \boldsymbol{\Lambda}^{-1} [\boldsymbol{I}_f, 0] \left( \boldsymbol{I}_{2f} - \tilde{\boldsymbol{Q}} \begin{bmatrix} 0 \\ \boldsymbol{I}_f \end{bmatrix} \left( \frac{1}{r} \boldsymbol{Q} \otimes p + \sigma_\iota^2 \boldsymbol{\mu} \boldsymbol{\mu}^{\mathrm{T}} \right)^{-1} [0, \boldsymbol{I}_f] \right) \tag{12.35}$$

Hence

$$\boldsymbol{q} = \left( \boldsymbol{\Gamma}^{\mathrm{T}} \tilde{\boldsymbol{Q}}^{-1} \boldsymbol{\Gamma} \right)^{-1} \boldsymbol{\Gamma}^{\mathrm{T}} \tilde{\boldsymbol{Q}}^{-1} e_{2f}$$

$$= \boldsymbol{\Lambda}^{-1} \left( e_f + \sigma_\iota^2 \boldsymbol{\mu} \boldsymbol{\mu}^{\mathrm{T}} \left( \frac{1}{r} \boldsymbol{Q}_p + \sigma_\iota^2 \boldsymbol{\mu} \boldsymbol{\mu}^{\mathrm{T}} \right)^{-1} e_f \right) \tag{12.36}$$

It is not difficult to verify that

$$\boldsymbol{I}_f + \sigma_\iota^2 \boldsymbol{\mu} \boldsymbol{\mu}^{\mathrm{T}} \left( \frac{1}{r} \boldsymbol{Q}_p + \sigma_\iota^2 \boldsymbol{\mu} \boldsymbol{\mu}^{\mathrm{T}} \right)^{-1} = \boldsymbol{I}_f + \alpha \boldsymbol{\mu} \boldsymbol{\mu}^{\mathrm{T}} \boldsymbol{Q}_p^{-1} \tag{12.37}$$

with $\alpha = \left[ (r\sigma_\iota^2)^{-1} + \boldsymbol{\mu}^{\mathrm{T}} \boldsymbol{Q}_p^{-1} \boldsymbol{\mu} \right]^{-1}$. Hence, for $\boldsymbol{q}$ we find

$$\boldsymbol{q} = \boldsymbol{\Lambda}^{-1} \left( \boldsymbol{I}_f + \alpha \boldsymbol{\mu} \boldsymbol{\mu}^{\mathrm{T}} \boldsymbol{Q}_p^{-1} \right) e_f \tag{12.38}$$

It is rather easy to prove:

$$\left( \boldsymbol{\Gamma}^{\mathrm{T}} \tilde{\boldsymbol{Q}}^{-1} \boldsymbol{\Gamma} \right)^{-1} \otimes \frac{1}{k} \overline{\boldsymbol{W}}^{-1} = \boldsymbol{Q}_{\hat{a}\hat{a}}^{(\text{fixed})} \tag{12.39}$$

To specify the time variation of troposphere design matrix $\boldsymbol{g}$ and elevation-dependent weight matrix, we assign the epoch index $t$ to $\boldsymbol{g}$ and $\boldsymbol{W}$. The normal matrix of $k$ epochs reads

$$\begin{bmatrix} \left( \boldsymbol{\Gamma}^{\mathrm{T}} \tilde{\boldsymbol{Q}}^{-1} \boldsymbol{\Gamma} \right) \otimes k \overline{\boldsymbol{W}} & \left( \boldsymbol{\Gamma}^{\mathrm{T}} \tilde{\boldsymbol{Q}}^{-1} e_{2f} \right) \otimes k \overline{\boldsymbol{W} \boldsymbol{g}} \\ \left( e_{2f}^{\mathrm{T}} \tilde{\boldsymbol{Q}}^{-1} \boldsymbol{\Gamma} \right) \otimes k \overline{\boldsymbol{g}_t^{\mathrm{T}} \boldsymbol{W}} & c_{\hat{\tau}|a}^{-2} \sum_{t=1}^{k} \boldsymbol{g}_t^{\mathrm{T}} \boldsymbol{W}_t \boldsymbol{g}_t \end{bmatrix} \tag{12.40}$$

with $\overline{W} = \frac{1}{k}\sum_{t=1}^{k} W_t$, $\overline{g} = \left(\sum_{t=1}^{k} W_t\right)^{-1}\sum_{t=1}^{k} W_t g_t$, $\sum_{t=1}^{k} W_t g_t = k\overline{W}\overline{g}$ and

$$c_{\hat{\tau}|a}^{-2} = e_{2f}^{\mathrm{T}} \tilde{Q}^{-1} e_{2f} = e_{2f}^{\mathrm{T}} \left(\frac{1}{r}Q_f + \sigma_t^2 v v^{\mathrm{T}}\right)^{-1} e_{2f} \tag{12.41}$$

Reducing the ZTD parameter, the normal matrix of ambiguities over $k$ epochs is

$$\left(Q_{\hat{a}\hat{a}}^{\text{(fixed)}}\right)^{-1} - \frac{\left(\Gamma^{\mathrm{T}}\tilde{Q}^{-1}e_{2f} \otimes \overline{W}\overline{g}\right)\left(e_{2f}^{\mathrm{T}}\tilde{Q}^{-1}\Gamma \otimes \overline{g}^{\mathrm{T}}\overline{W}\right)}{k^{-2}c_{\hat{\tau}|a}^{-2}\sum_{t=1}^{k} g_t^{\mathrm{T}} W_t g_t} \tag{12.42}$$

Using the matrix inversion lemma, we obtain the variance matrix (12.43) of ambiguities

$$Q_{\hat{a}\hat{a}} = Q_{\hat{a}\hat{a}}^{\text{(fixed)}}$$
$$+ \frac{k^2 c_{\hat{\tau}|a}^2 Q_{\hat{a}\hat{a}}^{\text{(fixed)}}\left(\Gamma^{\mathrm{T}}\tilde{Q}^{-1}e_{2f}e_{2f}^{\mathrm{T}}\tilde{Q}^{-1}\Gamma \otimes \overline{W}\tilde{g}\overline{g}^{\mathrm{T}}\overline{W}\right)Q_{\hat{a}\hat{a}}^{\text{(fixed)}}}{\sum_{t=1}^{k} g_t^{\mathrm{T}} W_t g_t - \left(e_{2f}^{\mathrm{T}}\tilde{Q}^{-1}\Gamma \otimes \tilde{g}^{\mathrm{T}}\overline{W}\right)Q_{\hat{a}\hat{a}}^{\text{(fixed)}}\left(\Gamma^{\mathrm{T}}\tilde{Q}^{-1}e_{2f} \otimes \overline{W}\overline{g}\right)k^2 c_{\hat{\tau}|a}^2}$$
$$= Q_{\hat{a}\hat{a}}^{\text{(fixed)}} + \frac{U}{v} \tag{12.43}$$

Let us now focus on the fraction $U/v$ of the second term only. We first simplify its numerator $U$. Substituting (12.39) into it yields:

$$U = c_{\hat{\tau}|a}^2 \left(\Gamma^{\mathrm{T}}\tilde{Q}^{-1}\Gamma\right)^{-1}\Gamma^{\mathrm{T}}\tilde{Q}^{-1}e_{2f} \times e_{2f}^{\mathrm{T}}\tilde{Q}^{-1}\Gamma\left(\Gamma^{\mathrm{T}}\tilde{Q}^{-1}\Gamma\right)^{-1} \otimes \overline{g}\overline{g}^{\mathrm{T}}$$
$$= c_{\hat{\tau}|a}^2 q q^{\mathrm{T}} \otimes \overline{g}\overline{g}^{\mathrm{T}} \tag{12.44}$$

Substituting (12.39) into the denominator $v$ of fraction yields:

$$v = \sum_{t=1}^{k} g_t^{\mathrm{T}} W_t g_t - \frac{k}{c_{\hat{\tau}|a}^{-2}} e_{2f}^{\mathrm{T}}\tilde{Q}^{-1}\Gamma\left(\Gamma^{\mathrm{T}}\tilde{Q}^{-1}\Gamma\right)^{-1} \times \Gamma^{\mathrm{T}}\tilde{Q}^{-1}e_{2f}\overline{g}^{\mathrm{T}}\overline{W}\overline{g}$$
$$= \sum_{t=1}^{k} g_t^{\mathrm{T}} W_t g_t - \frac{k}{c_{\hat{\tau}|a}^{-2}}\left(c_{\hat{\tau}|a}^{-2} - c_{\hat{\tau}_{\text{free}}}^{-2}\right)\overline{g}^{\mathrm{T}}\overline{W}\overline{g} \tag{12.45}$$

where use is made of (12.33) and (12.41). Further substituting the identity

$$\sum_{t=1}^{k}\left(g_t^{\mathrm{T}} W_t g_t\right) = \sum_{t=1}^{k}\left(g_t - \overline{g}\right)^{\mathrm{T}} W_t\left(g_t - \overline{g}\right) + k\overline{g}^{\mathrm{T}}\overline{W}\overline{g} \tag{12.46}$$

into (12.46) gives

$$v = \sum_{t=1}^{k} (g_t - \bar{g})^{\mathrm{T}} W_t (g_t - \bar{g}) + k \frac{c_{\hat{r}|a}^2}{c_{\hat{r}_{\mathrm{free}}}^2} \bar{g}^{\mathrm{T}} \overline{W} \bar{g} \tag{12.47}$$

Therefore,

$$\frac{U}{v} = \frac{1}{k} c_{\hat{\tau}}^2 q q^{\mathrm{T}} \otimes P_{\bar{g}} \overline{W}^{-1} \tag{12.48}$$

where

$$c_{\bar{\tau}}^2 = c_{\hat{\tau}_{\mathrm{free}}}^2 \left[ 1 + \frac{c_{\hat{\tau}_{\mathrm{free}}}^2}{c_{\hat{\tau}|a}^2} \frac{\sum_{t=1}^{k} (g_t - \bar{g})^{\mathrm{T}} W_t (g_t - \bar{g})}{k \bar{g}^{\mathrm{T}} \overline{W} \bar{g}} \right]^{-1} \tag{12.49}$$

$$P_{\bar{g}} = \bar{g} (\bar{g}^{\mathrm{T}} \overline{W} \bar{g})^{-1} \bar{g}^{\mathrm{T}} \overline{W}$$

## 12.4   ARTK Model

For the relative positioning between two platforms using the reduced observations from array antennas, we introduce the subscripts $b$ and $u$ to specify the array platform. The equations of reduced observations for two platforms read

$$\mathrm{E}(\bar{y}_b) = M_b x_b + e_{2f} \otimes \tilde{\tau}_b + v \otimes \tilde{\iota}_b + N\tilde{a}_{1,b} - \theta_1 \tag{12.50}$$

$$\mathrm{E}(\bar{y}_u) = M_u x_u + e_{2f} \otimes \tilde{\tau}_u + v \otimes \tilde{\iota}_u + N\tilde{a}_{1,u} - \theta_1 \tag{12.51}$$

The variance matrix is

$$\mathrm{D}\left( [\bar{y}_b^{\mathrm{T}}, \bar{y}_u^{\mathrm{T}}]^{\mathrm{T}} \right) = \frac{1}{r} \mathrm{blkdiag}(Q_{y_b}, Q_{y_u}) \tag{12.52}$$

It is noticed here that $\bar{y}_b$ and $\bar{y}_u$ are the reduced observations already corrected by the DD integer ambiguities within the platform. The corresponding DD equations

$$\mathrm{E}(\bar{y}_{bu}) = M_u x_u - M_b x_b + e_{2f} \otimes \tilde{\tau}_{bu} + v \otimes \tilde{\iota}_{bu} + Nz_{1,bu} \tag{12.53}$$

with the variance matrix as

$$\mathrm{D}(\bar{y}_{bu}) = \frac{1}{r} (Q_{y_b} + Q_{y_u}) \tag{12.54}$$

where $b$ is taken as a reference station. $\bar{y}_{bu} = \bar{y}_u - \bar{y}_b$ is DD observations. $\tilde{\tau}_{bu}$ and $\tilde{\iota}_{bu}$ are the DD tropospheric and ionospheric delays. $z_{1,bu}$ is the DD integer vector formed by the first antenna between two platforms. $\boldsymbol{Q}_{y_b}$ and $\boldsymbol{Q}_{y_u}$ are the variance matrix of observations from the single antenna on platform $b$ and $u$, respectively. In relative positioning $\boldsymbol{M}_b$ is usually very approximate to $\boldsymbol{M}_u$ and the coordinate of station $b$ is precisely known or computed using single point positioning (SPP), then the coordinate correction $\boldsymbol{x}_b$ is near to 0 and $\boldsymbol{M}_u\boldsymbol{x}_u - \boldsymbol{M}_b\boldsymbol{x}_b = \boldsymbol{M}_u\boldsymbol{x}_{bu} + \boldsymbol{M}_{bu}\boldsymbol{x}_b \approx \boldsymbol{M}_u\boldsymbol{x}_{bu}$.

For long baselines, the atmospheric effects cannot be ignored if the precise positioning is anticipated. Traditionally, the tropospheric delay is compensated by parameterizing it as a function of relative zenith tropospheric delay $\tau^z$ and mapping function as $\tilde{\tau}_{bu} = \boldsymbol{m}_{bu}\tau^z$ with $\boldsymbol{m}_{bu}$ mapping function. The ionospheric delays, as dominant and complicated systematic errors, are modeled by estimating all DD ionospheric delays as parameters. Then (12.53) becomes

$$\mathrm{E}(\bar{\boldsymbol{y}}_{bu}) = \boldsymbol{M}\boldsymbol{x}_{bu} + \left(\boldsymbol{e}_{2f} \otimes \boldsymbol{m}_{bu}\right)\tau^z + \boldsymbol{v} \otimes \tilde{\boldsymbol{\imath}}_{bu} + \boldsymbol{N}\boldsymbol{z}_{1,bu} \tag{12.55}$$

It is referred to as ionosphere-float model also equivalent to the ionosphere-free (IF) model. As well-known, the ionosphere-float model is too weak and one can then enhance the model strength by imposing the constraints to ionospheric parameters

$$\mathrm{E}(\tilde{\boldsymbol{\imath}}_{bu}) = \tilde{\boldsymbol{\imath}}_{bu}^0, \quad \text{with } \mathrm{D}(\tilde{\boldsymbol{\imath}}_{bu}) = \sigma_{\tilde{\imath}_{bu}}^2 \tag{12.56}$$

It is referred to as an ionosphere-weighted model. Usually, $\tilde{\imath}_{bu}^0 = 0$ for baselines even as long as several hundred kilometers. In RTK, the parameters $\boldsymbol{x}_{bu}$ and $\tilde{\boldsymbol{\imath}}_{bu}$ are various epoch by epoch; the ambiguity $\boldsymbol{z}_{1,bu}$ is constant and $\tau^z$ can be constant for a period as well. Then the least squares criterion is employed to solve the ionosphere-weighted model realizing RTK solution.

### 12.4.1  Experiment and Analysis

Total 3-h real GPS dual-frequency data was collected on an 80 km baseline with sampling interval of 1 s in Perth area, West Australia. The platforms were equipped with Sokkia (receiver type: GSR2700ISX, antenna type: Internal Pinwheel™) and Javad (receiver type: Javad Delta, antenna type: GrAnt-G3T) receivers. Any tricks of code smoothing were switched off for all receivers to cancel the time correlation in observations. The sky-plot of all 13 tracked satellites in 3-h observation span is also analyzed.

The satellite PRN 11 with the highest elevation at the first epoch is taken as reference satellite to form 12 pairs of DD observation series. In the computations, the cut-off elevation was set to $10°$. We reduce array data according to the implementation steps and output them in RINEX format. In the following, "ARTK" denotes the solution obtained using the reduced observations with 4 array antennas while "CRTK" with 1 antenna no matter in static or kinematic scenarios.

### *12.4.2  Estimation of Observation Standard Deviation*

As well known in the geodetic community, the only correct stochastic model can be used to compute the optimal estimator in the sense of least squares. Consequently, we need firstly to determine the stochastic model of (12.54) for precise RTK between-platform, which is specified by the stochastic characteristics of two types of antennas. The GNSS observation precision and time correlations may be different from the different antennas/receivers and observation types [8]. In our case, the code smoothing technique was switched off and time correlation is absent. Hence, we need only to examine the observation precisions of Sokkia and Javad receivers.

During the observation reduction on the array platform, the DD ambiguities are fixed and then we can retrieve the residuals

$$v_{\tilde{Y}} = \left(I_{r-1} \otimes R_y\right)\mathrm{vec}\left(\tilde{Y} - N\check{Z}\right) \tag{12.57}$$

with $R_y = Q_y \tilde{P}_y$. The one can estimate the standard deviations of all observation types with the residuals $v_{\tilde{Y}}$ by employing the variance component estimation theory [8, 9]. It is noticed here that the fixed ambiguities $\check{Z}$ are deemed as deterministic values. In theory, this assumption holds true only when the success-rate of ambiguity estimation is 100% It is however impossible practically, because the ambiguity is computed from the noisy observation. Fortunately, in our case, the baselines on the platform are so short that the IAR success-rate is always nearly equal to 100% even the rounding method is applied.

Figure 12.2 illustrates the estimated standard deviations as a function of a number of data epochs for all observation types. With the processing ongoing, the more data epochs are involved in the estimation and then the more stable estimates are obtained. The standard deviations of all observation types of Sokkia and Javad antennas are summarized in Table 12.1. In the experiment, we would like to explore the IAR capability using multiple antennas. Therefore, if the $r$ antennas are used on the platform for data reduction, the standard deviations of the reduced observations are the ones of single-antenna observations divided by $\sqrt{r}$.

### *12.4.3  Static Processing*

First of all, to get insight into the quality of reduced observations, we process data using Trimble Geomatics Office (TGO) commercial software in static mode. For such long baseline, TGO compulsively specifies the IF model with two cascading steps of IAR, i.e., widelane followed by narrowlane. In the processing, the hourly ZTD parameters are set up to absorb the tropospheric effects. The baseline is solved in ARTK and CRTK modes, respectively. It is emphasized here that ARTK and CRTK are both the static processing modes and the only difference is that ARTK uses the reduced data of 4 array antennas while CRTK the raw data of 1 antenna. The baseline

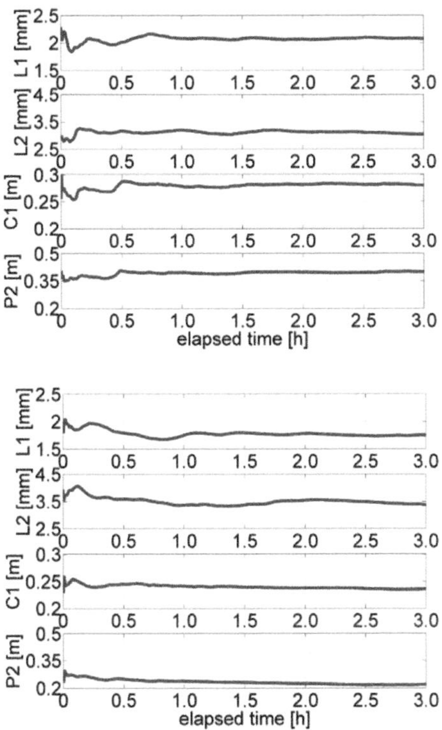

**Fig. 12.2** Standard deviations of all observation types of Sokkia (top) and Javad (bottom) receivers

**Table 12.1** Standard deviations of all observation types of two receivers

| Receiver | L1 (mm) | L2 (mm) | C1 (cm) | P2 (cm) |
| --- | --- | --- | --- | --- |
| Sokkia | 2.3 | 3.2 | 30 | 42 |
| Javad | 2.1 | 3.3 | 25 | 22 |

formal precisions of the north, east and up components are 0.4, 1.2 and 3.1 mm for ARTK and 1.1, 1.4 and 5.2 mm for CRTK.

After baseline resolved, the IF phase residuals are computed for all 12 pairs of DD satellites. Their means and standard deviations (STDs) are illustrated in Fig. 12.3. The means and STD of ARTK for all DD satellites are much closer to zero than their counterparts of CRTK. It means that ARTK indeed improves the measurement precision. The mean ratio of 12 STDs between CRTK and ARK is 1.7 which is close to the theoretical value 2. The difference 0.3 could be induced by the inadequately modeled systematic errors, like multipath, tropospheric errors. It is also observed that the means of satellites 3, 4 and 7 and the STDs of satellites 7 and 17 are apparently larger than the others, which is possibly attributed to their low elevations and then the misspecified systematic errors.

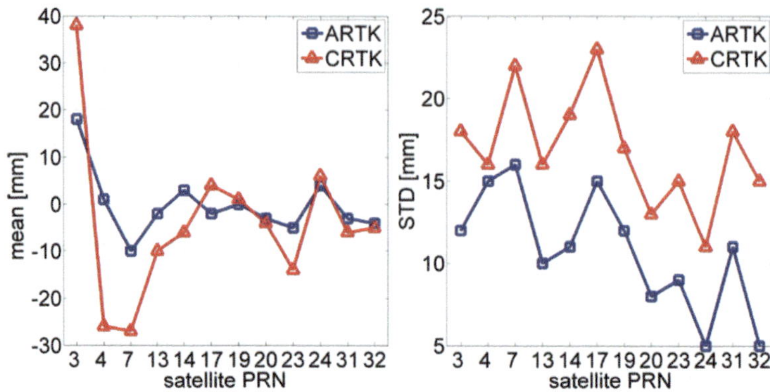

**Fig. 12.3** The means and STDs of DD IF phase residuals obtained with TGO static baseline processing for ARTK and CRTK, respectively

### 12.4.4   Kinematic Processing

In this subsection, we examine the superiority of ARTK against CRTK by processing data in kinematic modes based on the ionosphere-weighted model of Sect. 12.4. The STD of DD ionospheric constraint is set to $\sigma_{\tilde{I}_{bu}} = 15\,\text{cm}$. The Kalman filtering recursive processing is carried out with the dynamic noise as $1\,\text{cm}^2/\text{h}$ for ZTD and infinity for coordinates to identify the truly kinematic scenario.

First of all, we compare the IAR between ARTK and CRTK. The bootstrapped success rate is a good measure to indicate how much probability the successful IAR can be done [10]. It is defined as

$$P_B = \prod_{i=1}^{m} \left( 2\Phi\left(\frac{1}{2\sigma_{\hat{z}_{i|I}}}\right) - 1 \right) \tag{12.58}$$

with $\Phi(x) = \int_{-\infty}^{x} \frac{1}{\sqrt{2\pi}} e^{-t^2/2} \mathrm{d}t$. $\sigma_{\hat{z}_{i|I}}$ is the conditional standard deviation of the $i$th decorrelated float ambiguity $\hat{z}_i$ on the float ambiguities from $(i+1)$ to the total number ambiguities $m$, which is the $i$th diagonal element of $D$ computed from the Cholesky decomposition on the decorrelated ambiguity variance matrix $Q_{\hat{z}\hat{z}} = L^T D L$. In addition, one can also compute the empirical success rate $P_E$ defined as

$$P_E = \frac{\#\,\text{correct integer solutions}}{\#\,\text{total integer solutions}} \tag{12.59}$$

where the "correct integer solution" is evaluated by comparing with the "true" integer solution computed with all data in advance.

In the RTK campaigns, we expect to resolve ambiguities using short observation span (only a few epoch data). In such case, the geometry strength is too weak to fix all ambiguities, in particular for the ambiguities with larger STDs generally corresponding to lower elevations. Thus, we prefer to partial ambiguity resolution (PAR) in RTK, i.e., only fixing a subset of ambiguities that can be reliably fixed. The key of PAR is how to determine the optimal ambiguity subset for fixing. So far, several practical approaches have been developed but the optimal method is still under-developing [11–13]. Here we evaluate the ARTK performance using two PAR scenarios. One is to only fix widelane ambiguities, while the other is to fix ambiguities with elevations larger than a given threshold $\theta_0$, like 20°. In Fig. 12.4, the empirical success rates of widelane PAR of ARTK and CRTK are compared for the different number of data epochs. Figure 12.5 shows the empirical success rates of both RTK modes with ambiguity subset thresholds $\theta_0 = 20°$ and 30° under a varying number of data epochs, respectively. Both PAR results show that the ARTK success rates are much larger than CRTK counterparts for all scenarios, especially, for the cases with fewer data epoch, which indicates that the underlying model strength of IAR is indeed significantly enhanced in ARTK mode. As a byproduct, it is noticed that the empirical success rates of $\theta_0 = 30°$ are larger than those of $\theta_0 = 20°$ for both ARTK and CRTK, whereas the results of $\theta_0 = 40°$ are smaller than $\theta_0 = 30°$ although they are not shown here. This again highlights the open problem how to determine the optimal ambiguity subset for partial fixing but beyond the scope of this contribution.

Now, we analyze the RTK solutions in both ARTK and CRTK modes. The scatter-plot of horizontal positional errors and the vertical positional errors are shown in Fig. 12.6. The scatter-plot of ARTK is much more concentrated than that of CRTK and the vertical positional errors of ARTK are completely smaller than the counterparts of CRTK for the whole position series although they are both relatively larger at the beginning short span due to the severe systematic errors. The statistics of RTK solutions are listed in Table 12.2. Regarding the mean of positional error, the ARTK north is slightly worse than the CRTK north, while the ARTK east and up are much closer to 0 than CRTK counterparts. Particularly for up component, the mean of ARTK is just 1.25 cm against 8.27 cm of CRTK. The exciting result observed from

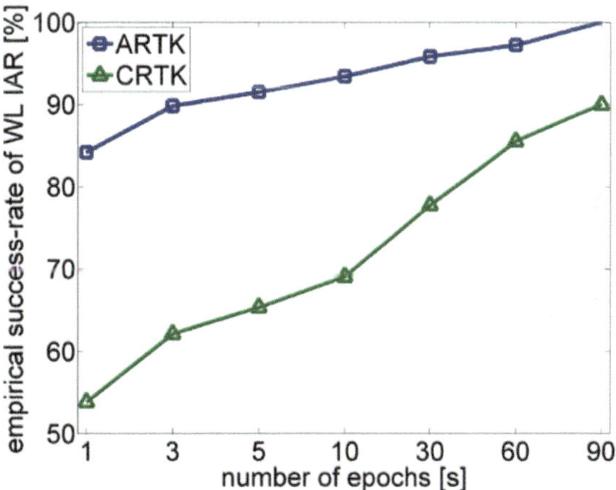


**Fig. 12.4** Empirical success-rate of widelane IAR for ARTK and CRTK, respectively

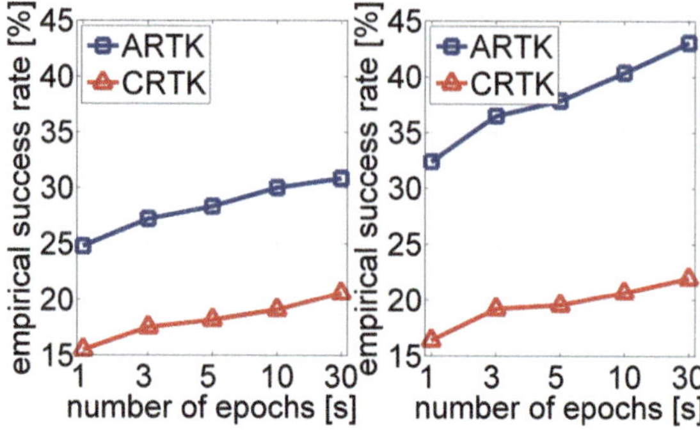

**Fig. 12.5** Empirical IAR success-rate for ARTK and CRTK as a function of data epochs with ambiguity subset thresholds $\theta_0 = 20°$ (left) and $\theta_0 = 30°$ (right), respectively

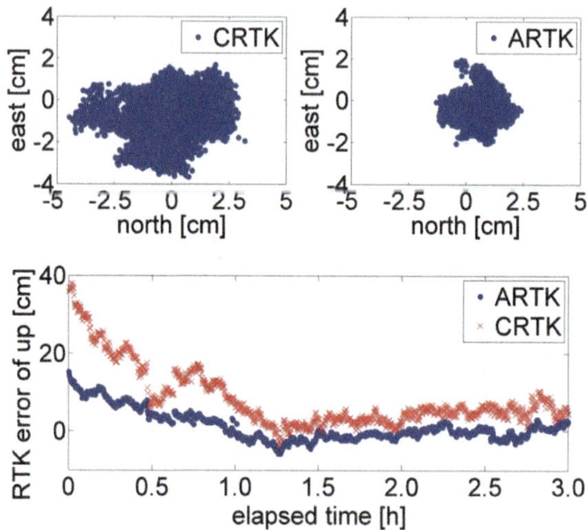

**Fig. 12.6** Scatter-plot of horizontal positional errors (top) and the vertical positional errors (bottom) for ARTK and CRTK, respectively

the standard deviations shows that the RTK solution is significantly improved in ARTK mode by factors of 2.2, 1.7 and 2.0 respectively to the north, east and up components. The mean of improved factors for three coordinate components is 1.97, which is rather consistent with the theoretical value 2 in our case study of 4 antennas being used on the platform.

With RTK solution resolved at each epoch, the residuals of DD IF phase and code are computed for all 12 pairs of DD satellites of 3-h observation span. To clearly illustrate the residual differences between ARTK and CRTK, the scalars of 200 and 2 are multiplied to the residuals of phase and code, respectively. It means that $1°$ variation in azimuth of sky-plot corresponds to 0.5 cm in phase residual while 0.5 m

established. The results from an 80 km baseline experiment with 4-antenna equipped stations suggested that besides the fast IAR with larger success rates and the improved RTK solutions with higher precision (half standard deviation), the ARTK can also improve the reliability of RTK solutions with doubled confidence. The reduced data is completely same as the raw data from individual antenna except its higher precision. Therefore, as one of ARTK benefits, the existent GNSS software can be immediately used to handle this reduced data without any modification. Because the receiver is generally much expensive than the antenna, considering the economic cost in practice, one may connect multiple antennas with one receiver on the platform. In addition, the proposed technique allows one to use multiple low-cost antennas, like very cheap ublox antennas, to realize the comparable IAR, positioning and attitude determination performance instead of the high-quality antennas, e.g., its attitude determination performance was demonstrated in [15, 16]. Moreover, such low-cost antennas have very small size and then make the platform portable in field surveying.

# References


1. Teunissen PJG (2012) A-PPP: array-aided precise point positioning with global navigation satellite systems. IEEE Trans Signal Process 60:2870–2881
2. Teunissen PJG (1998) Minimal detectable biases of GPS data. J Geod 72:236–244
3. Ray JK, Canon ME, Fenton P (2000) GPS code and carrier multipath mitigation using a multi-antenna system. IEEE Trans Aerosp Electron Syst 37:183–195
4. Li B, Teunissen PJG (2012) Real-time kinematic positioning using fused data from multiple GNSS antennas. Proceedings of th FUSION, pp 933–938
5. Giorgi G, Teunissen PJG, Verhagen S, Buist P (2010) Testing a new multivariate GNSS carrier phase attitude determination method for remote sensing platforms. Adv Space Res 46:118–129
6. Rao C (1973) Linear statistical inference and its applications. Wiley, New York
7. Teunissen PJG (2003) Adjustment theory: an introduction. In: Series on mathematical geodesy and positioning. Delft University Press, Delft
8. Li B, Shen Y, Xu P (2008) Assessment of stochastic models for GPS measurements with different types of receivers. Chin Sci Bull 53:3219–3225
9. Teunissen PJG, Amiri-Simkooei AR (2008) A comparison of the least-squares and the maximum likelihood estimation approaches for GNSS ambiguity resolution. J Geod 82:449–461
10. Verhagen S, Li B, Teunissen PJG (2013) Ps-LAMBDA: ambiguity success rate evaluation software for interferometric applications. Comput Geosci 54:361–376
11. Teunissen PJG, Joosten P, Tiberius CCJM (1999) Geometry-free ambiguity success rates in case of partial fixing. Proceedings of ION GPS 99, pp 201–207
12. Verhagen S, Teunissen PJG, van der Marel H, Li B (2011) GNSS ambiguity resolution: which subset to fix? Proceedings of the IGNSS symposium
13. Li B, Verhagen S, Teunissen PJG (2013) Robustness of GNSS integer ambiguity resolution in the presence of atmospheric biases. GPS Solut 18:283–296
14. Kaplan ED, Hegarty CJ (2006) Understanding GPS principles and applications. Artech House, Massachusetts
15. Nadarajah N, Teunissen PJG, Odijk D (2011) GNSS-based attitude determination for remote sensing platforms, Proceedings of ACRS-2011
16. Teunissen PJG, Nadarajah N, Giorgi G, Buist PJ (2011) Low-complexity instantaneous GNSS attitude determination with multiple low-cost antennas. Proceedings of ION GNSS-2011, pp 3874–3880

# Chapter 13
# CRTK: Cost-Effective RTK

## 13.1  Introduction of CRTK

In 'Google I/O of 2016', the Global Navigation Satellite Systems (GNSSs) raw data (including pseudorange, phase and doppler) is released to developers from smart devices with Android-N operating system [1]. The availability of raw data provides more opportunities in the booming location-based service (LBS) markets, allowing the users to carry out their positioning campaigns with flexible positioning modes in terms of their specific accuracy demands [2]. For instance, one can implement the pseudorange-based single point positioning [3], phase-based real-time kinematic (RTK) [4] or precise point positioning (PPP) [5] for the meter- to centimeter-accuracy LBS. With the growing demands for smartphone precise positioning, the researchers started to focus more on phase-based precise positioning techniques [6]. The correct integer ambiguity resolution (IAR) is the key issue to achieving precise positioning with carrier phase observations. In a short-baseline positioning mode where the atmospheric biases are basically eliminated, the success of IAR depends mainly on two factors. One is the integer property of ambiguity that is a prerequisite for IAR, while the other is the data quality that affects the ambiguity precision.

For the phase observations in smartphones, the integer property of ambiguity has been investigated for the different smartphone chips with embedded or external antennas. In the case of embedded antennas, the IAR is rather difficult or even impossible for some brands of smartphones, for instance, Nexus9, Huawei P10 and Galaxy S5 [7]. The reason is that their ambiguity fractions are time-variant dramatically from satellite to satellite. However, for Huawei Mate20X and P30 as well as Xiaomi Mi8 (Mi8), their ambiguities are of integer nature at some frequency signals, like Global Positioning System (GPS) L1 frequency [8]. In the case of external antennas, the results from [9] showed that with the Android system the constant offsets exist in the ambiguities for Nexus9 and Mi8, and thus their ambiguities can be fixed if these offsets are pre-calibrated. Moreover, they found that such property of constant

© The Author(s) 2025

B. Li et al., *GNSS Real-Time Kinematic Positioning*, Navigation: Science and
Technology 17, https://doi.org/10.1007/978-981-96-9116-6_13

offsets is not available for Mate20. Note, the above analyses for the integer property of ambiguity were mainly based on a given mobile operating system. In fact, the power-saving modes differ from the mobile operating systems [10], which may affect the smartphone chip to process the received GNSS signals. This leaves a question that whether the mobile operating system affects the integer property of ambiguity. In addition, with release of new chips of Huawei Kirin980, Huawei Mate20, Mate20X and P30 are all updated by these new chips. As a result, it is interesting to understand the ambiguity fixation for updated Huawei smartphones.

Regarding GNSS data quality, the previous studies indicated that the embedded antenna of smartphones is the key factor. The linearly polarized antennas and low-cost GNSS chipsets are generally used in smartphones [11], which together derive the GNSS signals featured by the lower and highly-variated carrier-to-noise density ratio (C/N0), the non-uniform signal strengths and low C/N0 at high elevations, the high noise in the order of tens of meters and frequent outliers for pseudoranges, as well as the Duty-cycle. However, the existing studies mainly concentrated on the data quality and its impacts on IAR at a given attitude. In fact, the smartphone attitude would frequently change in real applications. Since the smartphone antennas are generally omnidirectional rather than hemispherical, it is insufficient to understand the data quality of smartphones only at a given attitude. Instead, one needs to accurately understand the data quality at different attitudes so as to improve the IAR for smartphone positioning.

Different from the existing literatures where only the smartphone brands are analyzed for IAR, this chapter will address three factors hindering the smartphone IAR and thus the precise positioning, including the mobile operating systems and smartphone attitudes besides the smartphone brands. We comprehensively analyze their effects on the integer nature of ambiguities and data quality. The observations from the smartphones of Mate20 and Mi8 with embedded and external antennas and the geodetic receivers with external antennas are comparably analyzed.

## 13.2   Formulae of Precision Estimation

To study the integer property and noise characteristics of the observations from a smartphone, we will form the short baseline between a geodetic-grade receiver and a smartphone. For the between-receiver short baseline single-differenced (SD) observations, the systematic errors, e.g., satellite orbit and clock errors, satellite hardware delays and atmospheric effects, can be basically eliminated. Then the single-epoch, SD observation equations on frequency $j$ read [12]

$$\boldsymbol{\Phi}_j = \boldsymbol{B}b + e_s \delta t_j + \lambda_j a_j + \lambda_j e_s \varphi_j + \boldsymbol{\epsilon}_{\Phi_j}$$
$$\boldsymbol{P}_j = \boldsymbol{B}b + e_s \mathrm{d}t_j + \boldsymbol{\epsilon}_{P_j} \tag{13.1}$$

where $\boldsymbol{\Phi}_j$ is the SD observation vector of $s$ satellites for phase on frequency $j$, and $\boldsymbol{P}_j$ for code has the same structure as $\boldsymbol{\Phi}_j$. $\boldsymbol{B}$ is the design matrix to the baseline vector $\boldsymbol{b}$. $\boldsymbol{a}_j$ is the SD ambiguity vector with wavelength $\lambda_j$. $\varphi_j$ is the SD initial phase bias of receiver. $\delta t_j$ and $\mathrm{d}t_j$ are the SD receiver clock errors for phase and code. $\boldsymbol{\epsilon}_{\boldsymbol{\Phi}_j}$ and $\boldsymbol{\epsilon}_{\boldsymbol{P}_j}$ contain the measurement noise and multipath for the SD phase and code, respectively. The symbol $\boldsymbol{e}_s$ is the $s$-column vector with all elements of ones.

Obviously, the parameters $\delta t_j$ and $\varphi_j$ are fully dependent, and they are further dependent on parameter $\boldsymbol{a}_j$ with rank-deficiency of 1. In terms of [12], the full-rank single-epoch SD observation equations of phase and code on frequency $j$ read

$$\begin{bmatrix} \boldsymbol{\Phi}_j \\ \boldsymbol{P}_j \end{bmatrix} = \begin{bmatrix} \boldsymbol{B} & \lambda_j \boldsymbol{\Lambda} & \boldsymbol{e}_s & \boldsymbol{0} \\ \boldsymbol{B} & \boldsymbol{0} & \boldsymbol{0} & \boldsymbol{e}_s \end{bmatrix} \begin{bmatrix} \boldsymbol{b} \\ \boldsymbol{a}_j \\ \delta t_j \\ \mathrm{d}t_j \end{bmatrix} + \begin{bmatrix} \boldsymbol{\epsilon}_{\boldsymbol{\Phi}_j} \\ \boldsymbol{\epsilon}_{\boldsymbol{P}_j} \end{bmatrix} \tag{13.2}$$

where $\boldsymbol{\Lambda} = \begin{bmatrix} \boldsymbol{0}_{(s-1)\times 1}, \boldsymbol{I}_{s-1} \end{bmatrix}^{\mathrm{T}}$. Importantly, $\delta t_j$ is the nominated phase receiver clock error redefined as $\delta t_j = \delta t_j + \lambda_j \varphi_j + \lambda_j a_j^1$, which includes the receiver initial phase biases and the pivot ambiguity. $\boldsymbol{a}_j = \begin{bmatrix} -\boldsymbol{e}_{s-1} & \boldsymbol{I}_{s-1} \end{bmatrix} \boldsymbol{a}_j$ is the vector of double-differenced (DD) ambiguities, which must be integers for the geodetic grade receivers. However, it is not the case for phase observations of some smartphones. In such a case, the DD ambiguity can be deemed as a lumped variable of an integer and a real-valued between-satellite DD phase bias. As a result, the DD phase bias in $\boldsymbol{a}_j$ is responsible for smartphone IAR.

To analyze the stochastic characteristics of smartphone observations, we must first recover their noises. To be specific, once the DD ambiguities are correctly fixed by calibrating their phase biases, and the baseline is precisely known externally, Eq. (13.2) is written as:

$$\begin{bmatrix} \overline{\boldsymbol{\Phi}}_j \\ \overline{\boldsymbol{P}}_j \end{bmatrix} = \begin{bmatrix} \boldsymbol{e}_s & \boldsymbol{0} \\ \boldsymbol{0} & \boldsymbol{e}_s \end{bmatrix} \begin{bmatrix} \delta t_j \\ \mathrm{d}t_j \end{bmatrix} + \begin{bmatrix} \boldsymbol{\epsilon}_{\boldsymbol{\Phi}_j} \\ \boldsymbol{\epsilon}_{\boldsymbol{P}_j} \end{bmatrix} \tag{13.3}$$

where $\overline{\boldsymbol{\Phi}}_j$ and $\overline{\boldsymbol{P}}_j$ indicate the phase and code observations corrected with baseline and integer ambiguities. After single-epoch least-squares adjustment, the SD phase and code residuals are mainly affected by random noises and multipath. For smartphones with external antennas, the standard deviation (STD) of code and phase observations at frequency $j$ can be estimated by

$$\sigma_{O_j} = \sqrt{\frac{\boldsymbol{v}_{O_j}^{\mathrm{T}} \boldsymbol{v}_{O_j}}{2(s-1)}} \tag{13.4}$$

where $\boldsymbol{v}_j$ is the residual vector at frequency $j$. $\boldsymbol{O}$ stands for the code or phase observations. Note that (13.4) has a prerequisite that the baseline is precisely known. However, for smartphones with embedded antennas, the antenna phase center cannot be precisely measured and its variation is unclear. Therefore, a triple-difference in time-domain is applied to calculate the precisions of phase and code observations. The SD phase or code observations on frequency $j$ at adjacent epochs $k$, $k + 1$, $k + 2$ and $k + 3$ are denoted as $\overline{O}_{j,k}$, $\overline{O}_{j,k+1}$, $\overline{O}_{j,k+2}$ and $\overline{O}_{j,k+3}$, respectively. First, the between-epoch single-difference equations for SD observations read

$$\begin{cases} \overline{O}_{j,1} = \overline{O}_{j,k+1} - \overline{O}_{j,k} \\ \overline{O}_{j,2} = \overline{O}_{j,k+2} - \overline{O}_{j,k+1} \\ \overline{O}_{j,3} = \overline{O}_{j,k+3} - \overline{O}_{j,k+2} \end{cases} \tag{13.5}$$

Then, the between-epoch double-difference equations for SD observations read

$$\begin{cases} \overline{O}_{j,12} = \overline{O}_{j,k+2} - 2\overline{O}_{j,k+1} + \overline{O}_{j,k} \\ \overline{O}_{j,23} = \overline{O}_{j,k+3} - 2\overline{O}_{j,k+2} + \overline{O}_{j,k+1} \end{cases} \tag{13.6}$$

Finally, the between-epoch triple-difference equation for SD observations is formulated as [13]

$$\dddot{\overline{O}}_{j,k} = \overline{O}_{j,k+3} - 3\overline{O}_{j,k+2} + 3\overline{O}_{j,k+1} - \overline{O}_{j,k} \tag{13.7}$$

where $\dddot{(\cdot)}$ denotes the between-epoch triple-difference operator. In case of a short time duration (e.g., several minutes) where the satellite elevations are hardly changed, it is adequate to assume that the observation STDs are constant for each satellite. Let the STD of undifference code or phase observation as $\sigma_{O_j}$, it follows by using error propagation law in case of ignoring the time-correlations as

$$\sigma^2_{\dddot{\overline{O}}_{j,k}} = 2\sigma^2_{o_j} + 9 \times 2\sigma^2_{o_j} + 9 \times 2\sigma^2_{o_j} + 2\sigma^2_{o_j} = 40\sigma^2_{o_j} \tag{13.8}$$

By using the observations of total $K$ triple-difference epochs, the STD of undifference code or phase observation is estimated as

$$\sigma_{O_j} = \sqrt{\frac{\sum_{k=1}^{K} \dddot{\overline{O}}_{j,k}^{\mathsf{T}} \dddot{\overline{O}}_{j,k}}{40K}} \tag{13.9}$$

For more details, one can also refer to [14].

## 13.3   Integer Properties of Phase Ambiguities

In this section, we investigate the effects of smartphone brands and operating systems on ambiguity fixation. To suppress the multipath effects, the embedded antenna is replaced by an external geodetic-grade antenna. The ambiguity property of Mate20 is analyzed, and for Mi8 one can refer to [9]. Note that the operating system of Mate20 used in this study is EMUI 9.0.1. For the ultra-short baselines, the baseline-corrected DD phase observations can fully reflect the receiver-inherent phase offsets and variations besides the multipath and random noises. To be specific, the DD ambiguities are estimated epoch by epoch for the smartphone observations with an external antenna, and the fractional parts of those DD ambiguities are separated through a rounding operation. Since the reference station is a geodetic receiver without any phase offsets, the offsets of DD ambiguities are attributed to the smartphone observations.

### 13.3.1 Data Description

The static datasets were collected on the rooftop of a building at Tongji campus, and were employed to elucidate the impacts of smartphone brands, operating systems and attitudes on IAR. Two smartphones, Mate20 and Mi8, were placed next to each other in upward, horizontal and downward attitudes, respectively. Away from them by approximately 1 m, another Mate20 was located with an external SinoGNSS AT340 geodetic antenna powered by the Mate20 smartphone via a splitter. Through two outlets of the splitter, the L1 and L5 signals are transmitted to each of their own feeding points in the embedded antenna. In addition, two types of GNSS receivers, Trimble Alloy and u-blox ZED-F9T, are used, of which the u-blox ZED-F9T is a representative of the low-cost receivers. The u-blox ZED-F9T receiver tracks the L1 and L2 signals of GPS and the B1I and B2I signals of BeiDou Navigation Satellite Systems (BDS), while the smartphones track L1 and L5 signals of GPS and B1I signals of BDS. Thus in analysis of data quality, only GPS L1 signals and BDS B1I signals of u-blox are used for comparison; But in analysis of IAR, the dual-frequency GPS and BDS signals of u-blox are used. The detailed information of smartphones and receivers is presented in Table 13.1. The observation duration of static datasets for each attitude is in Table 13.2.

In the following, we define a combination set (denoted by U-T) that includes L1, B1I and E1 observations of u-blox, and L5 and E5a observations of Trimble, to comprehensively compare with dual-frequency smartphone signals.

The data quality of Mi8 observations is analyzed for comparison with Mate20. The GNSS chip of Huawei Kirin980 is embedded in Mate20 while the Broadcom

**Table 13.1** The information of data collecting devices

| Device | Antenna | Systems and frequencies |
| --- | --- | --- |
| Huawei Mate20 | Embedded | G:L1/L5; E:E1/E5a; C:B1I; J:L1/L5 |
| Huawei Mate20 | AT340 | G:L1/L5; E:E1/E5a; C:B1I; J:L1/L5 |
| Mi8 | Embedded | G:L1/L5; E:E1/E5a; C:B1I; J:L1/L5 |
| u-blox ZED-F9T | AT340 | G:L1/L2; E:E1; C:B1I/B2I; J:L1 |
| Trimble Alloy | TRM59800.00 | G:L1/L5; E:E1/E5a; C:B1I; J:L1/L5 |

**Table 13.2** Details of static datasets where the Trimble alloy is used as a reference for all devices

| Device | Baseline length (m) | Attitude | Duration (UTC time) (min) |
| --- | --- | --- | --- |
| Huawei Mate20 | 27.2 | Upward | 23 min (7:07–7:30, May 27, 2019) |
| | | Horizontal | 21 min (7:39–8:00, May 27, 2019) |
| | | Downward | 39 min (8:40–9:19, May 27, 2019) |
| Mi8 | 26.4 | Upward | 23 min (3:40–4:03, Oct. 3, 2020) |
| | | Horizontal | 21 min (4:09–4:30, Oct. 3, 2020) |
| | | Downward | 20 min (4:45–5:05, Oct. 3, 2020) |
| Huawei Mate20 | 24.1 | External | 48 min (0:16–1:04, Sept. 26, 2020) |
| u-blox ZED-F9T | 24.1 | External | 50 min (0:00–0:50, July 6, 2020) |

BCM47755 in Mi8. The smartphones are updated with the Android P operating system to provide observations and navigation messages of GPS, Galileo, QZSS and BDS. We developed an Android app Tongji GNSS RINEX Logger (TJGRL) to extract the observations with a sampling interval of 1 s through an application programming interface (API) provided by Android developers. It has been extensively tested by cooperating with Huawei Company and is freely available to the third parties upon required for academic usage at this stage. TJGRL can store the data in both RINEX 3.04 format and raw log format. The indicator of cycle slip pertained to the phase observation is set to 1 when the duty cycle occurs.

### 13.3.2   Temporal Properties of Ambiguity Fractions

We computed the time-series and histograms for the fractions of baseline-corrected DD phase observations between Alloy receiver and Mate20 with operating system EMUI 9.0.1. The fractions have constant offsets for all satellite systems, and the offsets differ from satellite systems and frequency bands. For example, the offset is about 0.5 cycles for L1/L5 signals of GPS and QZSS, while $-$ 0.5 cycles for B1/E1 signals of BDS and Galileo. With stable ambiguity offsets of Mate20, the IAR is expected if these offsets are corrected. Considering the result from [9] that the ambiguity fractions of Mate20 with EMUI 9.0 vary dramatically over time such that the ambiguities cannot be fixed, we conclude that the operating system is responsible for the time stability of ambiguity fractions. From this point of view, we can say that Huawei has solved the variations of ambiguity fractions for their smartphone GNSS chipsets in the operating system EMUI 9.0.1. In addition, [9] showed that ambiguity fixation is expected for Mi8, which further confirms that the integer nature of ambiguity depends highly on the operating systems and smartphone brands.

To show the efficiency of offset corrections, we first correct the DD observations with their corresponding offsets. We calculated the mean offsets for each frequency of each system over the entire observation period using the single-epoch offset estimates. The estimated mean offsets are as follows: 0.5 cycles for GPS L1/L5 and QZSS L1, − 0.5 cycles for BDS B1 and Galileo E1, 0.7 cycles for QZSS L5, and − 0.4 cycles, for Galileo E5a, respectively. By using these offsets, we correct the ambiguity fractions. After, the offset-corrected fractions are of zero-mean with magnitudes of ± 0.3 cycles for all frequencies. For the Mate20 observations with an embedded antenna, we also apply their ambiguities by using the estimated offsets. The results indicate that the ambiguity fractions with embedded antenna are of zero-mean for all frequencies but their magnitudes are larger than those with the external antenna.

### 13.3.3  *Offset-Calibrated Ambiguity Resolution*

After pre-calibrating the DD phase observations of Mate20 by using the mean offsets, in this section we investigate the IAR performance of different smartphone brands with embedded antennas. To make a comparison, the IAR performance of the smartphone with an external antenna and the survey-grade receiver are examined. Since the phase center of an embedded antenna cannot be precisely measured, we use the antenna reference point (ARP) as a truth benchmark to gauge their relative positions.

Multi-frequency multi-system real-time kinematic (MRTK) software developed by the GNSS group in Tongji University is used for IAR, which is able to process the data of each GNSS system or their combinations with the sequential least squares and extended Kalman filter (EKF) algorithms. In this study, we employ the EKF algorithm, where an elevation-dependent weighting function is applied [15]. The float ambiguities are continuously estimated and they are tried to be fixed at each epoch by using the partial ambiguity resolution (PAR) strategy [16] where the ambiguities with tracking duration of shorter than 30 epochs are excluded for fixing. Furthermore, an ambiguity-fixed epoch is obtained only when at least three ambiguities are successfully fixed [17] and the ratio is larger than the threshold of 3.0 [18]. Once the ambiguity-fixed epoch is reached, the time-to-first-fix (TTFF) is obtained. The fixing rate is defined as the proportion of the number of ambiguity-fixed epochs relative to the number of total epochs.

Figure 13.1 and Table 13.3 show the ambiguity fixing rate and positioning results of static datasets for Mate20 and Mi8 with embedded antennas, where root mean square (RMS) stands for the root mean square accuracy. Besides the positioning errors, the cumulative distribution function (CDF) of 3D positioning errors is illustrated as well. To compare, the results of u-blox ZED-F9T and Mate20 with external antenna are shown in Fig. 13.2 and Table 13.4. The ambiguity fixing rate of Mate20 with embedded antenna is 98.6%, which is higher than that of Mi8 by 9.2%. While if the external antenna is applied, the fixing rate can be further improved to 99.7% and the TTFF is shortened from 40 to 35 epochs. Regarding positioning results, the 3D positioning errors in the confidence of 95% are 3 cm and 5 cm for Mate20 and Mi8 with an embedded antenna, respectively. The results of u-blox are better. Its fixing rate reaches 100% with the TTFF of 30 epochs and the 3D errors of 2 cm in the confidence of 95%. From these results, the external antenna is an important factor to improve the IAR of smartphone.

The average number of satellites presented in Tables 13.3 and 13.4 indicates that the smartphone with an external antenna can track more satellites, which gives a quick understanding why the external antennas can obtain better results than embedded antennas, respectively. In fact, the high quality of phase observations with external antenna is even more important for better IAR and positioning. The phase residuals of Mate20 are smaller than those of Mi8, thus the ambiguity fixing rate is higher for Mate20. However, the phase residuals of Mate20 with external antenna are larger than those of u-blox, the ambiguity fixing rate is lower for Mate20. Therefore, we concluded that the external antenna affects the data quality and then the IAR.

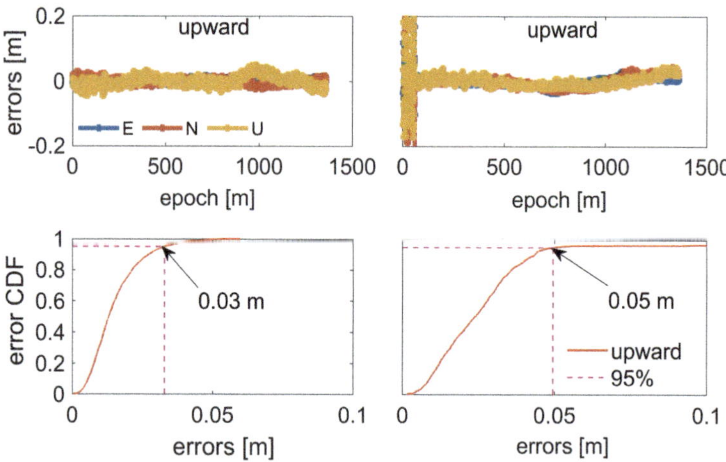

**Fig. 13.1** Positioning errors and their 3D CDFs for Mate20 (left) and Mi8 (right) with embedded antennas

**Table 13.3** Positioning and IAR statistics of Mate20 and Mi8 with embedded antennas in upward attitude

| Devices | | E | N | U |
|---|---|---|---|---|
| Mate20 | RMS (cm) | 2.3 | 2.6 | 3.6 |
| | Fix rate (%) | 98.6 | | |
| | TTFF (s) | 40 | | |
| | Average number of satellites | 18.7 [G:6.7; C:7.8; E:2.4; J:1.8] | | |
| Mi8 | RMS (cm) | 3.2 | 3.4 | 4.8 |
| | Fix rate (%) | 89.4 | | |
| | TTFF (s) | 56 | | |
| | Average number of satellites | 14.3 [G:5.6; C:4.6; E:2.3; J:1.8] | | |

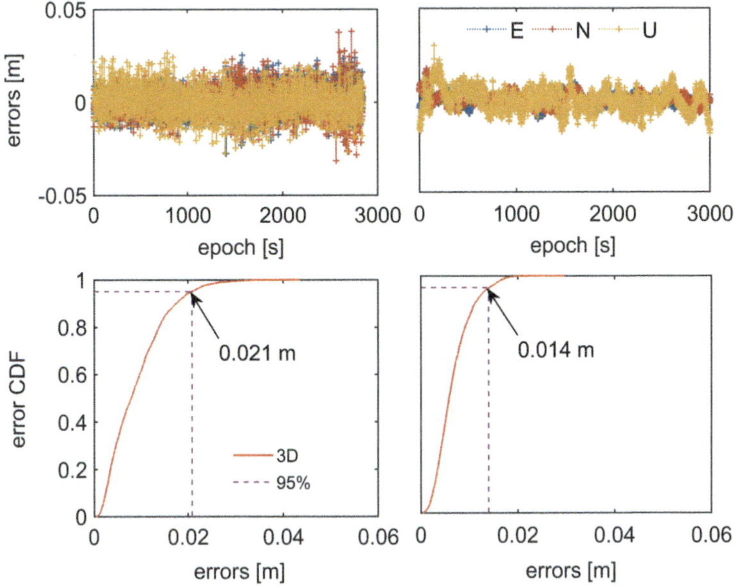

**Fig. 13.2** Positioning errors and their CDFs for Mate20 (left) and u-blox ZED-F9T (right) with external antennas

**Table 13.4** Positioning and IAR statistics of Mate20 and u-blox ZED-F9T with external antennas

|  | Mate20 | | | u-blox | | |
|---|---|---|---|---|---|---|
|  | E | N | U | E | N | U |
| RMS [cm] | 0.6 | 0.6 | 0.7 | 0.3 | 0.3 | 0.5 |
| Fix rate [%] | 99.7 | | | 100 | | |
| TTFF [s] | 35 | | | 30 | | |
| Average number of satellites | 24.5 [G:8.3; C:10.9; E:3.3; J:2.0] | | | 26.2 [G:9.9; C:6.5; E:6.8; J:3.0] | | |

## 13.4  Data Quality and Its Effects Under Different Situations

In this section, we examine the effects of the smartphone attitude on the observation noises and then on the IAR. Three indicators are defined to reflect the data quality at the different attitudes: (1) the data availability and data gap rate; (2) the relationship between C/N0 values and the satellite elevations; (3) the code and phase precisions with embedded and external antennas.

### 13.4.1   Data Availability and Data Gap Rates

The data availability rate (DAR) is defined as the proportion of number of real tracking satellites (NRTS), $n_a$, to the number of theoretical tracking satellites, $n_t$, at a given epoch, i.e., DAR $= n_a/n_t$, where the theoretical tracking satellites are defined as the satellites with elevations calculated by broadcast ephemeris higher than $0°$. DAR can overall reflect the signal acquisition ability of smartphone chips.

The DAR results of Mate20 and Mi8 are computed. In general, with an embedded antenna, the DAR of Mate20 is larger than Mi8, but both smaller than using an external antenna. The DAR of U-T is larger than Mate20 with an embedded antenna, but they are comparable if the external antenna is applied. Regarding Mate20 with an embedded antenna, when the antenna faces upwards or downwards, the DAR of L1/E1 is 92.5% on average, larger than that of L5/E5a, while when the antenna is horizontal, the DAR of L1/E1 is 87.3% on average. For Mi8 with an embedded antenna, the DAR of L1/E1 is 81.8% on average, larger than that of L5/E5a at different attitudes. It can be seen that the DAR gets minimal when the embedded antenna of Mi8 is placed downward and when the embedded antenna of Mate20 is horizontal. In addition, with an embedded antenna, the difference of DAR between different attitudes is smaller for Mate20 than for Mi8. An explanation for this phenomenon can be found in some studies [19]. Reference [20, 21] demonstrated the discrepancy between the antenna phase centers of Mi8 and Mate20. The antenna phase center of Mate20 is closer to the geometric center than Mi8, thus the data quality of Mate20 seems less attitude-dependent. This is in agreement with our results.

In summary, although the embedded antenna is omnidirectional, the number of tracking satellites varies dramatically with the antenna attitudes. The upward attitude is generally conducive to the observation reception.

For precise positioning, the continuous phase observations are rather important. Once a new ambiguity is introduced, it often needs a certain period of continuous phase observations to make its float solution converge. If the frequent interruptions occur, they will badly hamper or even be useless to the success of IAR. Therefore, to address the quality of phase data related to this issue, we define another indicator of data gap rate (DGR). As shown in Fig. 13.3, given a threshold $e_s$ (i.e., minimum continuous tracking epochs), for instance, $e_s = 30$ s, if the number of continuous epochs, $e_d$, for a satellite that is smaller than this threshold, the phase observations of these $e_d$ epochs are considered useless for ambiguity resolution and they are taken as an interruption. Then the DGR is defined as the ratio of the number of interrupted epochs $e_d$ to the total number of epochs $e_t$, i.e., DGR $= e_d/e_t$. The DGR gets larger for larger $e_s$. With $e_s = 30$ s, the DGRs of dual-frequency observations are 19.1, 23.7, and 35.7% on average for Mate20 with an embedded antenna in upward, horizontal, and downward directions, while they are 34.1, 46.1 and 35.9% for Mi8 with an embedded antenna. The DGR is maximum when the embedded antenna of Mi8 is placed downward and when the embedded antenna of Mate20 is horizontal. This implies that the tracking ability of phase observations varies between smartphones with different attitudes. However, with an external antenna, the DGRs of Mate20

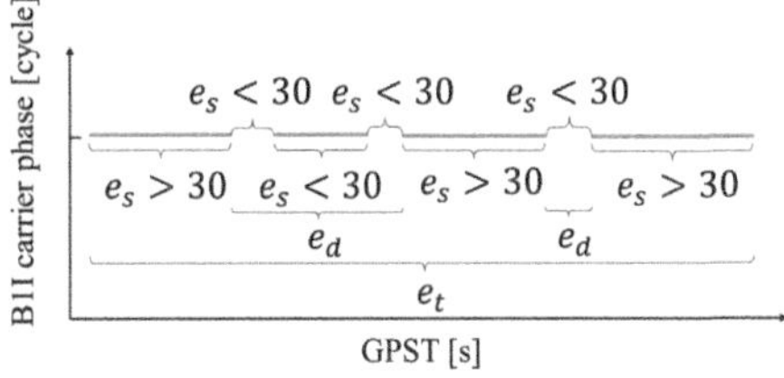



**Fig. 13.3** Graphical illustration of DGR definition with BI1 phase observations of C01 satellite as an example. The green line denotes the availability of phase observations

dual-frequency observations are reduced to 7.4, 7.4, 7.6 and 7.6% for $e_s = 30$, 60, 300 and 600 s, while they are nearly zero for U-T. From the above results we can conclude that, in addition to the effect of the smartphone antenna, the attitude of the smartphone does affect the continuity of the phase observations.

### 13.4.2   Elevation-Dependent C/N0 Values

The C/N0 is defined as the ratio of the signal power to the noise power in 1-Hz bandwidth, which reflects the quality of received signals from the energy aspect. In this section, we calculate the mean C/N0 values in an interval of 5° for each frequency observation with different attitudes. The results showed that the positive elevation-dependence is apparent for U-T receivers. For smartphones with either embedded or external antenna, the dependence is not clear and some fluctuations exist. Moreover, the C/N0 values of Mate20 with an external antenna are about 7 dB larger than those with an embedded antenna, and close to the U-T values. The reason is that the linearly polarized GNSS antennas employed in smartphones cannot compensate for the 3 dB signal power loss caused by polarization mismatch. In addition, we found that for the Mate20 and Mi8 with embedded antennas, the effects of attitude variations on C/N0 are up to 4dB and 11dB, respectively, in all GNSS systems. The observations with lower C/N0 may be outliers, which affects the IAR and positioning. Therefore, the effect of smartphone attitude on C/N0 values and thereby on the ambiguity resolution must be carefully considered in the actual data processing.

### 13.4.3   Observation Precisions

For test devices with external antennas, such as u-blox, Trimble Alloy and Mate20, the precisions of GNSS observations were evaluated based on the ultra-short base-lines with precisely known baseline coordinates, as shown in (13.4). Moreover, we quantified the precisions of smartphone observations with embedded antennas using a triple-difference method in the time-domain, as shown in (13.9). Note that the time-independent assumption is applied in (13.9). This is because this correlation can only cause a limited impact on the observation precision (1–2 mm) for all conditions in our experiments. Such small difference can hardly affect the cm-level positioning accuracy, and therefore the assumption of time-independent observations employed in (13.9) is acceptable.

The code and phase STDs of Mate20 and Mi8 are computed. The STDs of embedded antennas are generally similar for the different attitudes. When the external antenna is used, due to its high gain and low noise, the code STDs become smaller. For the B1/E1 signals of two smartphones, the code and phase STDs of BDS and Galileo satellites are smaller than those of GPS and QZSS satellites. However, the code and phase STDs of all frequencies in smartphones are larger than those of U-T. For the L5/E5a signals of two smartphones, the code STDs of all systems are significantly smaller than those of the L1/E1 signals. This shows that the L5/E5a signals have advantages in signal modulation and better anti-multipath ability in different scenes. To verify this point, the probabilities of the code outliers as a function of C/N0 values are calculated for the different smartphone attitudes. Here the code outlier is defined for the observation with its residual being three times larger than its STD. It is well known that the larger C/N0 value has generally a smaller noise influence. The result showed that the probability of code outliers decreases with the increase of C/N0 values. When the C/N0 value is larger than 30 dB-Hz, the probability of L1/E1 code outliers seems slightly higher than that of L5/E5a signals. It means that L5/E5a signals have the better anti-multipath capability in a smartphone. The code outliers occur more frequently when the observations have C/N0 values smaller than 30 dB-Hz for two smartphones, which gives experience in real data processing for setting the minimum C/N0 threshold in the actual data processing.

### *13.4.4  Ambiguity Resolution Under Different Attitudes*

The previous results show that the data quality differs from the smartphone attitudes, in this section, we will further study the effects of attitudes on the IAR. In terms of the afore-analysis, the code observations with C/N0-values lower than 30 dB-Hz would be outliers with high probability. Thus the observations only with C/N0 values larger than 30 dB-Hz are used. In addition, considering the effect of DGR on ambiguity fixation, the ambiguities with a tracking time shorter than 30 epochs will not be fixed in the data processing. Finally, the offsets obtained in Sect. 13.4 will be used for ambiguity fraction calibration for Mate20.

Table 13.5 shows the positioning results of statistic datasets for Mate20 and Mi8 in three antenna attitudes (i.e., upward, horizontal and downward). In general, the positioning performance of Mate20 is overall better than Mi8. The 3D errors of Mate20 are all smaller than 10 cm by 95% for three antenna attitudes, and even smaller than 5 cm for upward and downward attitudes. For Mi8, the 3D errors are in centimeters only for upward attitude, and reach 0.2 m and 0.4 m for downward and horizontal attitudes, respectively. The ambiguity fixing rates of Mate20 are larger than those of Mi8 with much shorter TTFF for all antenna attitudes. Moreover, the results of upward attitude are best with the highest accuracies, largest fix-rates and

**Table 13.5**  Positioning statistics of Mate20 and Mi8 with different antenna attitudes

| | | Upward | | | Horizontal | | | Downward | | |
|---|---|---|---|---|---|---|---|---|---|---|
| | | E | N | U | E | N | U | E | N | U |
| Mate20 | RMS (cm) | 2.3 | 2.6 | 3.6 | 5.3 | 5.6 | 6.1 | 4.3 | 4.6 | 5.5 |
| | Fix rate (%) | 98.6 | | | 80.2 | | | 81.1 | | |
| | TTFF (s) | 40 | | | 57 | | | 160 | | |
| Mi8 | RMS (cm) | 3.2 | 3.4 | 4.8 | 5.3 | 8.1 | 11.5 | 6.2 | 7.4 | 9.3 |
| | Fix rate (%) | 89.4 | | | 75.9 | | | 78.4 | | |
| | TTFF (s) | 56 | | | 102 | | | 213 | | |

shortest TTFF for both Mate20 and Mi8. Then the results of downward attitude are better than those of horizontal attitude. Therefore, the antenna attitude is indeed an important factor for smartphone positioning with an embedded antenna.

### *13.4.5  Kinematic Positioning*

The IAR and positioning have been investigated for smartphones with embedded antennas under different attitudes by using static data. However, most real smartphone positioning applications are in kinematic situations. In this section, we analyze the IAR and positioning of smartphones at upward attitude in two real kinematic experiments, aiming to provide the reference of quantitative accuracy for mass-market users. Table 13.6 summarizes the error characteristics and corresponding processing strategies for smartphones in real data processing.

Two kinematic datasets were collected on the playground of Tongji campus (denoted by Kin#1) and on the highway of Shanghai city (denoted by Kin#2). Note that for two kinematic experiments, the embedded antennas of the Mate20 and Mi8 face upwards. In Kin#1 dataset, all smartphones with embedded antennas were equipped on a kart and a Trimble receiver is used for comparison purposes. In Kin#2, two Mate20 smartphones were placed inside the windshield. A splitter was applied to separate the radio frequency signals from an external SinoGNSS AT340 antenna into a SinoGNSST30 receiver and one of two smartphones. In other words, one smartphone used an external geodetic antenna sharing with a geodetic receiver, while the other used its own embedded antenna. Kin#1 suffers from the

**Table 13.6**  The error characteristics and corresponding processing strategies for smartphones

| Device | C/N0 | DGR | Ambiguity fractions | Attitudes | Operating system |
|---|---|---|---|---|---|
| Mate20 | Larger than 30 dB-Hz are used | Longer than 30 epochs are used | Pre-calibrating | Upward | EMUI 9.0.1 |
| Mi8 | | | / | Upward | EMUI 9.0 |

semi-shielding surroundings by trees and buildings, while Kin#2 includes the open sky and semi-shielding surroundings. To identify the motion complexity, the varying velocities are included with the maximum speed of 7 km/h and 85 km/h for two experiments, respectively. In the following analysis, the epochs with ambiguity-fixed solutions from Trimble in Kin#1 and SinoGNSST30 receivers in Kin#2 are used as the true values to evaluate the solutions of smartphones.

For Kin#1, the ambiguity fixing rate and positioning statistics of Mate20 and Mi8 are shown in Fig. 13.4 and Table 13.7, respectively. The RMS accuracies of Mate20 and Mi8 are all smaller than 5 cm in three directions, and their ambiguity fixing rates of them are all above 90%. It means that with Mate20 and Mi8 smartphones the centimeter-level location-based services are achievable in such an environment. The TTFF of Mate20 is more than 2 times shorter than Mi8. The CDF results of positioning errors indicate smaller errors for Mate20 compared to Mi8. In conclusion, centimeter-accuracy positioning can be achieved in a semi-shaded environment using a smartphone with an embedded antenna placed upwards.

For Kin#2, the positioning errors and statistics of Mate20 with embedded or external antennas are shown in Fig. 13.5 and Table 13.8, respectively. Compared to Kin#1, the C/N0 values with embedded antenna are 7.2 dB-Hz lower for L1/B1/E1 signals and 2.3 dB-Hz lower for L5/E5a signals in Kin#2 due to the effect of the car front windshield. When the external antenna is used, the C/N0-values are significantly improved. In such a high-dynamic and obstruction environment, the ambiguity fixing rates are reduced by 62.1% for embedded antenna and still by 22.6%

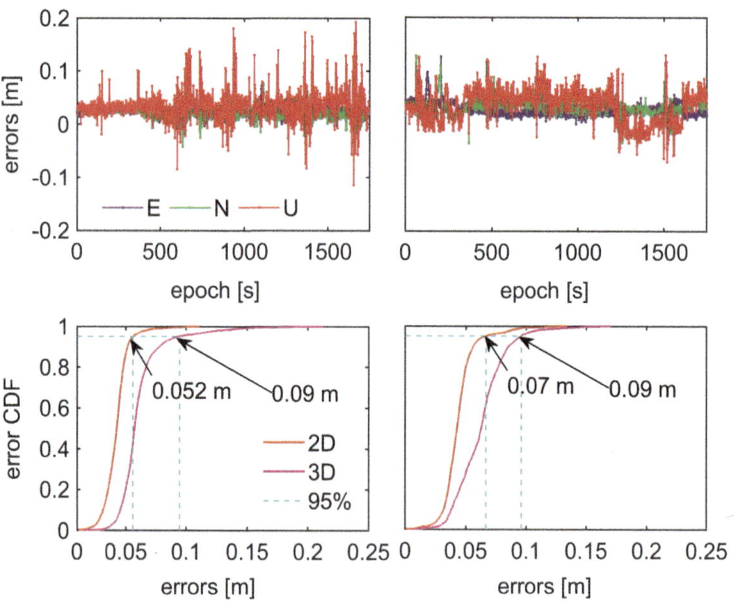

**Fig. 13.4** Positioning errors and their CDFs of 2D and 3D errors for Mate20 (left) and Mi8 (right) in Kin#1

**Table 13.7** Positioning statistics for Mate20 and Mi8 both with embedded antennas in Kin#1

| | Mate20 | | | Mi8 | | |
|---|---|---|---|---|---|---|
| | E | N | U | E | N | U |
| RMS (cm) | 2.5 | 2.6 | 4.4 | 3.1 | 3.5 | 4.7 |
| Fix rate (%) | 98.3 | | | 90.6 | | |
| TTFF (s) | 33 | | | 10 | | |

even for external antenna. In general, only the meter-level accuracy can be obtained in such a complicated city environment with an embedded antenna. The horizontal 2D errors are about 1.3 m and 3D errors 2.4 m in a percentage of 95%. However, once the external antenna is used, the accuracies of each coordinate component are improved to as high as centimeters, and the 3D and horizontal errors are about 20 cm and 10 cm by 95%, respectively. Such accurate positioning is very promising, allowing the variety of high-precision location-based services in the city environment, for instance, the vehicular-lane accurate positioning for intelligent transportation.

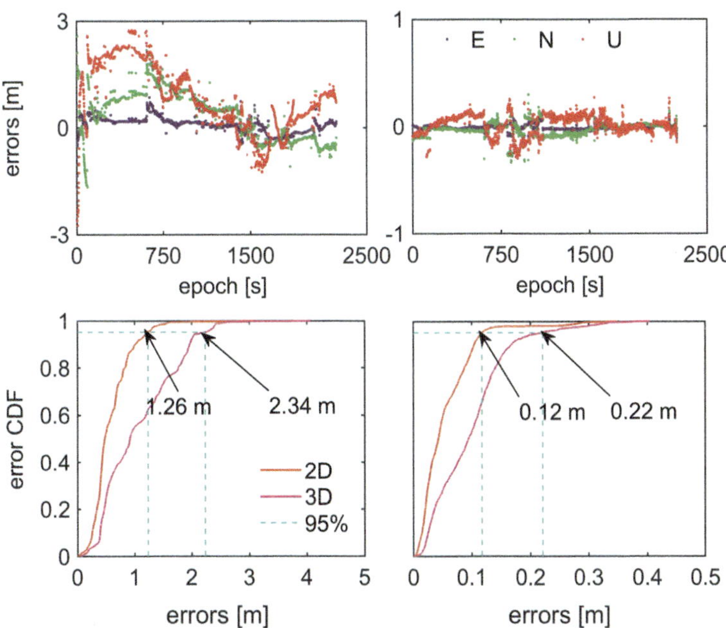


**Fig. 13.5** Positioning errors and their CDFs for Mate20 with embedded (left) or external (right) antenna in Kin#2

**Table 13.8** Positioning statistics of Mate20 with embedded and external antennas in Kin#2

| | Embedded antenna | | | External antenna | | |
|---|---|---|---|---|---|---|
| | E | N | U | E | N | U |
| RMS (m) | 0.36 | 0.79 | 0.99 | 0.03 | 0.07 | 0.09 |
| Fix rate (%) | 36.2 | | | 75.7 | | |
| TTFF (s) | 111 | | | 30 | | |

## 13.5   Conclusion

This contribution investigated three factors hindering smartphone IAR, including the smartphone brands, operating systems and smartphone attitudes. The success of IAR and positioning capability were assessed by using static and kinematic datasets. During the whole analysis, the geodetic-grade antenna was used to evaluate the impacts brought by the smartphone antennas. The research findings are summarized as follows.

The embedded antenna of smartphone is an important factor affecting the data quality. The data gap rates of Mate20 are larger than 20% and can be reduced to about 7% once the external antenna is applied. The C/N0 values are about 35 dB-Hz and smaller by 7 dB-Hz than the external antenna.

The antenna attitude also affects the data quality and ambiguity fixing rate. The upward attitude for both Mate20 and Mi8 achieves the best data quality with the smallest data gaps and largest data availability and then the highest ambiguity fixing rate.

The integer properties of phase ambiguities are related not only to smartphone brands but also to mobile operating systems. The phase ambiguities of Mate20 under Android 9.0.1 can be successfully fixed once the frequency-related constant offsets are properly calibrated. The fixing rate exceeds 90% in static scenarios and is higher than that of Mi8.

For a static dataset with an open-sky environment, the centimeter-accurate positioning solutions are achievable with 3D positioning errors smaller than 10 cm by 95%; while for city-environment with complicated obstructions, only the meter-level accuracy is obtained, which however can be significantly improved to centimeter to decimeter-level with positioning errors are smaller than 0.22 m by 95% if an external antenna is employed instead of embedded antenna. Such results are promising to satisfy with a lot of location-based services, such as the vehicular-lane accurate positioning for intelligent transportation.

## References


1. European GNSS Agency (2017) Using GNSS raw measurements on Android devices, Technical report

2. Banville S, Van Diggelen F (2016) Precision GNSS for everyone: precise positioning using raw GPS measurements from Android smartphones. GPS World 27:43–48
3. Shin D, Lim C, Park B, Yun Y, Kim E, Kee C (2017) Single-frequency divergence-free hatch filter for the Android N GNSS raw measurements. Proceedings of ION GNSS+ 2017, pp 188–225
4. Realini E, Caldera S, Pertusini L, Sampietro D (2017) Precise GNSS positioning using smart devices. Sensors 17:2434
5. Gill M, Bisnath S, Aggrey J, Seepersad G (2017) Precise point positioning (PPP) using low-cost and ultra-low-cost GNSS receivers. Proceedings of ION GNSS+ 2017, pp 226–236
6. European GNSS Agency (2019) PPP-RTK market and technology report, Technical report
7. Håkansson (2019) Characterization of GNSS observations from a Nexus 9 Android tablet. GPS Solut 23:21
8. Wanninger L, Heßelbarth A (2020) GNSS code and carrier phase observations of a Huawei P30 smartphone: quality assessment and centimeter-accurate positioning. GPS Solut 24:64
9. Geng J, Li G (2019) On the feasibility of resolving Android GNSS carrier-phase ambiguities. J Geod 93:2621–2635
10. Humphreys TE, Murrian M, van Diggelen F, Podshivalov S, Pesyna KM (2016) On the feasibility of cm-accurate positioning via a smartphone's antenna and GNSS chip. Proceedings of the IEEE/ION PLANS, pp 232–242
11. Zhang Y, Yao Y, Yu J, Chen X, Zeng Y, He N (2013) Design of a novel quad-band circularly polarized handset antenna. Proceedings of the ISAP, pp 146–148
12. Li B (2016) Stochastic modeling of triple-frequency BeiDou signals: estimation, assessment, and impact analysis. J Geod 90:593–610
13. Li B, Shen Y, Lou L (2010) Analysis of the stochastic characteristics for medium and long baseline GPS measurements. Geom Inf Sci Wuhan Univ 35:176–180
14. Leick A, Rapoport L, Tatarnikov D (2015) GPS satellite surveying. Wiley, New York
15. Li B, Li Z, Zhang Z, Tan Y (2017) ERTK: extra-wide-lane RTK of triple-frequency GNSS signals. J Geod 91:1031–1047
16. Nardo A, Li B, Teunissen PJG (2016) Partial ambiguity resolution for ground and space-based applications in a multi-GNSS scenario: a simulation study. Adv Space Res 57:30–45
17. Teunissen PJG (1995) The least squares ambiguity decorrelation adjustment: a method for fast GPS integer estimation. J Geod 70:65–82
18. Euler HJ, Schaffrin B (1990) On a measure of the discernibility between different ambiguity solutions in the static-kinematic GPS mode. Springer, New York
19. Yong C, Odolinski R, Zaminpardaz S, Moore M, Rubinov E, Er J, Denham M (2021) Instantaneous, dual-frequency, multi-GNSS precise RTK positioning using Google Pixel 4 and Samsung Galaxy S20 smartphones for zero and short baselines. Sensors 21:8318
20. Netthonglang C, Thongtan T, Satirapod C (2019) GNSS precise positioning determinations using smartphones. Proceedings of the APCCAS, pp 401–404
21. Darugna F, Wübbena J, Ito A, Wübbena T, Wübbena G, Schmitz M (2019) RTK and PPP-RTK using smartphones: from short-baseline to long-baseline applications. Proceedings of ION GNSS 2019, pp 3932–3945

# Chapter 14
# SSR-RTK: RTK with SSR Corrections

## 14.1 Introduction

In traditional real time kinematic positioning (RTK) models, precise positioning is usually performed using observations space representation (OSR) corrections generated by the reference station or reference station network. In recent years, precise positioning models based on precise point positioning (PPP)/PPP ambiguity resolution (PPP-AR)/PPP-RTK have tended to use state space representation (SSR) corrections for positioning solutions. In fact, there is a relationship between OSR corrections and SSR corrections showed in Fig. 14.1. Let $y_{OSR}$ represent the OSR correction. Based on the functional relationship between OSR and SSR corrections, we can establish the following function between OSR and SSR: $y_{OSR} = Ax_{SSR} + e$. By employing a suitable parameter estimation criterion, SSR corrections can be derived from OSR corrections. Conversely, OSR corrections can be obtained by combining the generated SSR corrections. However, it should be noted that when converting OSR to SSR corrections, the "information" in OSR is distributed between the SSR $x_{SSR}$ and the residual $e$. In general, it is assumed that the residual $e$ is white noise and is not considered in practical data processing. Under this assumption, the corrections expressed by OSR and SSR are equivalent. However, if the residual $e$ is not white noise, the information expressed by the corrections of OSR and SSR will no longer be equivalent.

The sources of SSR corrections for precise positioning in practical applications are diverse. In recent years, the B2b signal of the BeiDou Global Navigation Satellite System (BDS-3) system has provided an SSR correction for PPP/PPP-RTK positioning.

The characteristics, capability, and applications of PPP-B2b service have since attracted great attentions in both academic and engineering fields. The PPP-B2b products are resolved by the shanghai astronomical observatory (SHAO) using observations of 7 stations in mainland China and 30 globally distributed stations for BDS-3

© The Author(s) 2025

B. Li et al., *GNSS Real-Time Kinematic Positioning*, Navigation: Science and Technology 17, https://doi.org/10.1007/978-981-96-9116-6_14

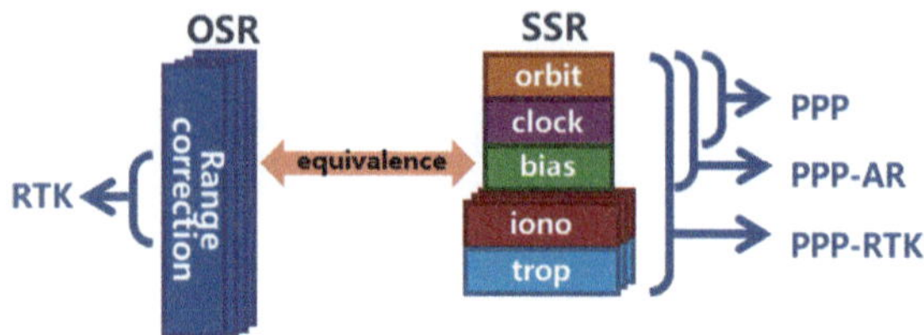


**Fig. 14.1** Relationship between different positioning methods

and Global Positioning System (GPS) respectively [1]. Limited by the station distribution and resolving strategies, the accuracy of satellite orbit and clock products provided by PPP-B2b is inferior to that of real-time service (RTS) provided by the Centre National d'Etudes Spatiales (CNES), and the PPP-B2b service is capable of offering the effective positioning assistance only to users in the Asian-Pacific region [2]. When using the PPP-B2b service, critical factors should be carefully handled, for instance, the satellite-specific bias in the PPP-B2b clock offset, outage of PPP-B2b products, mismatching problems, etc. [3]. Thus far, the innovative applications of PPP-B2b service have been extensively investigated for future requirements, including marine positioning, time transfer, etc. [4]. Meanwhile, compensation methods for existing problems in PPP-B2b products are also proposed to improve service performance [5].

Despite modifications that have been made, the PPP-B2b service is still limited in decimeter-level accuracy and long convergence time of real-time positioning [6]. Integer ambiguity resolution (IAR) is critical for precise positioning, which is, up to now, difficult to implement only with the PPP-B2b products. The primary reason is that ambiguities are contaminated by biases originating from receivers and satellites, and there is a lack of corrections for these biases. To enable IAR, the PPP is generally augmented by a network of continuously operating reference stations (CORS) [7, 8], where augmentation corrections in SSR format are provided to recover the integer property of ambiguities. Based on integer recovery strategies, the models can be categorized into the integer recovery clock (IRC) model [9], the uncalibrated phase delay (UPD) or the fractional cycle bias (FCB) model [10, 11] and the decoupled satellite clock (DSC) model [12] Accordingly, IAR-enabled precise point positioning (PPP-RTK) using PPP-B2b products is preliminarily investigated. With a sparse station network, extra atmosphere, phase bias, and satellite clock products are resolved to enhance the user positioning performance [13]. However, the network processing scheme relies heavily on a substantial number of stations and favorable communication conditions. Given that the PPP-B2b service is typically applied to specific scenarios like marine surveys, developing infrastructure for augmentation corrections is quite challenging. Instead, single-station PPP-RTK, a special case of the network-based PPP-RTK, outperforms due to its flexibility when only a few reference stations are available. Initial studies on single-station PPP-RTK are conducted based on the $S$-system theory [14, 15], demonstrating the feasibility of augmentation service on one reference station. Nonetheless, International GNSS Service (IGS) orbit/clock products are commonly used in the existing research, whereas investigations using

PPP-B2b products are insufficient. This raises the question of whether the PPP-B2b service is compatible with the single-station PPP-RTK scheme. Taking the significant differences between the IGS and PPP-B2b products into account, one needs to comprehensively understand the characteristics of PPP-B2b products as well as develop the single-station IAR-enabled PPP-B2b precise positioning (SSR-RTK) for particular applications.

To fulfill the centimeter-level real-time service, this contribution is devoted to offering a novel method of SSR-RTK, in which fast IAR is realized by PPP-B2b products and extra SSR corrections. Firstly, a full-rank version of the undifferenced and uncombined (UDUC) PPP-B2b model is formulated, considering the different reference signals of PPP-B2b clock products. The characteristics of PPP-B2b products are briefly analyzed to expound the impact on positioning. Then, SSR corrections, including satellite-specific phase biases and atmospheric corrections, are generated by using a single reference station to enhance the positioning. The single-station SSR-RTK is thus realized within the PPP-B2b service. Finally, experiments are carried out in kinematic mode to demonstrate the positioning performance with discussions on specific IAR and atmospheric augmentation methods.

The rest of the chapter is organized as follows. In Sect. 14.2, the full-rank PPP-B2b model is deduced with discussions on the characteristics of PPP-B2b products. On this basis, the single-station SSR-RTK model is presented in Sect. 14.3. Experiments are carried out in Sect. 14.4, demonstrating the PPP-B2b positioning performance based on raw observations. Finally, some conclusions are given in Sect. 14.5.

## 14.2  Full-Rank PPP-B2b Model and Product Characteristics

The PPP-B2b service currently provides satellite orbit and clock corrections for users to realize real-time PPP. Typically, two models are widely used in PPP, that is, the ionosphere-free (IF) combination model and the uncombined model [16]. The former is the most commonly used in PPP-B2b studies. Instead, this chapter prefers the latter, which retains all parameters and allows for flexible constraints to enhance the model [17]. In this section, we mainly deduce a full-rank PPP-B2b model considering the characteristics of PPP-B2b products.

### 14.2.1  PPP Model with PPP-B2b Products

The raw GNSS code and phase observation equations read

$$
\begin{aligned}
P_{j,r}^{s} &= \rho_r^s + T_r^s + \mu_j \iota_j^s + dt_r - dt^s + D_{j,r} - d_j^s + \varepsilon_{P_{j,r}^s} \\
\Phi_{j,r}^{s} &= \rho_r^s + T_r^s - \mu_j \iota_j^s + dt_r - dt^s + B_{j,r} - b_j^s - \lambda_j a_{j,r}^s + \varepsilon_{\Phi_{j,r}^s}
\end{aligned}
\tag{14.1}
$$

where the subscripts $j$, $r$, and $s$ denote the frequency, receiver, and satellite, respectively. $P_{j,r}^s$ and $\Phi_{j,r}^s$ are the GNSS code and phase observations, respectively. $\rho_r^s$ denotes the satellite-to-receiver distance. $T_r^s$ represents the slant tropospheric delay. $\iota_j^s$ represents the slant ionospheric delay and $\mu_j = f_1^2/f_j^2$ is the ionospheric factor, where $f_j$ is the value of the $j$th frequency. $dt_r$ and $dt^s$ are the clock offsets of the receiver and satellite, respectively. $D_{j,r}$ and $d_j^s$ are the receiver and satellite code hardware delays, respectively. $B_{j,r}$ and $b_j^s$ are the receiver and satellite phase hardware delays, respectively. $a_{j,r}^s$ denotes the integer ambiguity with wavelength $\lambda_j$. $\varepsilon_{P_{j,r}^s}$ and $\varepsilon_{\Phi_{j,r}^s}$ are the measurement noises of code and phase observations, respectively. In addition, the relativistic effect, tide displacement, phase windup, etc., have been corrected by corresponding models [18–20].

To deduce a full-rank PPP-B2b model, firstly, satellite products are applied according to the PPP-B2b protocol [21]. In contrast to precise clock products provided by IGS, the B3I signal is selected as the reference for BDS-3 in the PPP-B2b service, while the L1/L2 IF combination is still the reference for GPS. Considering the difference, the following equations are based on BDS-3 and GPS dual-frequency observations, and it is easy to extend to other systems or frequencies. Assuming that each system has $n$ satellites, the PPP-B2b precise clocks $dt_{\mathrm{B2b}}^s$ can be expressed as

$$dt_{\mathrm{B2b}}^s = dt^s + (\boldsymbol{\Lambda} \otimes \boldsymbol{I}_n)\boldsymbol{d}_{\mathrm{ref}}^s \tag{14.2}$$

where $\boldsymbol{dt}^s = \left[dt_C^{s\,\mathrm{T}}, dt_G^{s\,\mathrm{T}}\right]^{\mathrm{T}}$ and $\boldsymbol{dt}_{\mathrm{B2b}}^s$ are matrices composed of the original and the PPP-B2b clock products of BDS-3 and GPS satellites, respectively. $\boldsymbol{dt}_k^s = \left[dt_k^1, \ldots, dt_k^n\right]^{\mathrm{T}}$ represents the satellite clocks of the specific system $k$.
$\boldsymbol{\Lambda} = \begin{bmatrix} 1 & 0 & 0 \\ 0 & \lambda_{L2}^2/(\lambda_{L2}^2 - \lambda_{L1}^2) & -\lambda_{L1}^2/(\lambda_{L2}^2 - \lambda_{L1}^2) \end{bmatrix}$ represents the coefficient matrix.
$\boldsymbol{d}_{\mathrm{ref}}^s = \left[\boldsymbol{d}_{\mathrm{B3I}}^{s\,\mathrm{T}}, \boldsymbol{d}_{\mathrm{L1}}^{s\,\mathrm{T}}, \boldsymbol{d}_{\mathrm{L2}}^{s\,\mathrm{T}}\right]^{\mathrm{T}}$ denotes the satellite code hardware delays for BDS-3 B3I signal, GPS L1 and L2 signal with $\boldsymbol{d}_j^s = \left[d_j^1, \ldots, d_j^n\right]^{\mathrm{T}}, j = \mathrm{B3I, L1, L2}$, respectively. The symbol $\otimes$ denotes the Kronecker product. When applying PPP-B2b clock products, the corresponding code biases need to be corrected for code observations as

$$\boldsymbol{P}_I = \boldsymbol{P} - (\boldsymbol{e}_2 \otimes \boldsymbol{\Lambda} \otimes \boldsymbol{I}_n)\boldsymbol{d}_{\mathrm{ref}}^s + \boldsymbol{d}^s \tag{14.3}$$

where $\boldsymbol{P} = \left[\boldsymbol{P}_{1,C}^{s\,\mathrm{T}}, \boldsymbol{P}_{1,G}^{s\,\mathrm{T}}, \boldsymbol{P}_{2,C}^{s\,\mathrm{T}}, \boldsymbol{P}_{2,G}^{s\,\mathrm{T}}\right]^{\mathrm{T}}$ and $\boldsymbol{P}_I$ denote the dual-frequency code observation vectors before and after correcting, respectively. $\boldsymbol{P}_{j,k}^s = \left[P_{j,k}^1, \ldots, P_{j,k}^n\right]^{\mathrm{T}}$ represents code observations on frequency $j$ and system $k$. $\boldsymbol{d}^s = \left[\boldsymbol{d}_{1,C}^{s\,\mathrm{T}}, \boldsymbol{d}_{1,G}^{s\,\mathrm{T}}, \boldsymbol{d}_{2,C}^{s\,\mathrm{T}}, \boldsymbol{d}_{2,G}^{s\,\mathrm{T}}\right]^{\mathrm{T}}$ is the satellite code hardware delays of different systems and frequencies, which can be corrected by the code observable-specific signal biases (code OSBs). Substituting (14.2) and (14.3) into (14.1), they become

$$P_l = e_2 \otimes \rho + (e_2 \otimes I_2 \otimes e_n)dt_r - e_2 \otimes dt^s_{B2b}$$
$$+ (\mu \otimes I_n)\iota + e_2 \otimes T_r + (I_4 \otimes e_n)D_r + \varepsilon_{P_l}$$
$$\Phi = e_2 \otimes \rho + (e_2 \otimes I_2 \otimes e_n)dt_r - e_2 \otimes dt^s_{B2b} - (\mu \otimes I_n)\iota + e_2 \otimes T_r$$
$$- \lambda a + (I_4 \otimes e_n)B_r - b^s + (e_2 \otimes \Lambda \otimes I_n)d^s_{ref} + \varepsilon_\Phi \tag{14.4}$$

where $dt_r = \left[dt_{r,C}, dt_{r,G}\right]^{\mathrm{T}}$ denotes the receiver clock. $\rho = \left[\rho_C{}^{\mathrm{T}}, \rho_G{}^{\mathrm{T}}\right]^{\mathrm{T}}$, $\iota = \left[\iota_C{}^{\mathrm{T}}, \iota_G{}^{\mathrm{T}}\right]^{\mathrm{T}}$, $T_r = \left[T_{r,C}{}^{\mathrm{T}}, T_{r,G}{}^{\mathrm{T}}\right]^{\mathrm{T}}$, and $b^s = \left[b^s_{1,C}{}^{\mathrm{T}}, b^s_{1,G}{}^{\mathrm{T}}, b^s_{2,C}{}^{\mathrm{T}}, b^s_{2,G}{}^{\mathrm{T}}\right]^{\mathrm{T}}$ are the vectors of the satellite-specific distances, slant ionospheric delays, tropospheric delays, and phase hardware delays, respectively. $\mu = \begin{bmatrix} 1 & 0 & \mu_C & 0 \\ 0 & 1 & 0 & \mu_G \end{bmatrix}^{\mathrm{T}}$ is the coefficient matrix of slant ionospheric delays with $\mu_k = (f_{k,1}/f_{k,2})^2$. $D_r = \left[D_{1,C}, D_{1,G}, D_{2,C}, D_{2,G}\right]^{\mathrm{T}}$ and $B_r = \left[B_{1,C}, B_{1,G}, B_{2,C}, B_{2,G}\right]^{\mathrm{T}}$ are the receiver hardware delays of code and phase, respectively. $a$ is the integer ambiguity vector in the unit of cycles.

### 14.2.2  Full-Rank PPP-B2b Model

The observation equations formulated in Sect. 14.2.1 are unsolvable due to the rank-deficient problem. Therefore, we turn to $S$-system theory to construct the full-rank PPP-B2b model [22]. Although the satellite code biases are eliminated in (14.3), they are introduced into phase observations simultaneously. Considering the receiver code hardware biases will be absorbed by the receiver clock and slant ionosphere, these parameters are shown as

$$\tilde{dt}_r = dt_r + \mathbf{M}D_r \tag{14.5}$$

$$\tilde{\iota} = \iota + (K \otimes e_n)D_r \tag{14.6}$$

where $\tilde{dt}_r = \left[\tilde{dt}_{r,C}, \tilde{dt}_{r,G}\right]^{\mathrm{T}}$ is the estimated receiver clock. $\mathbf{M} = \begin{bmatrix} \alpha_C & 0 & \beta_C & 0 \\ 0 & \alpha_G & 0 & \beta_G \end{bmatrix}$ captures the coefficient matrix for system-specific receiver code biases. $\alpha_k = \lambda^2_{k,2}/(\lambda^2_{k,2} - \lambda^2_{k,1})$ and $\beta_k = -\lambda^2_{k,1}/(\lambda^2_{k,2} - \lambda^2_{k,1})$ are the factors for the dual-frequency of different systems that are distinguished by subscript $k$. $\tilde{\iota} = \left[\tilde{\iota}_C^{\mathrm{T}}, \tilde{\iota}_G^{\mathrm{T}}\right]^{\mathrm{T}}$ is the estimated slant ionospheric vector with $\tilde{\iota}_k = \left[\tilde{\iota}_k^1, \ldots, \tilde{\iota}_k^n\right]^{\mathrm{T}}$. $\mathbf{K} = \begin{bmatrix} \beta_C & 0 & -\beta_C & 0 \\ 0 & \beta_G & 0 & -\beta_G \end{bmatrix}$ denotes the coefficient matrix. Substituting (14.5) and (14.6) to the raw observation equations, the linearized equations using PPP-B2b products read

$$\tilde{P} = (e_2 \otimes A_x)x + (e_2 \otimes I_2 \otimes e_n)\tilde{dt}_r$$

$$+ (\boldsymbol{\mu} \otimes \boldsymbol{I}_n)\tilde{\boldsymbol{\iota}} + (\boldsymbol{e}_2 \otimes \boldsymbol{g})\tau + \boldsymbol{\varepsilon}_{\tilde{P}}$$

$$\tilde{\boldsymbol{\Phi}} = (\boldsymbol{e}_2 \otimes \boldsymbol{A}_x)\boldsymbol{x} + (\boldsymbol{e}_2 \otimes \boldsymbol{I}_2 \otimes \boldsymbol{e}_n)\tilde{\boldsymbol{dt}}_r$$

$$- (\boldsymbol{\mu} \otimes \boldsymbol{I}_n)\tilde{\boldsymbol{\iota}} + (\boldsymbol{e}_2 \otimes \boldsymbol{g})\tau - \lambda\tilde{\boldsymbol{a}} + \boldsymbol{\varepsilon}_{\tilde{\Phi}} \tag{14.7}$$

where $\tilde{\boldsymbol{P}}$ and $\tilde{\boldsymbol{\Phi}}$ represent observed-minus-computed code and phase observations, respectively. $\boldsymbol{A}_x$ is the design matrix for coordinate corrections $\boldsymbol{x}$. $\boldsymbol{g} = \left[\boldsymbol{g}_C{}^{\mathrm{T}}, \boldsymbol{g}_G{}^{\mathrm{T}}\right]^{\mathrm{T}}$ is the mapping function of residual zenith tropospheric delay with $\boldsymbol{g}_k = \left[g_k^1, \ldots, g_k^n\right]^{\mathrm{T}}$. $\tilde{\boldsymbol{a}}$ is the estimated ambiguity vector, which contains the original ambiguities $\boldsymbol{a}$ and bias terms. The float ambiguities $\tilde{\boldsymbol{a}}$ read

$$\tilde{\boldsymbol{a}} = \boldsymbol{a} - \left((\boldsymbol{e}_2 \otimes \boldsymbol{\Lambda} \otimes \boldsymbol{I}_n)\boldsymbol{d}_{\mathrm{ref}}^s + (\boldsymbol{\mu}\boldsymbol{K} \otimes \boldsymbol{e}_n - (\boldsymbol{e}_2 \otimes \boldsymbol{I}_2 \otimes \boldsymbol{e}_n)\mathbf{M})\boldsymbol{D}_r\right)/\lambda$$
$$- \left((\boldsymbol{I}_4 \otimes \boldsymbol{e}_n)\boldsymbol{B}_r - \boldsymbol{b}^s\right)/\lambda \tag{14.8}$$

To accurately describe the weight of code and phase observations, the stochastic model of (14.7) can be expressed as

$$\boldsymbol{Q} = \mathrm{blkdiag}\left(\left[\sigma_P^2, \sigma_\Phi^2\right]\right) \otimes \boldsymbol{I}_2 \otimes \boldsymbol{Q}_0 \tag{14.9}$$

where $\sigma_P^2$ and $\sigma_\Phi^2$ capture the precisions of code and phase at the zenith direction, respectively. $\boldsymbol{Q}_0$ is the cofactor matrix with elevation-dependent dispersions [23].

### 14.2.3   *Characteristics of PPP-B2b Products*

According to the Interface Control Document (ICD) published by the China Satellite Navigation Office [21], products of 7 types are broadcast for BDS-3 and GPS satellites, as summarized in Table 14.1. The characteristics of these products determine how we establish a proper mathematical model for PPP-B2b precise positioning. In this section, we only summarize the characteristics of 31-day PPP-B2b products during Day of Year (DOY) 214–244 in 2020, while the accuracy assessments, which have been fully studied yet, are not discussed in this chapter.

- Satellite mask

The satellite mask defines whether the corrections of one satellite are broadcast or not. In one message, there are a total of 255 bits to identify the broadcast status, in which BDS-3 occupies 63 bits while GPS, GLONASS, and Galileo each occupy 37 bits. The corresponding position in a bit will be assigned to "1" if the corrections of one satellite are broadcast. Besides, to ensure the relevance between PPP-B2b corrections, the Issue Of Data, State Space Representation (IOD SSR) and Issue Of Data, PRN mask (IODP) information are also broadcast for matching. Shown in message type 1, corrections of 59 satellites, including 27 BDS-3 satellites and 32

**Table 14.1**  PPP-B2b information

| Message content | Message type | Update interval | Nominal validity |
| --- | --- | --- | --- |
| Satellite mask | 1 | 48 s | – |
| Orbit corrections | 2, 6, 7 | 48 s | 96 s |
| DCB corrections | 3 | 48 s | 86,400 s |
| Clock corrections | 4, 6, 7 | 6 s | 12 s |
| User range index | 5 | 48 s | 96 s |
| Reserved | 8–62 | – | – |
| Null | 63 | – | – |

GPS satellites, are broadcast by PPP-B2b service. Corrections for the Galileo and GLONASS systems have not been broadcast yet.

- Orbit correction

Satellite orbit corrections in radial, along-track, and cross-track are broadcast in message type 2. When applying orbit corrections, the coordinate transformation is required because the satellite position usually refers to the Earth Centered Earth Fixed (ECEF) coordinate system. Additionally, Issue Of Data, Navigation (IODN) and IOD Cor information are also broadcast for orbit corrections to match the ephemeris and clock corrections, respectively.

The time series of PPP-B2b orbit corrections for BDS-3 and GPS satellites on August 1st, 2020 are computed. Apparently, the magnitude of orbit corrections for BDS-3 and GPS satellites is 0.2 m and 3 m respectively. It is also manifested that corrections in radial are much smaller than those in along-track and cross-track directions. Moreover, the seemingly irregular jumps existing in orbit corrections are caused by two aspects. Associated with the update rate of broadcast ephemeris CNAV1 and LNAV, large jumps occur every hour for BDS-3 satellites and every two hours for GPS satellites. Concerning the small jumps, orbit renewal in less than one hour is the dominant factor [1].

- Clock correction

The PPP-B2b clock corrections in meters along with IODP and IOD Cor information are broadcast in message type 4. The results showed that the clock corrections for BDS-3 are within 2 m whereas it is larger for GPS. The clock product for the corresponding satellite is marked invalid if the value equals − 26.2128 m. Besides obvious jumps due to ephemeris update, a small magnitude of jumps exist and may be caused by the resource information update when estimating real-time satellite clock offsets. Moreover, it should be noted that jumps in clock offsets can be absorbed by undifferenced ambiguities.

- DCB correction

The DCB in message type 3 defines the code biases of code observations between the ranging signal and the reference signal. From decoded PPP-B2b messages, DCB corrections for only BDS-3 satellites are provided whereas others are not. Since the B3I signal is taken as the reference signal, several DCBs including B1I-B3I, B1C-B3I, B2a-B3I, and B2b-B3I are available to users. Furthermore, it is noted that the DCB corrections of a signal component for one satellite remain constant during DOY 214–244.

- PPP-B2b products availability

The availability of PPP-B2b products is an essential prerequisite for precise positioning. As mentioned in [21], the PPP-B2b service mainly serves users in China and surrounding areas. The valid service scope is within latitude (48.47° S, 58.40° N) and longitude (16.36° W, 102.99° W) according to the figure.

To visually represent the availability of PPP-B2b products, the average number of available satellites is counted by day. The average number of available orbit products for BDS-3 satellites is about 10, while the same indicator for clock products is about 11. The abnormal clock corrections of C19 continuously broadcast during this period can account for this difference. The average number of GPS satellites with available orbit and clock corrections is almost the same, with a maximum of 9.90 (DOY 217) and a minimum of 9.32 (DOY 244). Furthermore, it should be noted that the interruption of PPP-B2b products may occur on certain satellites, which can influence the positioning performance and needs to be carefully handled.

## 14.3  Single-Station Augmented SSR-RTK

As for the PPP-B2b service, the real-time high-accuracy positioning can be effectively improved by ambiguity fixing, which is currently hindered by the lack of augmentation corrections. Additionally, in particular scenarios like ocean, desert, etc., there are not enough reference stations for correction generation. Facing the above two problems, we propose the SSR-RTK model augmented by SSR corrections from a single reference station. Accordingly, a three-step process of single-station SSR-RTK is defined as follows.

### 14.3.1  *Generating the SSR Corrections*

To be compatible with the PPP-B2b service, a station-based computing mode is adopted to generate augmentation corrections. With the satellite orbit, clock, and station coordinates fixed, other parameters are estimated using the full-rank PPP-B2b observation equations mentioned in Sect. 14.2. Firstly, phase bias products are derived from estimated ambiguities as [24]

$$N_{\mathrm{pb}} = \tilde{a} - \mathrm{round}(\tilde{a}) \tag{14.10}$$

**Table 14.2** SSR corrections generated at a reference station

| Correction | Notation and interpretation |
|---|---|
| Ionospheric delay | $\tilde{\iota}_B = \iota_B + \beta(D_1 - D_2)$ |
| Tropospheric delay | $\tau_B$ |
| Phase bias | $N_{pb} = \tilde{a} - \text{round}(\tilde{a})$ |
| Atmospheric factor | $Q_{\tilde{\iota}_B}, Q_{\tau_B}$ |

where $\text{round}(*)$ and $N_{\mathrm{pb}}$ indicate the round operation and fractional part of estimated ambiguities, respectively. To enable fast IAR, atmospheric corrections are considered for dynamic constraints. The estimated ionosphere $\tilde{\iota}_B$ and residual zenith tropospheric delay $\tau_B$ at the reference station can be directly used for atmospheric constraints. Meanwhile, the variances of atmospheric corrections $Q_{\tilde{\iota}_B}$ and $Q_{\tau_B}$ are introduced to define the constraint reliability, minimizing the impacts of inaccuracy or interruption on user positioning, especially in the single-station case. The notation and interpretation of SSR corrections are summarized in Table 14.2.

In terms of broadcasting extra SSR corrections, the data format and transmission rate depend on the properties of corrections, and there are several ways to broadcast them via BDS-3 short message service, the Internet, radio, etc. for different scenarios. Since it is out of the scope of this chapter, the optimal broadcast strategy will be further studied in the future.

### 14.3.2 SSR-RTK Model

Based on the full-rank PPP-B2b model, phase bias corrections are applied to users to eliminate the effects of satellite phase biases. The single-difference equations between the user and the reference station are nominally formulated as

$$
\begin{aligned}
\tilde{P}_U &= \left(e_2 \otimes A_{x_U}\right)x_U + (e_2 \otimes I_2 \otimes e_n)\tilde{dt}_{r,U} + (\mu \otimes I_n)\tilde{\iota}_U \\
&\quad + \left(e_2 \otimes g_U\right)\tau_U + \varepsilon_{\tilde{P}_U} \\
\tilde{\Phi}_U + \lambda N_{\mathrm{pb}} &= \left(e_2 \otimes A_{x_U}\right)x_U + (e_2 \otimes I_2 \otimes e_n)\tilde{dt}_{r,U} - (\mu \otimes I_n)\tilde{\iota}_U \\
&\quad + \left(e_2 \otimes g_U\right)\tau_U - \lambda\tilde{a}_{U,\mathrm{sd}} + \varepsilon_{\tilde{\Phi}_U}
\end{aligned}
\tag{14.11}
$$

where subscript $U$ denotes the user side. $\tilde{a}_{U,\mathrm{sd}} = \tilde{\iota}_U - N_{\mathrm{pb}}$ is the estimated ambiguity vector that absorbs phase bias corrections. The interpretations of other parameters are the same as (14.7). For the sake of fast ambiguity resolution, atmospheric corrections are utilized to form virtual observations. However, in the single-station case, the geospatial correlation of ionospheric and tropospheric delays can decline dramatically as the baseline length increases. To mitigate this impact, compensation is carried out by using empirical atmospheric models [15], and the constraint equations are written as

$$y_{\text{ion}} = \tilde{\imath}_B - \text{ion}_B + \text{ion}_U$$
$$= \tilde{\imath}_U + \text{bias}_{\text{ion}} + \varepsilon_{y_{\text{ion}}} \tag{14.12}$$

$$y_{\text{trop}} = \tau_B$$
$$= \tau_U + \varepsilon_{y_{\text{trop}}} \tag{14.13}$$

where subscript $B$ denotes the base station. $y_{\text{ion}}$ and $y_{\text{trop}}$ are the compensated corrections for atmospheric constraints. $\text{ion}_B$ and $\text{ion}_U$ are the ionospheric delays calculated by the Klobuchar model [25]. The additional parameter $\text{bias}_{\text{ion}} = -\beta_k(D_{1,k,U} - D_{2,k,U}) + \beta_k(D_{1,k,B} - D_{2,k,B})$ contains receiver bias terms of user side and base station, which is system-specific and distinguished by subscript $k$. Considering that the unmodelled spatial errors enlarge with baseline extension, the distance-dependent variances for constraint equations can be empirically modeled as

$$Q_{y_{\text{ion}}} = Q_{\tilde{\imath}_B} + \frac{10^3\left(\left(\text{lat}_{\text{ipp},U} - \text{lat}_{\text{ipp},B}\right)^2 + \left(\text{lon}_{\text{ipp},U} - \text{lon}_{\text{ipp},B}\right)^2\right)}{\sin(\text{elev})^2} \tag{14.14}$$

$$Q_{y_{\text{trop}}} = Q_{\tau_B} + \frac{\text{Baseline}^2}{10^{12}} \tag{14.15}$$

where $Q(*)$ indicates the variances of parameters. $\text{lat}_{\text{ipp},U}$, $\text{lon}_{\text{ipp},U}$ and $\text{lat}_{\text{ipp},B}$, $\text{lon}_{\text{ipp},B}$ are the latitude and longitude of the ionosphere pierce point (IPP) at the user side and base station, respectively. elev is the satellite elevation angle. Baseline is the distance between the user and the base station.

### 14.3.3  Ambiguity Resolution

Regarding float ambiguities in the single-station SSR-RTK model is still contaminated by receiver hardware delays, the difference between satellites is further performed to recover the ambiguity integer property. The most qualified satellite of each system is chosen as the reference satellite. Thus, the integer ambiguity is given as

$$a_{j,UB}^{sv} = a_{j,U}^{sv} - a_{j,B}^{sv} + \text{round}(\tilde{a}_{j,B}^{sv}) \tag{14.16}$$

where $a_{j,r}^{sv}$ represents the single-difference ambiguity with respect to satellite $s$ and $v$. The integer ambiguity $a_{j,UB}^{sv}$ consists of original ambiguities and is additionally biased by integer term $\text{round}(\tilde{a}_{j,B}^{sv})$ after applying phase bias products. The integer ambiguities are partially fixed by the least-squares ambiguity decorrelation adjustment (LAMBDA) method [26] and other parameters are updated at the same time. The overall flowchart of single-station SSR-RTK is shown in Fig. 14.2.

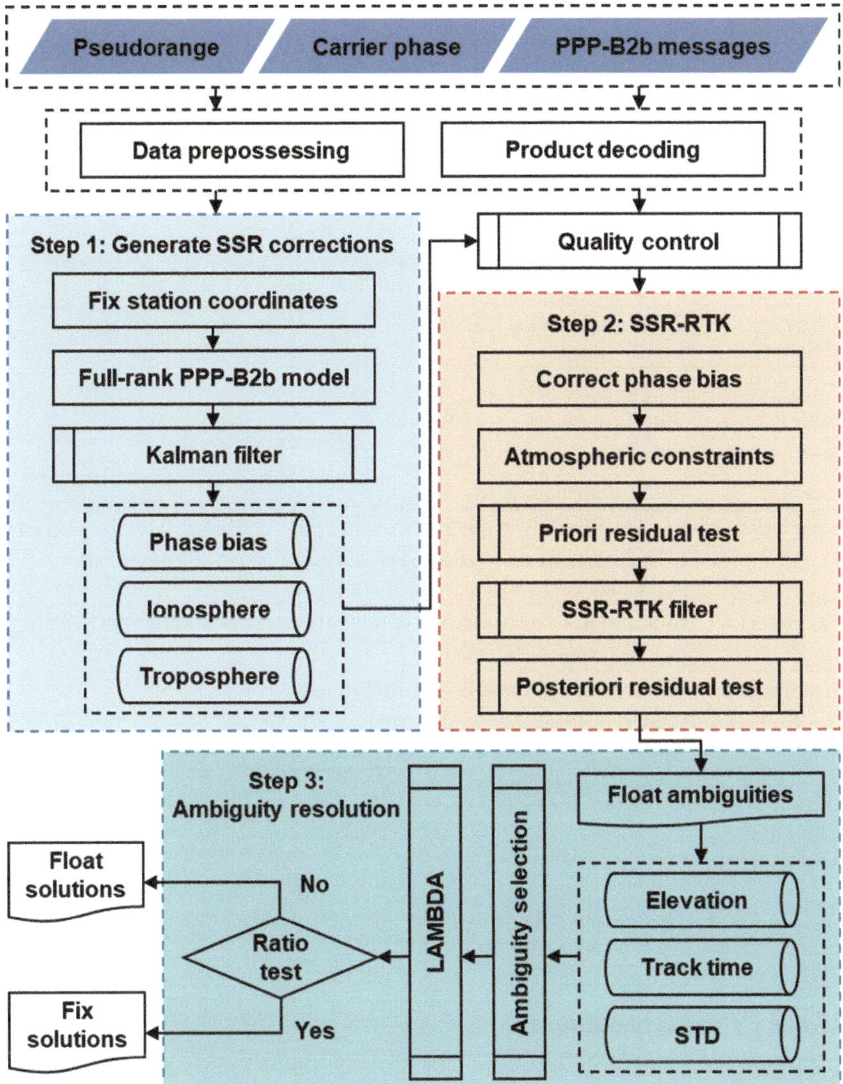



**Fig. 14.2**  Flowchart of single-station SSR-RTK

## 14.4  Experimental Analysis

Based on the proposed method in Sect. 14.3, experiments in kinematic mode are carried out with PPP-B2b service. The first one is the precise point positioning test using PPP-B2b corrections, where continuous observation data from 8 stations are used to demonstrate the positioning performance. The second is the single-station

**Table 14.3** Processing strategies with PPP-B2b service

| Items | Strategies |
|---|---|
| Systems | BDS-3 and GPS |
| Observations | BDS-3: B1I & B3I raw observations<br>GPS: L1 & L2 raw observations |
| Cut-off elevation | $10°$ |
| Satellite orbit and clock | Derived from CNAV1 and LNAV and corrected by PPP-B2b products during DOY 214–244, 2020 |
| Observation weighting | Elevation-dependent [23] |
| PCO/PCV | Corrected with igs14.atx |
| Solid tide | IERS Conventions 2010 [19] |
| Ocean loading | IERS Conventions 2010 [19] |
| Pole tide | IERS Conventions 2010 [19] |
| Coordinates | Estimated in a way of epoch-wise, with a prior value obtained from Standard Point Positioning (SPP) and the epoch noise of $60\ m/\sqrt{s}$ for kinematic positioning |
| Receiver clock | Estimated in a way of epoch-wise, with a prior value obtained from SPP and the epoch noise of $60\ m/\sqrt{s}$ |
| Ionosphere | Estimated as random walk ($4 \times 10^{-2}\ m/\sqrt{s}$) |
| Troposphere | Estimated as random walk ($1 \times 10^{-4}\ m/\sqrt{s}$) with a priori model [27] and GMF [28] |
| Ambiguity | Estimated as constant |

SSR-RTK experiment using extra SSR corrections generated by a single reference station. The processing strategies are listed in Table 14.3.

## 14.4.1 PPP Experiment

To investigate the PPP-B2b positioning performance based on raw observations and make preparations for augmentation generation, 30-s observations from 4 IGS stations and 4 International GNSS Monitoring and Assessment System (iGMAS) stations are selected for the experiment. The observation data duration is from DOY 214 of 2020 to DOY 244 of 2020. The reference coordinates are derived from SINEX files for IGS stations, while those of iGMAS stations are conducted with PPP using precise orbit and clock products provided by IGS.

During the 31-day test, one-day solutions of 3 stations are first selected for detailed comparison. The GPS-only (G), BDS-3-only (C), and combined GPS and BDS-3 (GC) solutions are resolved respectively to demonstrate the PPP-B2b positioning errors in east (E), north (N), and up (U) directions as well as position dilution of precision (PDOP) values of BJF1, SHA1, and CUSV on DOY 214, 2020, as shown in Fig. 14.3. Evidently, centimeter-level accuracy can be achieved by GC-combined PPP, and the results of BDS-only PPP are better than those of GPS-only PPP. Regarding the GPS-only PPP, the reduction in positioning is led by reduced satellites and poor geometry, especially within the period of 8:00 to 12:00 on the CUSV station. Among these stations, the positioning performance of CUSV is a little bit worse than others, which is related to the decrease of visible satellites with available PPP-B2b products. Taking 0.2 m for horizontal and 0.4 m for vertical as the convergence threshold, the averaged root mean square error (RMSE) of 31-day G, C, and GC solutions are computed. Statistically, the average RMSE values in E, N, and U directions are better than 7.93, 4.87, and 13.40 cm respectively for all stations, and the BDS-only solutions are superior to GPS-only solutions in most instances. Nevertheless, the positioning performance will be severely affected by the absence of observations and unavailable PPP-B2b products. Thus for POL2 and GUA1 stations, the BDS-only solutions are worse but explicable. For BDS and GPS dual-systems users, positioning accuracy of better than 4.75, 3.24, and 8.87 cm in E, N, and U directions is obtained with abundant observations and favorable geometry.

In terms of the convergence time for PPP-B2b positioning, a series of thresholds are set to reveal the convergence rate under different conditions. As shown in

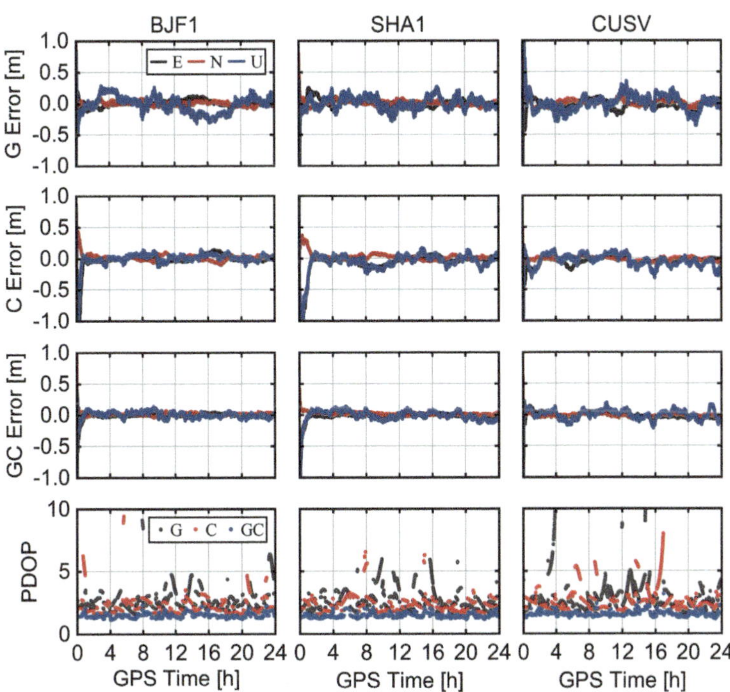

**Fig. 14.3** The PPP-B2b positioning errors in east, north, and up directions and PDOP values of BJF1, SHA1, and CUSV on DOY 214, 2020

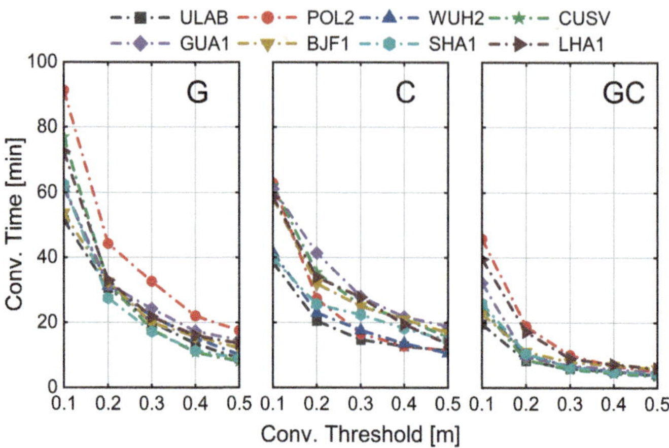



**Fig. 14.4** Averaged convergence (conv.) time of 8 stations with 31-day G, C, and GC solutions. The convergence thresholds vary from 0.1 to 0.5 m for horizontal, while it is double for vertical

Fig. 14.4, less time is needed for GC combined PPP to achieve targeted accuracy, and at least 20 min is required to converge to 0.1 m horizontally. Fast convergence of about 12 min is obtained by GPS-only PPP with an undemanding convergence threshold, whereas 15 min is required for BDS-only PPP. On the contrary, GPS-only PPP takes about 66 min to satisfy the 0.1 m threshold, while BDS-only PPP takes about 52 min to achieve the targeted accuracy.

### 14.4.2  SSR-RTK Experiment

To enhance the traditional PPP-B2b positioning, field tests of single-station SSR-RTK with different baselines are carried out to investigate the IAR and atmospheric augmentation. The station TJCH is selected as the reference station to provide extra SSR corrections. The other two rover stations TJJD and TJLG are set on different campuses of Tongji University with a baseline of about 27.6 km and 61.6 km respectively. All three stations are equipped with multi-frequency GNSS receiver Alloy from Trimble. Multi-frequency observations are collected by 1 Hz on 16 August 2020 at TJJD station and 14 November 2023 at TJLG station. Observations of BDS-3 and GPS are used for PPP-B2b positioning, and all qualified satellites are chosen for ambiguity resolution.

Float PPP with coordinates fixed is performed at the reference station TJCH using PPP-B2b products, and SSR corrections, such as phase bias, ionospheric delays, and residual zenith tropospheric delay, are generated for augmentation in the meantime. Actually, the corrections can be used even before PPP convergence [15], but a two-hour operation is conducted at the reference station in advance to ensure the accuracy and stability of SSR corrections. The results of PPP and single-station SSR-RTK in kinematic mode are computed. Positioning errors in east, north, and up directions are detailedly compared between float PPP and single-station SSR-RTK. The time series of positioning errors for the first hour is also depicted. Apparently, instant centimeter-level positioning accuracy can be achieved with the SSR corrections, and the positioning series are quite stable particularly in horizontal direction. For traditional PPP-B2b positioning, the results of TJJD and TJLG are better than 2.59, 1.70, and 5.32 cm in E, N, and U directions as illustrated in Table 14.4. A better solution, by contrast, is obtained by single-station SSR-RTK. The positioning accuracy is better than 1.05, 1.11, and 3.58 cm in E, N, and U directions with relative improvements of 59.29%, 34.71%, and 32.75% respectively. The fixing rate of single-station SSR-RTK exceeds 99% for two rover stations, though float solutions of a few epochs are conducted due to the missing PPP-B2b products. For single-station SSR-RTK in different baseline cases, the positioning accuracy may be severely affected by inaccurate atmospheric information and inappropriate stochastic models. However, users are still able to obtain centimeter-level positions as the baseline length grows longer, once the ambiguities are correctly fixed.

The atmospheric correlation between the reference and rover stations will degrade as the baseline length grows, affecting the rapid ambiguity resolution. To further explore the performance of single-station SSR-RTK, we reinitialize the positioning engine every hour and record the time to first fix (TTFF) of ambiguity resolution. The positioning errors with hourly reinitializing at rover station TJJD and TJLG are shown in Fig. 14.5, and the mean convergence time as well as the ambiguity fixing rate are simultaneously computed in Table 14.5. For a short baseline of 27.6 km, the ambiguities are instantly fixed within 8 s, in which only satellites continuously tracked for 5 epochs are considered for ambiguity fixing. When the baseline length extends to 61.6 km, the effect of atmospheric corrections is impaired for rapid ambiguity resolution. The convergence time, on average, is 71 s before ambiguities are correctly fixed. Meanwhile, the fixing rate is 99.74% and 97.45% for TJJD and TJLG respectively. Moreover, it should be noted that an extended period of fixing ambiguities is required due to the active ionosphere, especially during the GPS time 6:00–8:00 and 12:00–16:00 in our tests. Hence a more adaptive stochastic model is expected to improve the mean convergence time and fixing rate for single-station SSR-RTK.

For single-station SSR-RTK, the station-generated augmentations are quite effective at a specific distance. However, the atmospheric conditions at the base and rover

**Table 14.4** Mean RMSE of PPP and single-station SSR-RTK at station TJJD and TJLG

| Rover | Mean RMSE of PPP (cm) | | | Mean RMSE of single-station SSR-RTK (cm) | | |
|---|---|---|---|---|---|---|
| | East | North | Up | East | North | Up |
| TJJD | 1.77 | 1.70 | 5.32 | 1.00 | 0.93 | 3.08 |
| TJLG | 2.59 | 1.49 | 5.01 | 1.05 | 1.11 | 3.58 |

**Fig. 14.5** Positioning errors of single-station SSR-RTK with reinitializing every hour at rover station TJJD and TJLG

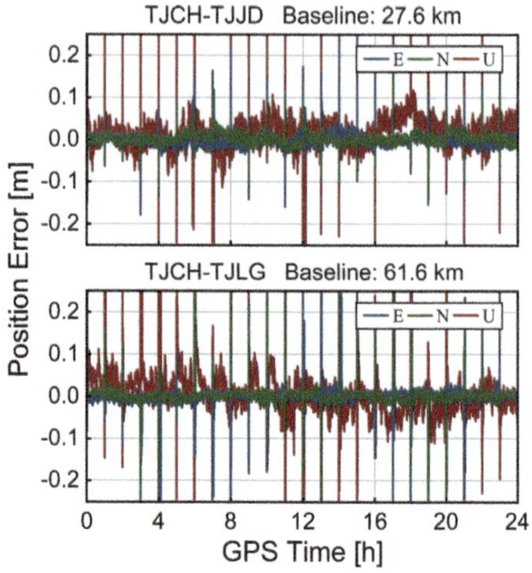



**Table 14.5** Mean convergence time and fixing rate of single-station SSR-RTK at rover station TJJD and TJLG

| Rover | Mean conv. time (s) | Fixing rate (%) |
| --- | --- | --- |
| TJJD | 8 | 99.74 |
| TJLG | 71 | 97.45 |

station can be entirely distinct when the baseline length is enough long. The atmospheric constraint, in the meantime, will have a small weight according to (14.14) and (14.15). Here we conduct an extra-long baseline case to discuss whether the phase bias products are still useful on this occasion. The IGS station JFNG is selected as the rover station, and SSR corrections generated by reference station TJCH are employed for single-station SSR-RTK. The test is performed in kinematic mode using 1-s observations from JFNG and TJCH stations on 14 November 2023. The baseline length is about 674.8 km, and the configuration is the same as above. For a simple demonstration, a set of 12-h data is continuously resolved with reinitializing the positioning engine every hour. The positioning error in time series and statistical results of station JFNG are shown in Fig. 14.6 and Table 14.6. The RMSE of fixed solutions are 0.90, 1.02, and 4.73 cm in east, north, and up directions. Compared to short baselines, the mean convergence time of this 674.8 km baseline soars to 673 s, mainly because the atmospheric augmentations do not work anymore. Nevertheless, there is strong evidence that phase bias products are still available for recovering the integer property of ambiguities, and the centimeter-level positions are obtained as the short baseline cases are.

**Fig. 14.6** Positioning errors of single-station SSR-RTK with reinitializing every hour at rover station JFNG

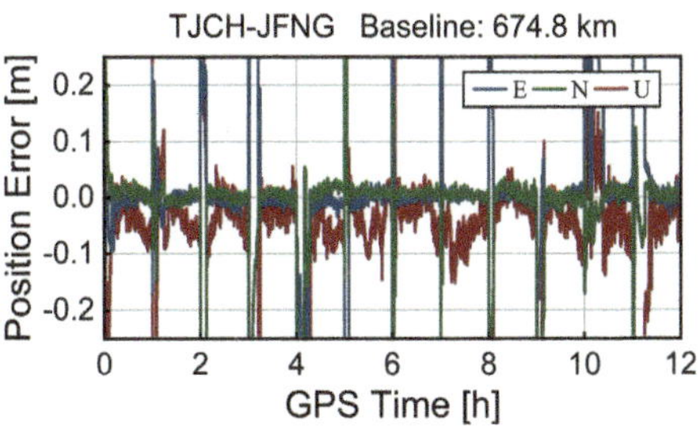



**Table 14.6** Mean convergence time and Mean RMSE of single-station SSR-RTK at station JFNG

| Rover | Mean conv. time (s) | Mean RMSE of single-station SSR-RTK (cm) | | |
|---|---|---|---|---|
| | | East | North | Up |
| JFNG | 673 | 0.90 | 1.02 | 4.73 |

## 14.5 Conclusion

This study investigates a feasible method of PPP-B2b positioning with IAR and atmospheric augmentation on the basis of PPP-B2b product characteristics as well as PPP analysis. The proposed station-based processing mode is tested with PPP-B2b products and real observation data. Augmented by one reference station, significant improvements have been made in positioning accuracy and convergence time, which is more flexible and conducive for users. The research conclusions are summarized as follows:

Regardless of the update of broadcast ephemeris, the discontinuity of PPP-B2b precise products is led by the real-time orbit and clock estimation strategies, resulting in small jumps in corrections. According to the characteristics of PPP-B2b products, the user algorithm is competitively optimized to enable IAR.

For kinematic PPP-B2b positioning based on raw observations, the accuracy of better than 4.75, 3.24, and 8.87 cm in E, N, and U directions is achieved after at least 20 min convergence. However, restricted by the serving area, positioning performances vary by location under different observation conditions.

The proposed single-station SSR-RTK is realized using PPP-B2b corrections and effective SSR corrections from one reference station. Compared to traditional PPP-B2b positioning, the accuracy of 1.05, 1.11, and 3.58 cm in E, N, and U directions is obtained with improvements of 59.29%, 34.71%, and 32.75%, respectively. The convergence time is tens of seconds rather than minutes. Furthermore, the results of a 674.8 km baseline positioning test prove that phase bias products are still available when the baseline grows extremely long, and centimeter-level positions are obtained after a convergence time of 673 s.

The ambiguity fixing rate exceeds 97% in our field test for single-station SSR-RTK, but advancements can still be made considering the adaptive modeling for the atmosphere. Additionally, the positioning performance is influenced by the latency of corrections, atmospheric correlations between different stations, length of baseline, etc., and further investigation is clearly warranted.

# References


1. Tang C, Hu X, Chen J, Liu L, Zhou S, Guo R, Li X, He F, Liu J, Yang J (2022) Orbit determination, clock estimation, and performance evaluation of BDS-3 PPP-B2b service. J Geod 96:60
2. Tao J, Liu J, Hu Z, Zhao Q, Chen G, Ju B (2021) Initial assessment of the BDS-3 PPP-B2b RTS compared with the CNES RTS. GPS Solut 25:131
3. Ouyang C, Shi J, Peng W, Dong X, Guo J, Yao Y (2023) Exploring characteristics of BDS-3 PPP-B2b augmentation messages by a three-step analysis procedure. GPS Solut 27:119
4. Geng T, Li Z, Xie X, Liu W, Li Y, Zhao Q (2022) Real-time ocean precise point positioning with BDS-3 service signal PPP-B2b. Measurement 203:111911
5. Sun S, Wang M, Liu C, Meng X, Ji R (2023) Long-term performance analysis of BDS-3 precise point positioning (PPP-B2b) service. GPS Solut 27:69
6. Xu X, Nie Z, Wang Z, Zhang Y, Dong L (2023) An improved BDS-3 PPP-B2b positioning approach by estimating signal in space range errors. GPS Solut 27:110
7. Wübbena G, Schmitz M, Bagge A (2005) PPP-RTK: precise point positioning using state-space representation in RTK networks. Proceedings of ION GNSS 2005, pp 2584–2594
8. Teunissen PJG, Odijk D, Zhang B (2010) PPP-RTK: Results of CORS network-based PPP with integer ambiguity resolution. J Aeronaut Astronaut Aviat Ser A 42:223–230
9. Laurichesse D, Mercier F, Berthias JP, Broca P, Cerri L (2007) Integer ambiguity resolution on undifferenced GPS phase measurements and its application to PPP and satellite precise orbit determination. Navigation 56:135–149
10. Ge M, Gendt G, Rothacher M, Shi C, Liu J (2008) Resolution of GPS carrier-phase ambiguities in precise point positioning (PPP) with daily observations. J Geod 82:389–399
11. Geng J, Shi C, Ge M, Dodson A, Lou Y, Zhao Q, Liu J (2012) Improving the estimation of fractional-cycle biases for ambiguity resolution in precise point positioning. J Geod 86:579–589
12. Collins P, Bisnath S, Lahaye F, Héroux P (2010) Undifferenced GPS ambiguity resolution using the decoupled clock model and ambiguity datum fixing. Navigation 57:123–135
13. Zha J, Zhang B, Liu T, Zhang X, Hou P, Yuan Y, Li Z (2023) Undifferenced and uncombined PPP-RTK aided by BDS-3 PPP-B2b precise orbits. Acta Geodaetica et Cartographica Sinica 52:1449–1459
14. Khodabandeh A (2021) Single-station PPP-RTK: correction latency and ambiguity resolution performance. J Geod 95:42
15. Lyu Z, Gao Y (2022) PPP-RTK with augmentation from a single reference station. J Geod 96:40
16. Li B, Ge H, Shen Y (2015) Comparison of ionosphere-free, Uofc, and uncombined PPP observation models. Acta Geodaetica et Cartographica Sinica 44:734–740
17. Zhang B, Chen Y, Yuan Y (2019) PPP-RTK based on undifferenced and uncombined observations: theoretical and practical aspects. J Geod 93:1011–1024
18. Ashby N (2003) Relativity in the global positioning system. Liv Rev Relativ 6:1
19. Petit G, Luzum B (2010) IERS Conventions 2010, Technical report
20. Wu J, Wu S, Hajj G, Bertiger W, Lichten S (1993) Effects of antenna orientation on GPS carrier phase. Manuscr Geodaet 18:91–98
21. CSNO (2020) BeiDou navigation satellite system signal in space interface control document precise point positioning service signal PPP-B2b, Version 1.0
22. Teunissen PJG (1985) Generalized inverses, adjustment, the datum problem and S-transformations. Proceedings of the optimization of geodetic networks, pp 253–267

23. Li B (2016) Stochastic modeling of triple-frequency BeiDou signals: estimation, assessment, and impact analysis. J Geod 90:593–610
24. Geng J, Meng X, Dodson A, Teferle F (2010) Integer ambiguity resolution in precise point positioning: method comparison. J Geod 84:569–581
25. Klobuchar J (1987) Ionospheric time-delay algorithm for single-frequency GPS users. IEEE Trans Aerosp Electron Syst AES-23:325–331
26. Teunissen PJG (1995) The least-squares ambiguity decorrelation adjustment: a method for fast GPS integer ambiguity estimation. J Geod 70:65–82
27. Saastamoinen J (1972) Atmospheric correction for the troposphere and stratosphere in radio ranging of satellites. Proceedings of the Geophysics Monograph Series 15, pp 247–251
28. Boehm J, Niell A, Tregoning P, Schuh H (2006) Global mapping function (GMF): a new empirical mapping function based on numerical weather model data. Geophys Res Lett 33:L07304